Air Pollution

This textbook covers the entire spectrum of topics required to completely understand air pollution. It emphasizes the atmospheric processes governing air pollution – emissions, atmospheric dispersion, chemical transformations, deposition on surfaces and ecosystems. Other areas of focus include air pollutant emission control technologies, health and environmental impacts, regulations and public policies, and interactions between climate change and air pollution. Topics are first presented conceptually, and then in terms of their fundamental aspects. Actual case studies are incorporated throughout to illustrate major air pollution phenomena, such as the dispersion of pollutants in the atmosphere, and the development of strategies to reduce urban air pollution, mitigate acid rain, and improve atmospheric visibility. Graduate students, researchers, and air quality professionals will find the full coverage of these important matters to be well suited to their needs.

Christian Seigneur is a professor at École des Ponts ParisTech, one of France's premier engineering graduate schools. He has both academic and private-sector experience in the field of air pollution. He was the director of the Atmospheric Environment Center (CEREA) for nine years. Previously, he was a vice-president at Atmospheric & Environmental Research, Inc. (AER), a research and consulting company, where he led the Air Quality Division located in California. He has published extensively in the scientific literature and has been an expert advisor to the U.S. Environmental Protection Agency and the French Agency for Health and Safety.

Air Pollution
Concepts, Theory, and Applications

CHRISTIAN SEIGNEUR
École des Ponts ParisTech

CAMBRIDGE
UNIVERSITY PRESS

University Printing House, Cambridge CB2 8BS, United Kingdom

One Liberty Plaza, 20th Floor, New York, NY 10006, USA

477 Williamstown Road, Port Melbourne, VIC 3207, Australia

314–321, 3rd Floor, Plot 3, Splendor Forum, Jasola District Centre, New Delhi – 110025, India

79 Anson Road, #06–04/06, Singapore 079906

Cambridge University Press is part of the University of Cambridge.

It furthers the University's mission by disseminating knowledge in the pursuit of education, learning, and research at the highest international levels of excellence.

www.cambridge.org
Information on this title: www.cambridge.org/9781108481632
DOI: 10.1017/9781108674614

© Christian Seigneur 2019
Originally published in France as: *Pollution atmosphérique. Concepts, théorie et application* by Christian Seigneur © Editions Belin / Humensis, 2018

This publication is in copyright. Subject to statutory exception and to the provisions of relevant collective licensing agreements, no reproduction of any part may take place without the written permission of Cambridge University Press.

This edition first published 2019

Printed in the United Kingdom by TJ International Ltd. Padstow Cornwall

A catalogue record for this publication is available from the British Library.

Library of Congress Cataloging-in-Publication Data
Names: Seigneur, Christian, 1952– author.
Title: Air pollution : concepts, theory, and applications / Christian Seigneur.
Description: Cambridge, United Kingdom ; New York, NY : Cambridge University Press, 2019. | Includes bibliographical references.
Identifiers: LCCN 2018055109 | ISBN 9781108481632
Subjects: LCSH: Air pollution.
Classification: LCC TD883 .S39 2019 | DDC 577.27/6–dc23
LC record available at https://lccn.loc.gov/2018055109

ISBN 978-1-108-48163-2 Hardback

Additional resources for this publication at www.cambridge.org/seigneur.

Cambridge University Press has no responsibility for the persistence or accuracy of URLs for external or third-party internet websites referred to in this publication and does not guarantee that any content on such websites is, or will remain, accurate or appropriate.

Contents

Preface	*page* ix
List of Main Notations	xi

1 Brief History of Air Pollution   1
    1.1 The Earth's Atmosphere   1
    1.2 Air Pollution   1
    References   4

2 Emissions of Air Pollutants and Emission Control Technologies   6
    2.1 Sources of Air Pollution   6
    2.2 Emission Inventories   10
    2.3 Emission Control Technologies   13
    2.4 Numerical Modeling of Air Pollutant Emissions   28
    References   30

3 Meteorology: General Circulation   33
    3.1 General Considerations on the Atmosphere   33
    3.2 General Atmospheric Circulation   40
    3.3 Meteorological Regimes   45
    References   49

4 Air Pollution Meteorology   51
    4.1 The Atmospheric Planetary Boundary Layer   51
    4.2 Vertical Motions of Air Parcels   52
    4.3 Winds and Turbulence   57
    4.4 Heat Transfer   66
    4.5 Local Circulations: Land-Sea Breezes and Mountain-Valley Winds   71
    4.6 Meteorological Modeling   72
    References   74

5 Atmospheric Radiative Transfer and Visibility   76
    5.1 General Considerations on Atmospheric Radiative Transfer   76
    5.2 Atmospheric Visibility   83
    5.3 Numerical Modeling of Atmospheric Radiation   92
    References   94

6 Atmospheric Dispersion — 95
6.1 General Considerations on Atmospheric Dispersion — 95
6.2 Lagrangian Representation of Turbulence — 96
6.3 Eulerian Representation of Turbulence — 99
6.4 Lagrangian Models of Atmospheric Dispersion — 103
6.5 Eulerian Models of Atmospheric Dispersion — 115
6.6 Street-canyon Models — 119
6.7 Numerical Modeling of Pollutant Transport in the Atmosphere — 121
References — 123

7 The Stratospheric Ozone Layer — 125
7.1 Fundamentals of Chemical Kinetics — 125
7.2 Chemistry of the Stratospheric Ozone Layer — 136
7.3 Numerical Modeling of Atmospheric Chemical Kinetics — 144
References — 145

8 Gaseous Pollutants — 146
8.1 General Considerations on Gaseous Pollutants — 146
8.2 Oxidizing Power of the Atmosphere and Chemical Reactivity — 149
8.3 Gas-phase Chemistry of Photochemical Air Pollution — 153
8.4 Chemical Kinetic Mechanisms of Photochemical Air Pollution — 169
8.5 Emission Control Strategies for Photochemical Air Pollution — 175
8.6 Numerical Modeling of the Gas-phase Chemistry of Photochemical Air Pollution — 185
References — 188

9 Atmospheric Particles — 190
9.1 General Considerations on Atmospheric Particles — 190
9.2 Dynamics of Atmospheric Particles — 192
9.3 Equilibrium Thermodynamics — 203
9.4 Secondary Inorganic Fraction of Particulate Matter — 204
9.5 Organic Fraction of Particulate Matter — 207
9.6 Emission Control Strategies for Atmospheric Particulate Matter — 217
9.7 Numerical Modeling of Atmospheric Particulate Matter — 227
References — 236

10 Clouds and Acid Rain — 239
10.1 General Considerations on Clouds and Fogs — 239
10.2 Aqueous-phase Chemistry — 240
10.3 Aqueous-phase Chemical Transformations — 249
10.4 Emission Control Strategies for Acid Rain — 255
10.5 Numerical Modeling of Aqueous-phase Chemistry — 257
References — 257

11 Transfer of Pollutants between the Atmosphere and Surfaces — 259
    11.1 Dry Deposition — 259
    11.2 Wet Deposition — 269
    11.3 Reemissions of Pollutants Deposited on Surfaces and Natural Emissions of Particles — 276
    11.4 Numerical Modeling of Atmospheric Deposition and Emissions — 283
    References — 284

12 Health Effects — 286
    12.1 Identification and Characterization of Health Effects — 286
    12.2 Health Risk Assessment — 297
    References — 302

13 Environmental Impacts — 304
    13.1 Ozone — 304
    13.2 Acid Rain — 306
    13.3 Eutrophication — 308
    13.4 Persistent Organic Pollutants — 309
    13.5 Heavy Metals — 311
    References — 316

14 Climate Change and Air Pollution — 318
    14.1 General Considerations on Climate Change — 318
    14.2 Effect of Air Pollution on Climate Change — 322
    14.3 Effect of Climate Change on Air Pollution — 324
    References — 325

15 Regulations and Public Policies — 327
    15.1 Regulations for Air Pollutant Concentrations — 327
    15.2 Public Policies — 334
    15.3 Regulations for Atmospheric Deposition and Global Pollution — 341
    References — 344

Index — 346

Preface

This book corresponds to graduate classes taught since 2009 in France at École des Ponts ParisTech ("Atmospheric environment" in the engineering curriculum, in French and English), at Université Paris-Diderot ("Air pollution impacts and regulations" and "Modeling of air pollution" in the Master program titled Environmental Science and Engineering, which is jointly organized by Université Paris-Est Créteil, Université Paris-Diderot, and École des Ponts ParisTech, in French), at École Centrale in Nantes ("Chemistry of air pollution" in the international Master program titled Science and Technology of the Urban Environment, in English), and at Institut Mines Télécom Atlantique in Nantes ("Dispersion of air pollutants" in the international Master program titled Project Management in Environmental and Energy Engineering, in English). However, the content of the book has been augmented so that it can serve as a scientific reference on the subject of air pollution for graduate students and scientists involved in air pollution research, as well as for air quality professionals.

This book covers air pollution from its sources to its health and environmental impacts. The various air pollution topics treated include gaseous air pollutants, such as ozone and nitrogen dioxide, particles, acid rain, heavy metals, such as mercury and chromium, persistent organic pollutants, and eutrophication via atmospheric deposition of nitrogenous species. The main technologies available to control air pollutant emissions are described, with special emphasis on vehicle emission control. Atmospheric processes are described in detail; they include pollutant dispersion, chemical transformation, particle dynamics, and atmospheric deposition on surfaces and ecosystems. The analysis of health impacts is presented mostly in terms of its use for the development of regulations of major air pollutants. Problems of a global nature, such as the destruction of the stratospheric ozone layer and climate change, are described more succinctly, and the emphasis is placed on the interactions between these global phenomena and air pollution. Indoor air pollution is not addressed explicitly, but the physico-chemical processes treated here for outdoor air are transferable to confined environments.

I want to thank a large number of colleagues who have been of tremendous help during the preparation of this book, through their reading of chapters, discussion of specific topics, adaptation of figures, etc., and have contributed to a much improved result. Among these colleagues, I wish to thank in particular Karine Sartelet, Hadjira Foudhil-Schmitt, Bernard Aumont, Yelva Roustan, David Pollack, Geoffrey Sunshine, Luc Musson-Genon, Michel André, Julien Cattiaux, Bertrand Carissimo, Ève Lecœur, Laëtitia Thouron, Ruiwei Chen, Youngseob Kim, Hélène Marfaing, Véronique Ghersi, Robert W. Bergstrom, and Carole

Legorgeu. In addition, this book has also benefited from the many interactions that I had throughout my career, in the United States as a consultant and in France as a professor and director of a research laboratory, with colleagues, clients, Ph.D. students, and post-doctoral students. Nevertheless, if there are some remaining errors, omissions or inaccuracies, those are my sole responsibility.

Main Notations

The units that are the most commonly used in this book are indicated in parentheses.

A	Surface area (m^2)
A_j	Activity of source j (activity unit s^{-1})
A_T	Parameter of the reaction rate constant
B_T	Parameter of the reaction rate constant
b_a	Absorption coefficient for atmospheric radiation (m^{-1})
b_e	Extinction coefficient for atmospheric radiation (m^{-1})
b_{Ray}	Rayleigh scattering coefficient for atmospheric radiation (m^{-1})
$b_{Ray,ref}$	Rayleigh scattering coefficient for atmospheric radiation at 1 atm and 0 °C (m^{-1})
b_s	Scattering coefficient for atmospheric radiation (m^{-1})
C	Concentration ($\mu g\ m^{-3}$, ppm, M)
C_0	Initial concentration ($\mu g\ m^{-3}$, ppm)
C_{aq}	Concentration in the aqueous phase (M)
C_b	Background concentration ($\mu g\ m^{-3}$)
C_g	Gas-phase concentration ($\mu g\ m^{-3}$, $g\ m^{-3}$, atm, ppm, ppb)
$C_{g,e}$	Gas-phase concentration at thermodynamic equilibrium (atm, $g\ m^{-3}$)
C_P	Molar heat capacity at constant pressure ($J\ mol^{-1}\ K^{-1}$)
C_u	Upwind concentration ($\mu g\ m^{-3}$)
c	Speed of light ($3 \times 10^8\ m\ s^{-1}$)
c_c	Cunningham correction factor
c_d	Drag coefficient
c_e	Correction factor for aeolian particle emissions
c_p	Specific heat capacity at constant pressure ($J\ kg^{-1}\ K^{-1}$)
$c_{p,dry}$	Specific heat capacity of dry air at constant pressure ($J\ kg^{-1}\ K^{-1}$)
$\bar{c}_t$	Mean thermal velocity ($cm\ s^{-1}$)
c_v	Specific heat capacity at constant volume ($J\ kg^{-1}\ K^{-1}$)
Co	Contrast of an object against the horizon
Co_0	Intrinsic contrast of an object (at distance zero from the object)
CSF	Cancer slope factor ($mg^{-1}\ kg\ day$)
D_E	Exposure duration (h, day, year)
D_H	Dose ($mg\ kg^{-1}\ day^{-1}$)
D_J	Julian day (day)

D_m	Molecular diffusion coefficient (cm^2 s^{-1})
D_p	Brownian diffusion coefficient (cm^2 s^{-1})
d_b	Displacement height (m)
d_f	Characteristic length of a surface for particle deposition via inertia or interception (μm)
d_p	Diameter of a particle (μm)
$d_{p,a}$	Aerodynamic diameter of a particle (μm)
$d_{p,St}$	Stokes diameter of a particle (μm)
d_r	Diameter of a raindrop (μm)
d_s	Diameter of a source (stack) (m)
DHI	Deciview haze index (dv)
E	Collision efficiency between a particle and an object (drop, surface)
E_a	Activation energy (J mole^{-1})
E_B	Collision efficiency for brownian motion
E_b	Chemical bond dissociation energy (J mole^{-1})
E_e	Irradiance (W m^{-2} or W m^{-2} nm^{-1})
E_{IM}	Collision efficiency for impact by inertia
E_{IN}	Collision efficiency by interception
E_k	Kinetic energy of the flow per unit mass (m^2 s^{-2})
E_p	Energy of a photon (J)
E_S	Solar constant (1,361 W m^{-2})
E_t	Mean thermal energy of a gas (J mole^{-1})
EF	Emission factor (g per activity unit)
EFR	Etiologic fraction of risk
ER	Excess relative risk
F_a	Kinetic energy term in the Briggs plume rise equation (m^4 s^{-2})
F_b	Buoyancy term in the Briggs plume rise equation (m^4 s^{-3})
F_c	Troe falloff parameter in the rate constant
F_d	Dry deposition flux (g m^{-2} s^{-1})
F_e	Aeolian particle emission flux (μg m^{-2} s^{-1})
$F_{l,r}$	Addition of particles to the roadway (g km^{-1} h^{-1})
F_m	Mass transfer flux (g m^{-2} s^{-1})
F_{Qs}	Sensible heat flux (J m^{-2} s^{-1})
F_s	Sea-salt particle emission flux (μg m^{-2} s^{-1})
F_w	Wet deposition flux (g m^{-2} s^{-1})
f	Coriolis parameter (s^{-1})
f_{aq}	Fraction of a chemical species present in the aqueous phase
f_{as}	Effect of relative humidity on inorganic fine particle size
f_{clay}	Clay fraction of the soil
f_{er}	Erodible fraction of the soil

f_p	Fraction of particles sticking to a surface after contact
f_{ss}	Effect of relative humidity on sea salt coarse particle size
f_v	Fraction of particles present on a roadway reemitted by traffic (vehicle^{-1})
Fo_{St}	Stokes frictional force (kg m s^{-2})
g	Earth's gravitational constant (9.81 m s^{-2})
g_{ec}	Variable in the equation of the Sun ecliptic longitude (°)
g_i, g_j	Coefficients used in the intermediate regime for the brownian coagulation coefficient (cm)
GR	Gas ratio, ratio of the available ammonia concentration and total inorganic nitrate concentration
H	Henry's law constant (M atm^{-1})
H_{eff}	Effective Henry's law constant (M atm^{-1})
H_i	Henry's law constant for species i (M atm^{-1})
h	Planck constant (6.626 × 10^{-34} J s)
h_a	Characteristic length of the atmosphere in the vertical direction (m, km)
h_b	Mean height of buildings (m)
h_c	Height of the air column under cloud base (m)
h_{ne}	Height of non-erodible elements (m)
HI	Hazard index
HI_t	Total hazard index
HQ	Hazard quotient
HQ_i	Hazard quotient for exposure pathway i
I	Intensity or radiance (W m^{-2} sr^{-1} or W m^{-2} sr^{-1} nm^{-1})
I_b	Background intensity or radiance (W m^{-2} sr^{-1} or W m^{-2} sr^{-1} nm^{-1})
I_H	Incidence rate of an adverse health effect
I_J	Actinic flux (photons cm^{-2} s^{-1})
I_o	Intrinsic intensity or radiance of the observed object (W m^{-2} sr^{-1} or W m^{-2} sr^{-1} nm^{-1})
I_p	Precipitation intensity (m h^{-1}, m s^{-1})
I_v	Growth law of particles due to condensation (μm^3 s^{-1})
J	Photolysis rate constant (s^{-1})
J_n	Nucleation rate (particles cm^{-3} s^{-1})
K_1, K_2	Ionic dissociation equilibrium constants (units depend on the equilibrium)
K_{eq}	Chemical equilibrium constant (units depend on the equilibrium)
K_H	Turbulent heat transfer coefficient (m^2 s^{-1})
K_h	Eulerian horizontal dispersion coefficient (m^2 s^{-1})
K_M	Turbulent kinematic viscosity (m^2 s^{-1})
$K_{om,i}$	Gas/particle partition coefficient for organic species SVOC$_i$ (m^3 g^{-1})
K_{xx}	Eulerian dispersion coefficient in the x direction (m^2 s^{-1})
K_{yy}, K_y	Eulerian dispersion coefficient in the y direction (m^2 s^{-1})

K_{zz}, K_z	Eulerian dispersion coefficient in the z direction (m^2 s^{-1})
k	Reaction rate constant (cm^6 molec^{-2} s^{-1}, cm^3 molec^{-1} s^{-1} or s^{-1} depending on the order of the reaction)
k'	Pseudo-first-order rate constant (s^{-1}, min^{-1})
k_0	Low-pressure limit rate constant (cm^3 molec^{-1} s^{-1} for unimolecular reactions, cm^6 molec^{-2} s^{-1} for trimolecular reactions)
k_∞	High-pressure limit rate constant (s^{-1} for unimolecular reactions, cm^3 molec^{-1} s^{-1} for trimolecular reactions)
k_a	Absorption efficiency for atmospheric radiation (m^2 g^{-1})
k_B	Boltzmann constant (1.381 × 10^{-23} J K^{-1})
k_f	Rate constant of the forward reaction (cm^3 molec^{-1} s^{-1}, ppb^{-1} min^{-1})
k_m	Mass transfer coefficient (m s^{-1})
k_{MKE}	Specific kinetic energy of the mean flow (m^2 s^{-2})
k_r	Rate constant of the backward reaction (cm^3 molec^{-1} s^{-1}, ppb^{-1} min^{-1})
k_s	Scattering efficiency for atmospheric radiation (m^2 g^{-1})
k_{TKE}	Specific turbulent kinetic energy (m^2 s^{-2})
Kn	Knudsen number
L	Liquid water content (g m^{-3})
L_{ec}	Variable in the equation of the Sun ecliptic longitude (°)
L_{MO}	Monin-Obukhov length (m)
L_s	Length of a street-canyon segment (m)
l_c	Characteristic length of the flow (m)
l_m	Characteristic mixing length, so-called Prandtl mixing length (m)
M	Pollutant mass (g)
M_A	Total mass of gases of the Earth's atmosphere (5.148 × 10^{18} kg)
M_{aq}	Pollutant mass in the aqueous phase (g)
$M_{l,r}$	Mass concentration of particles present on a roadway (g km^{-1})
M_o	Mass concentration of organic particulate matter (g m^{-3})
M_p	Mass concentration of particles (μg m^{-3})
m_{air}	Mass of air displaced by a particle or a droplet (g)
m_g	Mass of a gas molecule (g)
m_p	Mass of a particle (g)
MM	Molar mass of a chemical species (g mole^{-1})
MM_{air}	Mean molar mass of dry air (0.029 kg mole^{-1})
MM_{om}	Mean molar mass of organic particulate matter (g mole^{-1})
N	Avogadro's number (6.02 × 10^{23} molec mole^{-1})
N_a	Total number of particles in the accumulation mode (# m^{-3})
N_c	Total number of chemical pollutants included in a health risk assessment
N_{dh}	Number of total daylight hours used for the calculation of AOT40
N_H	Number of exposed individuals with an adverse health effect

List of Main Notations

N_{H2SO4}	Number concentration of gaseous sulfuric acid molecules (molec cm^{-3})
N_i	Total number concentration of particles in size section i (# m^{-3})
N_j	Variable used in the calculation of the ecliptic longitude
N_p	Number concentration of particles (# m^{-3})
N_r	Number of reflections taken into account in the gaussian plume equation
N_{rp}	Number concentration of raindrops for a precipitation event (# m^{-3})
N_T	Total number of exposed individuals
N_v	On-road traffic flow (vehicles h^{-1})
N_w	Number of days in a three-month period for the calculation of W126
n	Number of moles (#)
$n_p(v_p)$	Distribution of the number concentration of a population of particles as a function of particle volume v_p (# μm^{-3} cm^{-3}, # cm^{-6})
$n_{p,d}(d_p)$	Distribution of the number concentration of a population of particles as a function of particle diameter d_p (# μm^{-1} cm^{-3})
n_s	Emission rate in number of moles of gas (mole s^{-1})
OR	Odds ratio
P	Atmospheric pressure (atm, Pa (N m^{-2}), torr (mm Hg))
P_i	Partial pressure of gas i (atm)
$P_{s,H2O}$	Saturation vapor pressure of water (atm)
$P_{s,i}$	Saturation vapor pressure of gas i (atm)
$p(t)$	Probability distribution as a function of time
$p(x, t)$	Probability distribution as a function of space and time
$p(\Omega' \longrightarrow \Omega)$	Phase function of particle scattering of atmospheric radiation
Pr_t	Turbulent Prandtl number (0.95)
Q	Heat quantity per unit mass (J kg^{-1})
Q_h	Heat flux (J s^{-1})
q	Specific humidity (g of water per g of moist air)
R	Ideal gas law constant (8.206 × 10^{-5} atm m^3 mole^{-1} K^{-1}, 8.314 J mole^{-1} K^{-1}, 1.986 cal mole^{-1} K^{-1})
$R(\tau_L)$	Lagrangian correlation function
R_H	Health risk
$R_{H,i}$	Excess cancer risk for exposure pathway i
$R_{H,t}$	Total excess cancer risk
R_{ij}	Components of the tensor of the Reynolds stresses (m^2 s^{-2})
r	Correlation coefficient
r_a	Aerodynamic resistance to dry deposition (s m^{-1})
r_b	Quasi-laminar resistance to dry deposition (s m^{-1})
r_c	Surface resistance to dry deposition (s m^{-1})
r_r	Chemical reaction rate (molec cm^{-3} s^{-1})
r_T	Average radius of the Earth (Terra) (6,371 km)

r_t	Total resistance to dry deposition (s m^{-1})
Re	Reynolds number
RfC	Reference concentration (μg m^{-3})
RfD	Reference dose (mg kg^{-1} day^{-1})
RH	Relative humidity (%)
Ri_b	Bulk Richardson number
Ri_f	Flux Richardson number
Ri_g	Gradient Richardson number
RR	Relative risk
S	Continuous emission rate of a point source (g s^{-1})
S_0	Instantaneous emission rate (g)
S_{ij}	Emission rate of pollutant i from source j (g s^{-1})
S_l	Continuous emission rate of a line source (g km^{-1} s^{-1})
$S_{l,r}$	Loss rate of particles present on a roadway (g km^{-1} h^{-1})
S_p	Surface concentration of particles (μm^2 m^{-3})
S_Γ	Variable of K_h representing the stretching deformation of the wind field (s^{-1})
S_Λ	Variable of K_h representing the shearing deformation of the wind field (s^{-1})
s_r	Surface of a raindrop (m^2)
s_s	Sea water salinity (g kg^{-1})
Sc	Schmidt number
Sh	Sherwood number
St	Stokes number
T	Temperature (K)
T_N	Total exposure duration of a cohort (person h, person days)
T_s	Temperature of the effluent plume (K)
T_{ss}	Sea surface temperature (K)
$T_{ss,C}$	Sea surface temperature (°C)
T_v	Virtual temperature (K)
t	Time (s)
$t_{1/2}$	Half-life (s)
t_L	Lagrangian time scale (s)
t_l	Lifetime or residence time (s)
U	Internal energy per unit mass (J kg^{-1})
u	Wind speed in direction x (m s^{-1})
u_{10}	Wind speed in direction x at 10 m height (m s^{-1})
u_s	Average wind speed in a street-canyon (m s^{-1})
u_*	Friction velocity (m s^{-1})
u_{*s}	Surface friction velocity (m s^{-1})
u_{*t}	Threshold surface friction velocity (m s^{-1})
u_{*st}	Standard threshold surface friction velocity (m s^{-1})

$u_{*st,\,ref}$	Reference value of the standard surface threshold friction velocity (m s^{-1})
UR	Unit risk factor (µg^{-1} m^3)
V	Volume of air (m^3)
V'	Specific air volume (m^3 kg^{-1})
V_a	Total volume concentration of particles in the accumulation mode (µm^3 m^{-3})
V_i	Total volume concentration of particles in size section i (µm^3 m^{-3})
V_p	Volume concentration of particles (µm^3 m^{-3})
v	Wind speed in direction y (m s^{-1})
v_d	Dry deposition velocity (m s^{-1})
v_i	Initial velocity of effluent (m s^{-1})
v_m	Molecular volume (µm^3 molec^{-1})
v_p	Volume of a particle (µm^3)
v_r	Volume of a raindrop (m^3)
v_s	Sedimentation velocity (m s^{-1})
v_s^f	Terminal fall velocity (m s^{-1})
$v_{s,p}$	Particle fall velocity (m s^{-1})
$v_{s,r}$	Raindrop fall velocity (m s^{-1})
v_v	Speed of vehicles (km h^{-1})
VR	Visual range (m, km)
W	Quantity of work per unit mass (J kg^{-1})
$W_{v,w}$	Wear rate (g vehicle^{-1} km^{-1})
W_s	Average street width (m)
w	Wind speed in direction z (m s^{-1})
w_*	Convective wind speed
x, y, z	Cartesian coordinates (m)
x_i	Molar fraction of species i
x_s, y_s, z_s	Cartesian coordinates of a source (m)
x_t	Downwind distance for the calculation of the Briggs plume rise (m)
Y	SOA yield from the oxidation of an organic compound (fraction or %)
z_0	Roughness length (m)
z_i	Height of the planetary boundary layer or inversion layer (m)
z_r	Reference height (used to calculate dry deposition, m)
$z_{s,f}$	Height of the plume centerline following plume rise (m)
α	Angle or exponent in various formulas
$\alpha, \beta \ldots$	Parameter sets used for example in a generic chemical reaction or in the efficiency functions for dry deposition of particles
α_d	Correction term for dry deposition in the gaussian plume equation
β	Coagulation coefficient (cm^3 particle^{-1} s^{-1})
β_b	Empirical parameter characterizing the configuration of a built area

β_S	Regression slope between the relative health risk, $\ln(RR)$, and the pollutant concentration (m^3 µg^{-1})
ΔH_A	Enthalpy of dissolution (cal mole^{-1})
ΔH_R	Enthalpy of reaction (cal mole^{-1})
δ	Dirac function
δ_s	Declination of the Sun (°)
ε	Dissipation rate of the turbulent kinetic energy (m^2 s^{-3})
ε_{ob}	Inclination of the ecliptic (°)
Φ_H	Dimensionless number of the vertical temperature profile
Φ_{H0}	Value of Φ_H at $z = z_0$
Φ_M	Dimensionless number of the vertical wind shear
Φ_{M0}	Value of Φ_M at $z = z_0$
ϕ	Latitude (°)
ϕ_j	Quantum yield of a photolytic reaction (molec photon^{-1})
γ	Adiabatic index, also called the Laplace coefficient (1.4 for air)
γ_i	Activity coefficient of species i
η_p	Lognormal distribution of the number concentration of particles as a function of particle diameter or its logarithm (# m^{-3})
φ	Function relating L_{MO} and Ri_g
κ	von Kármán constant (0.4)
Λ	Scavenging coefficient (s^{-1})
λ	Radiation wavelength (nm, µm, m)
λ_b	Built fraction of an area
λ_{ec}	Ecliptic longitude (°)
$\lambda_{m,a}$	Mean free path in the air (cm)
μ_D	Dilatational viscosity (kg m^{-1} s^{-1})
μ_v	Dynamic viscosity (kg m^{-1} s^{-1})
$\mu_{v,a}$	Dynamic viscosity of air (1.84 × 10^{-5} kg m^{-1} s^{-1} at 1 atm and 25 °C for dry air)
$\mu_{v,w}$	Dynamic viscosity of water (kg m^{-1} s^{-1})
ν	Frequency of atmospheric radiation (s^{-1})
ν_v	Kinematic viscosity (m^2 s^{-1})
$\nu_{v,a}$	Kinematic viscosity of air (1.55 × 10^{-5} m^2 s^{-1} at 1 atm and 25 °C for dry air)
ρ	Density (g m^{-3}, kg m^{-3}, g cm^{-3})
ρ_a	Air density (1.225 kg m^{-3} at 1 atm and 15 °C for dry air)
ρ_p	Density of a particle (g cm^{-3})
ρ_w	Water density (1,000 kg m^{-3} at 4 °C)
σ_a	Standard deviation of the lognormal distribution of the number or volume concentration of particles for the accumulation mode
σ_p	Surface tension of a particle (erg cm^{-2})

List of Main Notations

σ_J	Absorption cross-section of a molecule (cm^2 $molec^{-1}$)
σ_u	Standard deviation of the distribution of the horizontal wind speed (m s^{-1})
σ_w	Standard deviation of the distribution of the vertical wind speed (m s^{-1})
$\sigma_x, \sigma_y, \sigma_z$	Standard deviations of the gaussian distribution of concentrations; gaussian dispersion coefficients (m)
σ_{SB}	Stefan-Boltzmann constant (5.67×10^{-8} W m^{-2} K^{-4})
τ	Optical depth
τ^*	Shear stress (kg m^{-1} s^{-2})
τ_{ij}	Components of the tensor of the shear stresses (kg m^{-1} s^{-2})
τ_L	Time variable in the lagrangian theory of dispersion
θ	Potential temperature (K)
θ_o	Angle between the Earth's radius toward the observer and that toward the point corresponding to the visual range (°)
θ_h	Angle of the solar hour (°)
θ_r	Angle between the incoming radiation and the normal direction to a surface (°)
θ_v	Virtual potential temperature (K)
θ_z	Zenith angle (°)
θ_*	Temperature scale (K)
θ_{*c}	Convective temperature scale (K)
$\upsilon_{p,a}$	Lognormal distribution of the volume concentration of particles as a function of the particle diameter or its logarithm for the accumulation mode (μm^3 m^{-3})
Ω	Direction of the atmospheric radiation
Ω_T	Angular velocity of the Earth (Terra) (rad s^{-1})
ω_a	Albedo
ω_T	Variable function of temperature used in the equation of the saturation vapor pressure of water
ξ_b	Empirical parameter of a built area
ξ	Dimensionless variable for altitude (z/L_{MO})

1 Brief History of Air Pollution

1.1 The Earth's Atmosphere

The Earth's atmosphere is composed mostly of molecular nitrogen (N_2, 78 % of dry air) and molecular oxygen (O_2, 21 % of dry air). It holds also a fair amount of water vapor (H_2O), which varies greatly in concentration (ranging from negligible in dry regions to a few % in humid regions) and leads to the formation of clouds and fogs in case of supersaturation. The Earth's atmosphere also contains carbon dioxide (CO_2), which has an average concentration of about 0.04 %. H_2O and CO_2 are gases that absorb infrared (IR) radiation, but let ultraviolet (UV) and visible solar radiation go through. Since they partially absorb IR radiation emitted by the Earth toward space, these species are called "greenhouse gases" (GHG). As a result, the temperature of the Earth's atmosphere is on average about 34 °C (61 °F) warmer than it would be without those GHG. However, an increase of GHG atmospheric concentrations leads to climate change and its associated consequences in terms of extreme meteorological events, changes in precipitation patterns, and increase of sea level.

Molecular oxygen undergoes photolysis by UV solar radiation, which leads in the upper atmosphere to the formation of oxygen atoms (O) and the formation of triatomic oxygen, i.e., ozone (O_3). Ozone absorbs part of UV solar radiation and, therefore, protects organisms on the Earth's surface from some of those harmful rays (which may lead to skin cancer).

In addition, gaseous chemical species and atmospheric liquid or solid particles are present in the atmosphere. They originate either from natural sources or from human activities (also called anthropogenic activities). Some of those gases and particles may be harmful to human health and the environment (e.g., ecosystems, buildings, atmospheric visibility). Those gases and particles are then considered to be air pollutants. Some of those chemical species are not necessarily harmful when present in the air, but may become harmful after their transfer to other environmental media and their possible transformation and/or bioaccumulation in those media (soil, water, and the food chain).

1.2 Air Pollution

One may distinguish several aspects of the perturbation of the atmospheric environment. On one hand, some phenomena are global in nature, because of the long lifetime (several years or decades) of the chemical species involved. This is the case in particular for:

- Climate change due to GHG
- The destruction of the stratospheric ozone layer

On the other hand, some phenomena pertain mostly to the lower atmosphere and their impacts occur at the Earth's surface. We group those phenomena here under the general term of air pollution. Air pollution may occur at spatial scales that are local, regional or global, and for time periods that may be short (on the order of an hour) or long (on the order of a year). Some air pollutants are emitted directly into the atmosphere, whereas others are formed in the atmosphere via chemical processes (chemical or photochemical reactions) and/or physical processes (gas-to-particle conversion). Therefore, one distinguishes:

- Primary air pollutants, which are emitted directly into the atmosphere
- Secondary air pollutants, which are formed in the atmosphere from other chemical species, called precursors

If all precursors of secondary air pollutants are not necessarily directly harmful to human health or the environment, they are nevertheless considered to be pollutants, because they contribute to air pollution.

As mentioned above, one may distinguish air pollutants that have adverse effects due directly to their concentrations in the air and those that lead to adverse effects following their deposition to the Earth's surface and subsequent transformation in other media, such as soil, water, and the food chain.

Historically, air pollution started with the discovery of fire and human exposure to the chemical species produced during biomass burning (combustion of wood and vegetation) in poorly ventilated locations where concentrations could reach levels that are harmful to human health (e.g., Hardy et al., 2012). During the 19th century, the Industrial Revolution led to the emission of significant amounts of air pollutants via the combustion of a variety of fossil fuels (coal, oil, and gas). In particular, the air pollution in London during the 19th and 20th centuries became particularly problematic. The most significant, documented London air pollution episode occurred during the winter of 1952 (Wilkins, 1954; Bell et al., 2004). Smoke from combustion mingled with London fog to become "smog," a term coined in the early 20th century by Harold Antoine des Vœux, a member of the Coal Smoke Abatement Society, at a conference of the American Medical Association in London (JAMA, 1905). The stagnant atmosphere during the 1952 event, which contained high concentrations of sulfur dioxide (SO_2), nitrogen oxides (NO_x), and atmospheric particles (including sulfate), contributed to the deaths of several thousands.

During the 1950s, a new type of air pollution appeared in the Los Angeles basin in California. The use of big cars to move around an area characterized by widespread urban planning, combined with industries (power plants, refineries, port activities, etc.) located on the Pacific coast, led to the formation of secondary pollutants such as ozone, oxygenated organic compounds (e.g., acrolein), and fine particles. This type of air pollution was the result of atmospheric chemical reactions. It was more troublesome during summer when solar radiation was intense. Arie Jan Haagen-Smit, a biochemistry professor at the California Institute of Technology (Caltech), proposed in 1952 that emissions of nitrogen oxides and volatile organic compounds (VOC) led, via chemical reactions in the presence of sunlight, to the formation of ozone. Solar radiation in the ultraviolet and visible range has sufficient energy to break the bonds of some molecules. This process, called photolysis, creates chemical species that are very reactive and can initiate a series of chemical reactions leading to the formation of the pollutants mentioned previously. Therefore, this form of air pollution was called photochemical smog, because sunlight was needed to initiate the photochemistry (chemical reactions involving the photolysis of molecules) of this type of air pollution.

In 1962, Rachel Carson published *Silent Spring*, a book that alerted readers to the harmful effects of pesticides, such as dichlorodiphenyltrichloroethane (DDT), on birds. Scientific studies followed, which confirmed the harmful effects of chlorinated organic compounds not only on the avian fauna, but also on other animals and aquatic ecosystems. Furthermore, their bioaccumulation in the food chain was shown to lead to potentially harmful effects (including carcinogenic ones) on humans. Since then, these organic compounds have been grouped under the term persistent organic pollutants (POP), because their lifetime in the environment is very long (on the order of several years).

During the 1970s, the destruction of forests in Europe and North America suggested the presence of another type of pollution. Atmospheric deposition of acidic species, such as sulfuric acid (H_2SO_4) and nitric acid (HNO_3), modified significantly the chemical equilibrium of soils and surface waters (e.g., Likens and Bormann, 1974; Likens et al., 1979). As a result, nutrients were mobilized and washed away, whereas some toxic metals became available to the roots, as well as to terrestrial and aquatic flora and fauna. This type of pollution was called "acid rain," because the load of those acidic species was associated in great part with rain events. Actually, dry deposition and snowfall also contribute to atmospheric deposition and the subsequent increase in acidity of soils and surface waters. Therefore, acid deposition is a better, more scientific and precise term than acid rain.

During the 1950s, the contamination of fish by mercury was identified in Minamata, a fishing village on the Island of Kyûshû in Japan, as the possible source of neurologic diseases affecting local people who consumed large amounts of fish (Smith and Smith, 1975). During the 1990s, atmospheric mercury was identified as a major source of contamination of the aquatic fauna, because it can be transformed by bacteria into organic mercury after entering an ecosystem. Next, organic mercury bioaccumulates in the aquatic food chain up to the higher trophic levels, where its concentrations may become harmful to humans who consume fish.

More recently, diesel particles have been listed as carcinogenic to humans by the International Agency for Research on Cancer (IARC) of the World Health Organization (WHO; Benbrahim-Tallaa et al., 2012). Later, IARC listed air pollution, as a whole, as carcinogenic to humans (Loomis et al., 2013).

Some pollutants, such as ozone, not only have harmful health effects, but may also impact vegetation, thereby leading to significant economic loss for agriculture (e.g., Rich, 1964). Furthermore, there are negative aesthetic impacts of air pollution. For example, buildings and statues may be deteriorated by acidic species, and soiled by soot particles (e.g., Kucera and Fitz, 1995). In addition, sunlight is scattered by fine particles and may be absorbed by some particles and gaseous pollutants. As a result, the human eye does not see as far during an air pollution episode (smog or photochemical smog) as under pristine conditions, because the visual range has been degraded by the presence of those fine particles and light-absorbing gases (e.g., Trijonis, 1979).

This brief historical summary highlights the fact that the term air pollution involves a large number of pollutants and adverse health and environmental effects. Furthermore, air pollution occurs at a variety of spatial and temporal scales. The objective of this book is to explain in technically simple, but fundamentally precise, terms the processes that lead to air pollution, to identify the main sources, and to describe its main health and environmental impacts. Possible solutions, both in terms of public policy and available air pollution control technologies, are also presented.

References

Benbrahim-Tallaa, L., R.A. Baan, Y. Grosse, B. Lauby-Secretan, F. El Ghissassi, V. Bouvard, N. Guha, D. Loomis, and K. Straif, on behalf of the International Agency for Research on Cancer Monograph Working Group, 2012. Carcinogenicity of diesel-engine and gasoline-engine exhausts and some nitroarenes, *Lancet Oncology*, **13**, 663–664.

Bell, M.L., D.L. Davis, and T. Fletcher, 2004. A retrospective assessment of mortality from the London smog episode of 1952: The role of influenza and pollution. *Environ. Health Perspect.* **112**, 6–8.

Carson, R., 1962. *Silent Spring*, 380 pp., Houghton Mifflin Company, Boston.

Haagen-Smit, A.J., 1952. Chemistry and physiology of Los Angeles smog, *Ind. Eng. Chem.*, **44**, 1342–1346.

Hardy, K., S. Buckley, M.J. Collins, A. Estalrrich, D. Brothwell, L. Copeland, A. García-Tabernero, S. García-Vargas, M. de la Rasilla, C. Lalueza-Fox, R. Huguet, M. Bastir, D. Santamaria, M. Madella, J. Wilson, Á.F. Cortés, and A. Rosas, 2012. Neanderthal medics? Evidence for food, cooking, and medicinal plants in dental calculus, *Naturwissenschaften*, **99**, 617.

JAMA, 1905. Smog, *J. Amer. Med. Assoc.*, **45**, 637.

Kucera, V. and S. Fitz, 1995. Direct and indirect air pollution effects on materials including cultural monuments, *Water Air Soil Pollut.*, **85**, 153–165.

Likens, G.E. and F.H. Bormann, 1974. Acid rain: A serious regional environmental problem, *Science*, **184**, 1176–1179.

Likens, G.E., R.F. Wright, J.N. Galloway, and T.J. Butler, 1979. Acid rain, *Scientific American*, **241**, 43–47.

Loomis, D., Y. Grosse, B. Lauby-Secretan, F. El Ghissassi, V. Bouvard, L. Benbrahim-Tallaa, N. Guha, R. Baan, H. Mattock, and K. Straif, on behalf of the International Agency for Research on Cancer Monograph Working Group, 2013. The carcinogenicity of outdoor air pollution, *Lancet Oncology*, **14**, 1262–1263.

References

Rich, S., 1964. Ozone damage to plants. *Annual Rev. Phytopathology*, **2**, 253–266.

Smith, W.E. and A.M. Smith, 1975. *Minamata, Words and Photographs*, Holt, Rinehart & Winston, New York.

Trijonis, J., 1979. Visibility in the southwest—An exploration of the historical data base. *Atmos. Environ.*, **13**, 833–843.

Wilkins, E.T., 1954. Air pollution aspects of the London fog of December 1952, *Q. J. R. Meteor. Soc.*, **80**, 267–271.

2 Emissions of Air Pollutants and Emission Control Technologies

Air pollution is due to emissions of pollutants in the atmosphere, which may be natural or of human origin. Thus, in order to understand air pollution, it is necessary to identify, characterize, and quantify those emissions. Furthermore, reducing air pollution requires either eliminating some of those emissions via a change in a product, process, or technology, or reducing those emissions using some control technologies. This chapter describes the main sources of air pollution and the technologies available to control those emissions. First, air pollutant sources are described. Next, the methods used to quantify the corresponding emissions and develop air pollutant emission inventories are presented. Finally, the main technologies used to control emissions of gaseous and particulate air pollutants are described.

2.1 Sources of Air Pollution

First, it is useful to recall the definitions of primary and secondary pollutants. A primary pollutant is a pollutant that is emitted directly in the atmosphere. A secondary pollutant is formed in the atmosphere via chemical reactions among other chemical species, which are called precursors. Some precursors may also be primary pollutants, and a chemical species may be both a primary pollutant and a secondary pollutant. Therefore, to understand air pollution, one must know not only the emissions of primary pollutants, but also those of precursors of secondary pollutants. Generally, precursors of secondary pollutants are considered to be an integral part of air pollution and are called air pollutants. It is the case, for example, in the United States (CFR, 2016) and in France (Code de l'Environnement, 2016). Therefore, this text will include both primary pollutants and precursors of secondary pollutants as air pollutant emissions.

Air pollutants may be emitted from anthropogenic sources (i.e., those sources related to human activities) and/or from natural sources. Examples of anthropogenic sources include transportation (on-road, rail, air, maritime, etc.), industry (fossil-fuel fired power plants, smelters, incinerators, refineries, etc.), agriculture (cattle, fertilizer use, etc.), and the residential, commercial, and institutional sector (heating, cleaning products, etc.). Examples of natural sources include emissions of volatile organic compounds (VOC) from vegetation and nitrogen compounds from soils, dust emissions due to wind erosion, ocean emissions, volcanic eruptions, geothermal sources, lightning (production of nitrogen oxides), and forest fires (however, those may also be due to human activities).

Four major categories of processes lead to air pollutant emissions: combustion, volatilization, mechanical processes (abrasion, resuspension, etc.), and natural processes that do not belong to one of the previous categories.

2.1.1 Combustion

Combustion may be the result of an anthropogenic activity (e.g., transportation, production of electricity, incineration of waste, heating) or a natural process (forest wildfires). Combustion leads to the production of heat, which can then be converted if needed into another form of energy (e.g., electrical, mechanical). The combustion process implies the presence of oxygen, which is available from the air, and carbon, which is the main component of fuels, such as coal, gasoline, diesel, and wood. This combustion occurs at high temperatures and leads to (1) the dissociation of oxygen (O_2) and nitrogen (N_2) molecules, both of which are present in the air, and (2) the oxidation of carbon. The dissociation of the O_2 and N_2 molecules leads to oxygen (O) and nitrogen (N) atoms, respectively. Then, the reactions among oxygen (O_2, O) and nitrogen (N_2, N) lead to the formation of nitrogen oxides (NO_x), mostly nitric oxide (NO), but also a fraction (<10 %) of nitrogen dioxide (NO_2). The complete oxidation of the fuel leads to carbon dioxide (CO_2), a greenhouse gas, and to water vapor (H_2O). However, combustion is generally not complete and carbonaceous compounds that are not completely oxidized are produced during combustion. Such compounds include, for example, carbon monoxide (CO), volatile organic compounds (VOC), soot particles (originating mostly from diesel engines and biomass fires), polycyclic aromatic hydrocarbons (PAH), and dioxins and furans. Some of those compounds are pollutants and some may even be carcinogenic (e.g., formaldehyde, soot particles from diesel engines, some PAH, dioxins, and furans). In addition, inorganic substances present in the fuel are released during combustion, often in their oxidized form due to chemical reactions occurring at high temperatures. Among substances present in coal, gasoline, and diesel, one may mention sulfur and mercury.

2.1.2 Volatilization

The volatilization of semi-volatile compounds consists in their transfer from a liquid phase to a gas phase, which may then be dispersed in the atmosphere. Volatilization affects, for example, hydrocarbons (e.g., oil, gasoline) during their storage and transfer and paints and solvents during their use. It also affects fuels contained in vehicles, and this volatilization process can contribute to a significant fraction of VOC emissions from vehicles when the ambient temperature is high or even moderate. Volatilization varies depending on the nature of the fuel, because it is a function of the physico-chemical properties of the hydrocarbons present in the fuel. Gasoline includes linear and branched alkanes (20 to 30 %), cycloalkanes (~5 %), alkenes (30 to 45 %), and aromatic compounds (30 to 45 %), as well as additives (such as ethanol). Laboratory chemical analyses lead to an average molecular formula for gasoline that is close to that of heptane (C_7H_{16}). Diesel includes mostly alkanes (linear, branched, and cyclic; >75 %) and aromatic compounds (<25 %). A theoretical average molecular formula of $C_{16}H_{29}$ may be used as representative of the ensemble of hydrocarbons present in diesel. (Note that although octane, C_8H_{18}, and cetane, $C_{16}H_{34}$, indices are used to characterize gasoline and diesel, respectively, these hydrocarbons represent only a small fraction of all the hydrocarbons present in these fuels and do not correspond to the average formula of these fuels.) Therefore, diesel is a fuel that includes VOC that are heavier, and therefore less volatile, than those of gasoline. As a result, the volatilization of VOC from vehicles pertains mostly to gasoline vehicles. Mercury may be emitted naturally from soils and oceans as elemental mercury, a form that is very volatile. In addition, reemission of semi-volatile pollutants (for example, some persistent organic pollutants, POP, such as PAH and pesticides) is a volatilization process.

2.1.3 Mechanical Processes

Among the mechanical processes leading to atmospheric emissions, one may mention anthropogenic activities such as construction activities, farming, and some industrial activities, as well as natural activities such as the emission of wind-blown dust and sea salt (aeolian emissions). Also, transportation is an important source of particles via processes such as braking (abrasion of the brake pads), driving (wear of tires, roads, metal wheels, railways, etc.), and the resuspension by traffic of particles present on roads.

2.1.4 Natural Processes

Natural processes other than the ones already mentioned include, for example, the metabolism of vegetation, which leads to the atmospheric emissions of VOC, and emissions associated with volcanic eruptions.

2.1.5 Summary of Global Emissions of Air Pollutants

Table 2.1 summarizes the global emissions of several major air pollutants: sulfur dioxide (SO_2), nitrogen oxides (NO_x), ammonia (NH_3), carbon monoxide (CO), volatile organic compounds (VOC), and particles. Particles are represented here by PM_{10} (particles with an aerodynamic diameter less than 10 µm) and $PM_{2.5}$ (particles with an aerodynamic diameter less than 2.5 µm, i.e., fine particles). Anthropogenic emissions were obtained from the Emission Database for Global Atmospheric Research (EDGAR) developed by the European Commission (EC, 2016) and are for the year 2010. Biomass fires may be from anthropogenic or natural origin and, accordingly, they are listed separately (Andreae and Marlet, 2001).

Natural sources vary depending on the pollutants. Natural emissions of SO_2 result mostly from volcanic activities (degasing and eruptions) and they vary significantly from one year to the next depending on eruptions from active volcanoes (Bates et al., 1992). Natural emissions of NO_x result from lightning and from soils, in similar proportions (Logan, 1983). Natural emission sources of ammonia include mostly the oceans, soils associated with natural vegetation, and the human population (Bouwman et al., 1997). Natural emissions of CO originate from vegetation and from the oceans (Khalil and Rasmussen, 1990). Natural emissions of VOC result from vegetation. They include isoprene for about

Table 2.1. Annual global emissions of selected major air pollutants (Tg/year). Data sources: EC (2016); Andreae and Marlet (2001); Bates et al. (1992); Bouwman et al. (1997); Gong et al. (2002); Guenther et al. (2012); Khalil and Rasmussen (1990); Logan (1983). (1 Tg = 10^{12} g.)

Sources	SO_2	NO_x[a]	NH_3	CO	VOC[b]	PM_{10}	$PM_{2.5}$
Electric power and heat production	49	31	0.1	6.5	0.7	4.9	3
Oil refineries	1.7	0.9	0	0.5	2.1	<0.1	<0.1
Other industrial sources	31	19.2	1.7	137.9	63.4	16.3	10.1
Waste and wastewater treatment	0.1	0.2	0.1	0.1	2.6	0.2	0.1
Residential, commercial, and institutional sector	8.2	6.2	5.1	232.5	36	30.2	18.2
Agriculture	0.4	6	47	75	4.4	10	8.3
On-road transportation	0.8	27	0.5	170	25.4	0.9	0.9
Aviation	0.3	2.9	0	0.5	0.1	<0.1	<0.1
Maritime shipping	11	17	0	5.3	1.2	1.9	1.8
Other modes of transportation	0.1	1.9	0	0.6	0.2	0.3	0.3
Sub-total of anthropogenic sources	103	112	54	629	136	65	43
Biomass fires[c]	3	25	6	413	251	49	36
Natural sources	19	53	16	140	1000	1690	460
Total	**125**	**190**	**76**	**1182**	**1387**	**1804**	**539**

(a) NO_x emissions are expressed as NO_2.
(b) VOC except methane, which is not chemically very reactive
(c) Except residential heating; biomass fires are mostly from anthropogenic activities, but there is also a natural contribution.

50 % (90 % from deciduous trees), monoterpenes for about 15 % (>80 % from deciduous trees), sesquiterpenes for about 3 % (from deciduous trees and evergreens), 2-methyl-3-buten-2-ol (MBO) for about 0.2 % (from evergreens), and other VOC, such as alcohols, aldehydes, ketones, alkenes, and carboxylic acids for the remainder (Guenther et al., 2012). Natural emissions of particles include aeolian soil erosion (mostly from deserts; Zender et al., 2003) and sea-salt emissions (Gong et al., 2002). Regarding sea salt, only fine particles are included here, because coarse particles have a lifetime of only a few hours and, therefore, do not have any long-range impacts. Volcanic eruptions are also a source of particles; this source is not included here because of its large interannual variability.

These global inventories do not include atmospheric chemical reactions. Some chemical reactions may be an important source for some of those pollutants, as described in Chapters 8, 9, and 10. For example, dimethyl sulfide (DMS) and hydrogen sulfide (H_2S) are oxidized into SO_2, VOC are oxidized and eventually form CO, and an important fraction of fine particles ($PM_{2.5}$) is formed in the atmosphere via chemical reactions that involve SO_2, NO_2, VOC, and NH_3.

Note that natural sources dominate the global inventory of VOC and particulate matter (PM_{10} and $PM_{2.5}$). However, in the case of VOC, chemical speciation is essential, because VOC differ significantly in terms of chemical reactivity and toxicity. In the case of particles, they are regulated in terms of mass rather their chemical composition; nevertheless, their health impacts could depend on their chemical composition. Furthermore, natural sources are distributed widely over the globe, whereas anthropogenic sources are generally concentrated in or near areas where people live.

These global emissions have been evaluated for some pollutants using satellite data and inverse modeling. Such evaluations have been performed, for example, for SO_2 emissions (Lee et al., 2011) and CO emissions (Kopacz et al., 2010). In both cases, the bottom-up global emission inventories were found to be consistent with the top-down emission estimates obtained from satellite data.

2.2 Emission Inventories

Emission inventories are needed to track the temporal evolution of air pollutant emissions. For example, countries in Europe and states in the United States must report their emissions on a regular basis to the European Union and the federal government, respectively. Greenhouse gas emissions may also be included in such emission reporting. In addition, numerical modeling of air pollution, which is conducted for air quality impact assessments, emission scenario simulations, and air quality forecasting, requires spatially distributed and temporally resolved emission inventories. Methods that are used to develop emission inventories are briefly described in this section.

The fundamental equation for the quantification of most air pollutant emissions is as follows:

$$S_{ij} = EF_{ij} \times A_j \tag{2.1}$$

where S_{ij} is the rate of emission of air pollutant i from source j (in g s^{-1}), EF_{ij} is the emission factor for air pollutant i emitted from a source category corresponding to source j (in g per activity unit), and A_j is the activity of source j (in activity units per second).

The activity of a source is defined in different ways depending on the source type. For example, it may be defined in terms of vehicle km per hour for on-road traffic, energy production per unit time (for example, MW) for power plants, and the amount of fuel used per year for residential heating.

Emission factors are expressed in units that are consistent with the unit of the corresponding activity. They may be obtained in several ways. For some sources, emission measurements may be performed at the source. The emission factor obtained for a specific source may then be used more generally for the source category (i.e., for other similar sources). For example, vehicle (or engine) emission measurements are performed on a dynamometer to obtain emission factors for on-road traffic (see Section 2.3.4). For some pollutants, a mass balance may be performed on the emission process. For example, in the case of sulfur present in a fuel (e.g., coal, gasoline, diesel), the sulfur content of the fuel can be used to estimate the emission of sulfur compounds (mostly sulfur dioxide and sulfuric acid), since the sulfur mass is conserved during the combustion process. In a few cases, a simulation of the process may be performed to obtain the chemical speciation of some pollutants (for example, the relative fractions of elemental and oxidized mercury emitted from coal-fired power plants).

For some emissions, the process may be more complex (it may depend, for example, on meteorology) and a parameterization must then be used. This is the case for biogenic VOC emissions from vegetation, which depend on ambient temperature and solar radiation, for VOC volatilization from gasoline vehicles, which depends on ambient temperature, and for wind-blown dust emissions from desert areas and sea-salt emissions from oceans, which depend on wind speed. Models have been developed to estimate those emissions as a function of meteorology. Some models used to estimate VOC emissions from vegetation are mentioned at the end of this chapter. Some models used to estimate aeolian emissions are presented in Chapter 11.

The development of an emission inventory typically requires some method to organize the various source categories. Examples are provided here for the United States and France.

In the United States, the National Emissions Inventory (NEI) is developed by the U.S. Environmental Protection Agency (EPA) from data reported by the states. The NEI uses several types of codes to classify source categories, industrial facilities, geographical regions, pollutants, and emission control equipment. Sources are classified according to Source Classification Codes (SCC). An SCC is specific to an item of equipment, an operation, or a practice that is a source of air pollutants. These codes include eight digits for large point sources (such as power plant stacks) and ten digits for other sources. The North American Industry Classification System (NAICS) is used to identify the primary activity of an industrial facility. The Federal Information Processing Standards (FIPS), state and county codes, and tribal codes are used to identify the state, county, territory, or tribe area where the source is located. Seven-digit numerical codes are used to identify specific pollutants. Finally, three-digit codes characterize the type of emission control equipment used on a specific source. Other codes are used to identify the emission calculation method and the type of reporting period (e.g., seasonal, annual). The use of such codes facilitates the retrieval of specific information on the method and data associated with the development of the emission inventory. More information on the U.S. EPA emission inventory system is available at www.epa.gov/air-emissions-inventories.

In France, CITEPA ("*Centre interprofessionnel technique d'études de la pollution atmosphérique*") is the organization responsible for the development of the national emission inventories for the French ministry in charge of the environment. CITEPA uses the SNAP 97 c (Selected nomenclature for air pollution 1997, corrected version) classification for source activities and NAPFUE 94 c (Nomenclature for air pollution and fuels 1994, corrected version) for fuels. SECTEN ("*Secteurs économiques et énergie*") and SNAP 97 c are generally used in France for the emission inventory output formats. Other formats, such as NFR (Nomenclature for reporting) and CRF (Common reporting format) are occasionally used for international reporting in Europe. Tables have been developed to convert emission inventories from one format to another (e.g., www.citepa.org).

In terms of a geographical coordinate system, several options are available, depending on the need of the user. For example, the Lambert, UTM (Universal Transverse Mercator), and latitude-longitude systems are widely used.

Once the emission rates have been calculated for all the identified sources of air pollution, these emission rates must be distributed spatially and temporally, if they are to be used in a numerical modeling study.

The spatial distribution is performed differently depending on the source type. Typically, sources are grouped as point sources (e.g., large stacks), area sources (lumping sources that are too small to be treated individually, such as residential heating), line sources (representing, for example, major roadways), and volume sources (used, for example, to represent industrial fugitive sources). In a standard air quality simulation model (see Chapter 6 for a discussion of different types of air quality models), emissions are only represented by means of point sources and area sources. However, these emissions are released in three-dimensional grid cells, and the corresponding sources are, therefore, equivalent to volume sources. In an air quality model that provides a multi-scale treatment of air pollution, it is possible to treat emission sources with greater detail, using point sources (e.g., for tall stacks), line sources (e.g., for major roadways), and volume sources (e.g., for fugitive emissions at industrial sites). Similarly, atmospheric dispersion models may treat individual sources of various types (see Chapter 6) and the four categories of sources may then be used.

The locations of point sources are identified exactly. Area sources represent a large amount of small sources, which cannot be identified individually exactly. Therefore, one must use a surrogate variable to distribute spatially the emissions of that source category. For example, population density may be used to treat residential heating so that the emissions can be distributed spatially over a city, a district, or a region. Line and volume sources correspond generally to specific sources (roadways, industrial sites), which can be localized precisely.

The temporal resolution of emissions is generally hourly for air quality simulations. In some cases, emissions are available with some temporal resolution. This is the case, for example, in Europe and North America for some industrial sources (e.g., power plants) that are required to monitor their emissions for some regulated pollutants (e.g., NO_x and SO_2). However, in most cases, no specific information is available. Then, one must use temporal distribution factors obtained from other databases. These factors may include different temporal scales. For example, in the case of on-road traffic, temporal distribution factors may include daily, weekly, monthly, and seasonal distribution factors. Daily factors may also vary depending on the day of the week.

Air pollutant emissions include generally gases, particulate matter, and greenhouse gases. For some gaseous pollutants, it is necessary to obtain a chemical speciation: this is the case for nitrogen oxides, which must be categorized as nitric oxide and nitrogen dioxide, and for VOC, which must be distributed among a large number of specific organic molecules. For particulate matter, the chemical composition is needed (black carbon, organic matter, sulfate, etc.). In addition, the particle size distribution is essential because (1) the regulations pertain to specific particle size ranges ($PM_{2.5}$ and PM_{10}, see Chapters 9, 12, and 15) and (2) the dynamics of particles in the atmosphere depends on their size.

Emission factors and methods available to calculate emission rates for major source categories may be obtained from the following list of selected organizations:

– In the United States: AP-42, Compilation of air pollution emission factors (www.epa.gov)
– In Europe: EMEP/EEA air pollutant emission inventory guidebook (www.eea.europa.eu/publications)

– In France: *"Organisation et méthodes des inventaires nationaux des émissions atmosphériques en France"* (OMINEA), available from CITEPA (www.citepa.org)

While air pollutant emission inventories are generally developed by international, national, or regional organizations, there are some countries and regions for which no emission inventory is available. In such cases, it is possible to develop an emission inventory using information on source activities available locally and emission factors available in the references provided above. The development of an emission inventory for Lebanon, and more specifically for its capital, Beirut, exemplifies such an approach (Waked et al., 2012).

In addition, one should note that some emissions are particularly difficult to estimate accurately. For example, biomass fires are highly variable from one year to the next. The use of satellite data for burning and burned areas may help develop emission inventories pertaining to biomass fires (e.g., Mieville et al., 2010).

2.3 Emission Control Technologies

2.3.1 Gaseous Pollutants

The technologies available to control gaseous pollutant emissions may be summarized according to the following major categories (Flagan and Seinfeld, 1988; Wang et al., 2005):

– Absorption in a liquid
– Adsorption on a solid
– Chemical transformation
– Incineration

Absorption in a Liquid

This approach may be used to reduce emissions of air pollutants that are very soluble, generally in water, for example hydrogen chloride (HCl) and hydrogen fluoride (HF). Dissolution in water occurs according to Henry's law, and the efficiency of this emission control process depends on the solubility of the pollutant in water. In some cases, it is possible to increase this efficiency by displacing the gas/water equilibrium toward the aqueous phase. For example, the gas/water equilibrium of sulfur dioxide (SO_2) may be displaced toward the liquid phase by using an alkaline solution (i.e., a basic solution with a pH greater than 7), because SO_2 is a weak acid and its dissolution increases with pH (see Chapter 10). The efficiency of the absorption process may also be increased by adding a chemical transformation to displace the transfer of the substance toward the liquid phase (see the section on chemical transformation).

Adsorption on a Solid

This approach is based on the formation of a bond between a gas molecule and the solid surface. This phenomenon includes various processes that may be categorized as follows:

- Adsorption on a non-polar solid, such as activated carbon, which is a carbonaceous substance with high porosity, thereby allowing a large surface area for interaction with the gas phase. This method is used, for example, to reduce the emissions of persistent organic pollutants (POP) and mercury from incinerators.
- Adsorption on a polar solid (alumina, silica ...); however, this type of solid will also adsorb water, which can significantly reduce the efficiency of the adsorption process when the emission effluent contains large amounts of water.
- Chemisorption, which corresponds to a chemical reaction of the adsorbed substance with the solid and which may lead to desorption of the (potentially less harmful) reaction product. Heterogeneous catalytic reactions (i.e., chemical reactions taking place on a solid catalyst) may be included in this category. For example, catalytic converters, which use a catalyst to convert carbon monoxide (CO) into carbon dioxide (CO_2) and unburned hydrocarbons into CO_2 (two-way catalytic converters), may be included in this category. Three-way catalytic converters also convert a fraction of nitrogen oxide emissions (NO_x) into molecular nitrogen (N_2) (see the description of emission control systems for on-road vehicles in Section 2.3.4).

In all these cases, the solid that becomes laden with the adsorbed pollutant must be disposed of in a safe and environmentally sound manner.

Chemical Transformation

Chemical transformation may be used to form a pollutant that is more easily controlled or to form a product that is not a pollutant (or at least a pollutant that is less harmful than the original pollutant). Two examples may be mentioned: the control of sulfur dioxide (SO_2) and nitrogen oxides (NO_x) emissions from coal-fired power plants. These emission control technologies have been implemented, for example, to reduce acid deposition in the United States (reduction of emissions of SO_2 and NO_x, which are precursors of sulfuric and nitric acids, respectively, see Chapters 10 and 13), as well as ozone levels (NO_x are precursors of ozone, see Chapter 8).

To reduce SO_2 emissions, SO_2 may be transformed into sulfate by reaction with calcium carbonate after absorption in a scrubber (i.e., partial dissolution in an aqueous phase):

$$SO_2 + CaCO_3 + 0.5\ H_2O \rightarrow CaSO_3 \cdot 0.5\ H_2O + CO_2 \quad (R2.1)$$

$$SO_2 + CaCO_3 + 0.5\ O_2 + 2\ H_2O \rightarrow CaSO_4 \cdot 2\ H_2O + CO_2 \quad (R2.2)$$

Calcium sulfite may be oxidized to sulfate:

$$CaSO_3 \cdot 0.5\ H_2O + 0.5\ O_2 + 1.5\ H_2O \rightarrow CaSO_4 \cdot 2\ H_2O \quad (R2.3)$$

The oxidation of SO_2 by calcium carbonate, in the form of limestone, leads to the formation of calcium sulfate ($CaSO_4$), also called gypsum, which precipitates as a solid. Gypsum must be removed from the scrubber to avoid its clogging. Typically, gypsum precipitation is minimized by maintaining the pH above 6. Gypsum may be sold as building material, if it does not contain too many toxic substances (metals …). The oxidation of SO_2 may alternatively be performed with calcium oxide, also called quicklime:

$$SO_2 + CaO + 0.5\, H_2O \rightarrow CaSO_3 \cdot 0.5\, H_2O \qquad (R2.4)$$

$$SO_2 + CaO + 0.5\, O_2 + 2\, H_2O \rightarrow CaSO_4 \cdot 2\, H_2O \qquad (R2.5)$$

The oxidation of SO_2 by quicklime is more efficient than that by calcium carbonate, but quicklime is more expensive. These reactions may take place in a scrubber where droplets are sprayed into the effluent that contains SO_2. Calcium sulfate particles are formed following the evaporation of the droplets, and those particles must be captured by filtration. These emission control systems for SO_2 are typically called flue gas desulfurization systems (FGD). The efficiency of an FGD is on the order of 75 to 95 % for SO_2 emission control. Dry FGD processes are also available; however, they are not as widely used as wet FGD (see Srivastava and Jozewicz, 2001, for a review of FGD technologies).

To reduce NO_x emissions, NO_x may be reduced to molecular nitrogen (N_2), which is the main constituent of the atmosphere, by using ammonia (NH_3). The corresponding chemical reactions are the following:

$$NO + NH_3 + 0.25\, O_2 \rightarrow N_2 + 1.5\, H_2O \qquad (R2.6)$$

$$NO + NO_2 + 2\, NH_3 \rightarrow 2\, N_2 + 3\, H_2O \qquad (R2.7)$$

Urea, $(NH_2)_2C=O$, may be used instead of ammonia. These reactions may occur with or without a catalyst, and one refers to selective catalytic reduction (SCR) and to selective non-catalytic reduction (SNCR). SNCR is less efficient (30 to 80 % efficient) than SCR (~90 %). The latter is of course more expensive due to the use of a catalyst. SNCR occurs in the gas phase (homogeneous reaction), whereas SCR occurs at the surface of the catalyst (heterogeneous reaction). The catalyst may be a titanium oxide (TiO_2) or a vanadium oxide (V_2O_5). The term "selective" characterizes the fact that ammonia reacts preferentially with NO or NO_2 rather than with oxygen (O_2) at temperatures between 1,200 and 1,300 K, because at $T > 1,370$ K, the following reaction tends to prevail:

$$4\, NH_3 + 5\, O_2 \rightarrow 4\, NO + 6\, H_2O \qquad (R2.8)$$

Srivastava et al. (2005) provide a review of such technologies for industrial NO_x emission control.

Incineration

Incineration is used to eliminate solid waste. This approach may also be used in a targeted manner to treat some air pollutants in some emission control systems (see Section 2.3.4 for the control of on-road vehicle emissions). By raising the temperature of the effluent significantly, it may be possible to oxidize soot particles and transform them into gaseous carbon dioxide (CO_2). In addition, high temperatures are conducive to the oxidation of some gases, such as volatile organic compounds, which are then more easily converted into CO_2.

2.3.2 Particulate Pollutants

The main technologies available to control industrial emissions of particulate pollutants are the following (Flagan and Seinfeld, 1988):

- Capture via aerodynamic inertia
- Filtration
- Electrostatic capture

The major physical processes that govern the deposition of particles on obstacles in a flowing fluid are the following:

- Sedimentation
- Impact by inertia or interception
- Brownian diffusion
- Migration in an electrical field

Some of these processes occur in the atmosphere. The first process is important for atmospheric dry deposition. The second and third processes are important for atmospheric deposition (dry and wet), as well as for deposition in the respiratory tract during inhalation. Therefore, the mathematical formulation of these processes for atmospheric deposition or deposition during inhalation applies to particulate emission control as well (see Chapter 11 for the mathematical formulation of these processes and the efficiency of these processes in the case of atmospheric deposition; see Chapter 12 for deposition of particles during inhalation). The physical processes involved in the major types of particulate emission control systems are described next.

Sedimentation Chamber

This process is efficient for particles with a size sufficiently large that their sedimentation velocity is significant, i.e., particles with a diameter of several tens of microns. The sedimentation velocity results from equilibrium between the gravitational force and the frictional force of the particle in the fluid (see Chapter 11). For particles with a diameter less than about 10 µm, the frictional force leads to a sedimentation velocity that is very low (<1 cm s^{-1}).

Cyclones

This type of device uses mostly the inertia of particles in the flow to capture them either on the wall of the device or on droplets sprayed into the device. Particle inertia is proportional to their mass, i.e., for a given density, to their volume (albeit with a correction factor that depends on their size, see Chapter 11). Therefore, large particles have more inertia than fine particles.

Filtration

Filtration captures particles via the same processes that govern atmospheric deposition or deposition within the respiratory tract during inhalation. Thus, particles will deposit on a filter via inertia, interception, and brownian diffusion. Brownian diffusion (i.e., the random motions of particles due to thermal agitation) becomes more efficient as the particle diameter decreases (see Chapter 11); therefore, this process is most efficient for ultrafine particles. Impact by inertia is roughly proportional to the particle volume and density. Impact by interception is roughly proportional to the cross-section area of the particle (albeit with a correction factor that depends on particle size, see Chapter 11). Therefore, the efficiency of these latter two processes increases with the diameter of the particle. As a result, particles of diameter between about 0.1 and 1 µm are the least subject to these processes, since they are too large to be significantly affected by brownian diffusion and they are too small to be significantly affected by inertia and interception. These particles are also those that are not efficiently removed from the atmosphere via wet or dry deposition and, therefore, have long atmospheric residence times. Filtration devices used at industrial sites are typically called baghouses or fabric filters.

Electrostatic Precipitators

In an electrostatic precipitator (ESP), an electrical discharge is sent into the effluent to create electrostatic charges on the particles. The device walls are electrically charged and, therefore, particles migrate toward those walls, which act as electrodes. At regular time intervals, the walls are shaken so that the particles may fall (i.e., precipitate) to the bottom of the device, where they can be collected. The electrostatic charge of a particle depends on its size: large particles (>1 µm about) are charged mostly via direct capture of gaseous ions, whereas fine particles (<1 µm about) are charged mostly by diffusion of gaseous ions toward those particles. The efficiency of an ESP is greater for ultrafine particles (<0.1 µm) and coarse particles. Therefore, fine particles (those with diameters in a range of about 0.1 to 1 µm) are the least efficiently captured by an ESP. ESP and baghouses are the most commonly used devices for controlling particulate emissions from coal-fired power plants.

Scrubbers

Scrubbers may be used to capture particles via their interactions with droplets. There exist several types of scrubbers (droplet sprays, venturi scrubbers, etc.). In a venturi scrubber, an acceleration of the effluent is created, which leads to an increase in the collision efficiency between particles and droplets. The inertia, interception, and brownian diffusion processes govern the efficiency of the capture of particles by the droplets (similarly to the wet scavenging of atmospheric particles by raindrops). The large particles are captured with the greatest efficiency (via inertia and interception). Ultrafine particles are captured via brownian diffusion. Thus, fine particles with diameters that are roughly between 0.1 and 1 µm are those that are the least efficiently captured by scrubbers.

Efficiency of Major Emission Control Devices for Particles

The efficiencies of different emission control technologies for particles are summarized in Table 2.2.

Table 2.2. Efficiency of particulate emission control technologies. Source: Flagan and Seinfeld (1988).

Control technology	Minimum diameter of the main particulate fraction captured (by mass)	Efficiency (percentage of particles of larger diameter captured)
Sedimentation chamber	50 μm	<50 %
Cyclone	5 to 25 μm	50 to 90 %
Venturi scrubber	0.5 μm	<99 %
Electrostatic precipitator	1 μm	95 to 99 %
Baghouse	1 μm	>99 %

2.3.3 Control of Mercury Emissions from Coal-fired Power Plants

Mercury (Hg) is emitted from coal-fired power plants in the form of elemental mercury and oxidized (divalent) mercury. The use of FGD and SCR (or SNCR) devices to reduce emissions of SO_2 and NO_x, respectively, favors the reduction of mercury emissions. The SCR equipment tends to oxidize elemental mercury into its divalent form, which is very soluble in water. The FGD equipment located downstream may then capture a large fraction of the oxidized mercury present in the effluent. If it is needed to further reduce the mercury emissions, then one must use a device with mercury adsorption on activated carbon. The injection of activated carbon into the effluent stream leads to the capture of gaseous mercury (elemental and oxidized). The mercury-laden carbon particles must then be captured, to avoid their release to the atmosphere. This capture may be performed downstream with an ESP or a baghouse. However, the captured particles will then need to be sent to a waste disposal site, because they will be contaminated by mercury. The other option consists of injecting the activated carbon downstream of the ESP or baghouse and capturing the contaminated carbon particles separately. Thus, the particles captured upstream may be sold for use, for example, as building/filling material.

2.3.4 Control of Air Pollutant Emissions from On-road Traffic

On-road traffic is an important source of gaseous and particulate air pollutants in urban areas and on major highways. The main air pollutants emitted by on-road traffic include carbon monoxide (CO), nitrogen oxides (i.e., nitric oxide, NO, and nitrogen dioxide, NO_2, grouped as NO_x), volatile and semi-volatile organic compounds (VOC and SVOC, respectively), sulfur dioxide (SO_2), and particulate matter (PM). PM may contain black carbon, organic compounds, inorganic compounds, and toxic metals. Atmospheric concentrations of CO, NO_2, SO_2, PM, and lead (Pb, a toxic metal) are regulated in the United States, in Europe, and in many other countries. Atmospheric concentrations of benzene (a carcinogenic VOC) are regulated in Europe. In addition, NO_x, VOC, and CO are precursors of ozone, a gaseous secondary pollutant, which is regulated in the United States, in Europe, and in many other countries, and NO_x, SO_2, VOC, and SVOC are precursors of the secondary fraction of PM. Furthermore, gaseous and particulate toxic compounds may be regulated via health risk assessments (e.g., in the United States). Therefore, it is essential to control on-road traffic emissions and regulations have been introduced, for example, in the United States and in Europe to reduce the emissions of some of those air pollutants.

Regulatory Standards

In the United States, regulations have historically been more stringent in California than at the federal level, because of the large air pollution problems identified in the Los Angeles basin and the Central Valley of California (in particular, the San Joaquin Valley). In addition, the following states have adopted California emission standards at various times (the corresponding vehicle model year is indicated in parentheses): Maine, Massachusetts, New York, and Vermont (2004), Connecticut, Pennsylvania, and Rhode Island (2008), New Jersey, Oregon, and Washington (2009), Maryland (2011), Delaware (2014), and New Mexico (2016). However, there has been some harmonization over the past recent years between the federal and California emission standards and the 2018 Tier 3 federal standards are very similar to the LEV III California standards. The emission standards are based on fleet-average values with objectives set for future years. Table 2.3 summarizes some of the California and U.S. mobile source exhaust emission standards for carbon monoxide (CO), nitrogen oxides (NO_x), non-methane organic gases (NMOG, equivalent here to non-methane VOC), formaldehyde (HCHO), and particulate matter (PM). NO_x and NMOG were regulated separately prior to 2014 for the federal standards and prior to 2012 for California, but a common emission standard was introduced in the Tier 3 and LEV III emission standard programs, respectively.

Table 2.3. U.S. federal and California emission standards for light-duty, medium-duty, and heavy-duty vehicles. LEV: low emission vehicle; ULEV: ultra-low emission vehicle; SULEV: super ultra-low emission vehicle; ZEV: zero emission vehicle. NMOG and NMHC correspond to non-methane organic gases and non-methane hydrocarbons, respectively.

(a) Light-duty vehicles (passenger cars and light-duty trucks, federal and California). Emissions are shown in mg mi^{-1} and in parentheses in mg km^{-1}. In California, fleet-average emission standards for (NO_x + NMOG) decrease from 86 mg mi^{-1} (54 mg km^{-1}) in 2017 to 30 mg mi^{-1} (19 mg km^{-1}) in 2025.

Category	LEV160	ULEV125	ULEV70	ULEV50	SULEV30	SULEV20	ZEV
CO	4,200 (2,625)	2,100 (1,312)	1,700 (1,062)	1,700 (1,062)	1,000 (625)	1,000 (625)	0 (0)
NO_x + NMOG	160 (100)	125 (78)	70 (44)	50 (31)	30 (19)	20 (12.5)	0 (0)
HCHO	4 (2.5)	4 (2.5)	4 (2.5)	4 (2.5)	4 (2.5)	4 (2.5)	0 (0)
PM*	3 (1.9)	3 (1.9)	3 (1.9)	3 (1.9)	3 (1.9)	3 (1.9)	0 (0)

* 10 mg mi^{-1} (6 mg km^{-1}) in California; however, the PM fleet-average emission standards are actually lower and must decrease from 3 mg mi^{-1} (1.9 mg km^{-1}) in 2021 to 1 mg mi^{-1} (0.6 mg km^{-1}) in 2028.

Table 2.3. (cont.)

(b) Medium-duty vehicles of gross vehicle weight rating (GVWR) between 8,501 and 10,000 lbs (California). Emissions are shown in mg mi^{-1} and in parentheses in mg km^{-1}.

Category	LEV395	ULEV340	ULEV250	ULEV200	SULEV170	SULEV150	ZEV
CO	6,400 (4,000)	6,400 (4,000)	6,400 (4,000)	4,200 (2,625)	4,200 (2,625)	3,200 (2,000)	0
NO$_x$ + NMOG	395 (247)	340 (212)	250 (156)	200 (125)	170 (106)	150 (94)	0
HCHO	6 (4)	6 (4)	6 (4)	6 (4)	6 (4)	6 (4)	0
PM*	120 (75)	60 (37.5)	60 (37.5)	60 (37.5)	60 (37.5)	60 (37.5)	0

* PM fleet-average emission standards are actually lower and must decrease to 8 mg mi^{-1} (5 mg km^{-1}) in 2021.

(c) Medium-duty vehicles of GVWR between 10,000 and 14,000 lbs (California). Emissions are shown in mg mi^{-1} and in parentheses in mg km^{-1}.

Category	LEV630	ULEV570	ULEV400	ULEV270	SULEV230	SULEV200	ZEV
CO	7,300 (4,562)	7,300 (4,562)	7,300 (4,562)	4,200 (2,625)	4,200 (2,625)	3,700 (2,312)	0
NO$_x$ + NMOG	630 (394)	570 (356)	400 (250)	270 (169)	230 (144)	200 (125)	0
HCHO	6 (4)	6 (4)	6 (4)	6 (4)	6 (4)	6 (4)	0
PM*	120 (75)	60 (37.5)	60 (37.5)	60 (37.5)	60 (37.5)	60 (37.5)	0

* PM fleet-average emission standards are actually lower and must decrease to 10 mg mi^{-1} (6 mg km^{-1}) in 2021.

(d) Heavy-duty vehicles of GVWR greater than 8,500 lbs (federal) or 14,000 lbs (California) for selected model years. Emissions are shown in g bhp-h^{-1} and in parentheses in g kW-h^{-1}.

Model year	1974	1985	1988	1990	1993	1998	2010
CO	40 (54)	15.5 (20.8)	15.5 (20.8)	15.5 (20.8)	15.5 (20.8)	15.5 (20.8)	15.5 (20.8)
NO$_x$	16 (21)*	10.7 (14.3)	10.7 (14.3)$^£$	6 (8)	5 (6.7)	4 (5.4)	0.2 (0.3)
HC		1.3 (1.7)	1.3 (1.7)	1.3 (1.7)	1.3 (1.7)	1.3 (1.7)	0.14[#] (0.19)
PM	–	–	0.6 (0.8)	0.6 (0.8)	0.25 (0.33)$^$$	0.1 (0.13)$^$$	0.01 (0.013)

*NO$_x$ + HC; $^£$ 6 (8) for California; [#] NMHC; $^$$ 0.1 (0.13) in 1993 and 0.05 (0.067) in 1998 for urban buses.

The emission standards are presented in both U.S. units (mg mi^{-1} or g (bhp h)$^{-1}$) and SI units (mg km^{-1} or g (W h)$^{-1}$) to facilitate the comparison with the European standards presented later in this section. The U.S. federal and California approaches offer some flexibility to car manufacturers since they may produce cars of various emission standard categories (low, ultra-low, super ultra-low, and zero emissions) as long as their sales-weighted car fleet meets the car fleet emission standard. These car-fleet average emission standards become more stringent with time, which requires car manufacturers to introduce vehicles with lower emissions or even zero emissions. For example, in 2018, the sales-weighted car fleet includes mostly ultra-low, super

Table 2.4. European emission standards for passenger cars and heavy-duty vehicles. The date corresponds to the model year (first registration date).

(a) Passenger cars using gasoline or liquefied natural gas (LNG)/liquefied petroleum gas (LPG). Emissions are in mg km^{-1}, except for particle numbers (PN), which are in number of particles per km. NMHC corresponds to non-methane hydrocarbons.

Standard	Euro 1	Euro 2	Euro 3	Euro 4	Euro 5	Euro 6
Date	01/01/1993	07/01/1996	01/01/2001	01/01/2006	01/01/2011	09/01/2015
CO	2,720	2,200	2,200	1,000	1,000	1,000
NO$_x$	–	–	150	80	60	60
HC	–	–	200	100	100	100
NMHC	–	–	–	–	68	68
PM	–	–	–	–	5	4.5
PN (number)	–	–	–	–	–	6×10^{12}

(b) Diesel passenger cars. Emissions are in mg km^{-1}, except for particle numbers (PN), which are in number of particles per km.

Standard	Euro 1	Euro 2	Euro 3	Euro 4	Euro 5	Euro 6
Date	01/01/1993	07/01/1996	01/01/2001	01/01/2006	01/01/2011	09/01/2015
CO	2,720	1,000	640	500	500	500
NO$_x$	–	–	500	250	180	80
NMHC + NO$_x$	970	900	560	300	230	170
PM	140	100	50	25	5	4.5
PN (number)	–	–	–	–	6×10^{11}	6×10^{11}

(c) Heavy-duty vehicles. Emissions are in g (kW h)$^{-1}$, except for particle numbers (PN), which are in number of particles per kW h. Emissions are expressed in pollutant mass or particle number per kW h, i.e., per amount of energy used.

Standard	Euro 0	Euro I	Euro II	Euro III	Euro IV	Euro V	Euro VI
Date	10/01/1990	10/01/1993	10/01/1996	10/01/2001	10/01/2006	10/01/2009	01/01/2014
CO	12.3	4.9	4	2.1	1.5	1.5	1.5
NO$_x$	15.8	9.0	7.0	5.0	3.5	2.0	0.4
NMHC	2.6	1.23	1.1	0.66	0.46	0.46	0.13
PM	–	0.4	0.15	0.1	0.02	0.02	0.01
PN (number)							6×10^{11}

To convert the data in this table into g km^{-1}, one must estimate the quantity of energy needed for a heavy-duty vehicle to drive 1 km. This conversion depends on the heavy-duty vehicle load, its speed, and the road slope. For example, one may estimate as a first approximation that to drive 1 km, it takes 10 kW h for a 15 t vehicle and 20 kW h for a 30 t vehicle. Therefore, to convert the data in this table into g km^{-1} or # km^{-1}, they must be multiplied by about a factor of 10 for a 15 t vehicle and by a factor of 20 for a 30 t vehicle.

ultra-low, and zero emission vehicles and, in 2021, it will include mostly super ultra-low and zero emission vehicles. California requires that 10 % of vehicles sold in 2025 be zero emission vehicles (ZEV). Emission standards have also been implemented for evaporative emissions of NMOG.

In Europe, emission standards were introduced as early as 1970 for passenger cars and 1988 for trucks (Hugrel and Joumard, 2006). Since then, European standards have been augmented and updated to reinforce emission controls for on-road vehicles. These European standards, called "Euro" standards (Euro 1 to 6 for passenger cars and Euro 0 to VI for heavy-duty vehicles), are summarized in Table 2.4. They apply to all vehicles sold after the standard comes into effect. Comparing the U.S./California and European light-duty vehicle emission standards, one notes that they are similar for CO. For gasoline passenger cars (the great majority of U.S. passenger cars run on gasoline), the European Euro 6 emission standard for (NO_x + NMHC) is similar to the U.S. LEV160 emission standard. However, it is higher than the fleet-averaged U.S. emission standard for 2017, which is less than half the Euro 6 value (54 versus 128 mg km^{-1}). For heavy-duty vehicles, the European Euro VI standards of 2014 are similar to the 2010 U.S. standards for NO_x, hydrocarbons (HC), and PM. Emissions of formaldehyde (HCHO), which is a carcinogenic compound and a precursor of ozone, are regulated in the U.S., but not in Europe. Note that caution is advised when comparing different emission standards, because they depend on the testing procedure (driving cycle and testing conditions). For example, the U.S. federal testing procedure for light-duty vehicles (FTP-75) includes three phases (cold-start, stabilized, and hot-start phases) with speed varying from 0 to 91 km h^{-1} and an average speed of 34 km h^{-1}. In addition, a supplemental FTP is used to address high-speed driving and the use of air conditioning. Different driving cycles are used in Europe that may lead to slightly different vehicle emissions.

These emission standards do not necessarily correspond to the actual emissions from vehicles, which can be greater or lower than these standards. These standards are set for new vehicles and are evaluated according to a driving cycle and experimental testing conditions (ambient temperature, for example) that may differ from actual driving conditions. In addition, some car manufacturers tend to optimize the emissions for the testing conditions. In some extreme cases, as was the case for some Volkswagen diesel vehicles in the United States, actual driving emissions significantly exceeded the emission standards (by factors of 10 or more), because the NO_x emission control system was modified outside of the testing conditions. However, the emission factors that are developed for emission inventories are not based on the emission standards, but are derived from emission tests conducted on dynamometers with used vehicles for driving cycles that better reflect actual driving conditions than the emission standard driving cycles (e.g., André et al., 2006; Franco et al., 2013). In particular, both cold-start and hot-start conditions are tested. This is important because CO and VOC emissions are much greater under cold-start conditions. In the U.S., emission factors are available in the emissions models to calculate emissions from on-road vehicles. For all states except California, the U.S. EPA recommends using the Motor Vehicle Emission Simulator (MOVES) model. For California, the U.S. EPA recommends using the latest EPA-approved version of the EMFAC model of the California Air Resources Board. In Europe, the Copert 4 database (Copert, 2006) provides emission factors by vehicle type as a function of driving conditions. NO_x emission factors under hot-start conditions are illustrated in Figure 2.1 for European gasoline and diesel passenger cars.

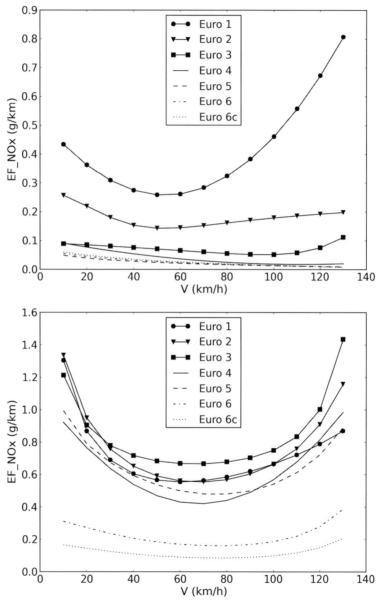

Figure 2.1. NO_x emission factors (g km^{-1}). Emission factors are for European passenger cars (1.4 to 2 liter engine; hot start) as a function of speed (km h^{-1}) based on the Copert 4 European database. Top figure: gasoline vehicles; bottom figure: diesel vehicles. Source: Chen et al. (2017).

Emission measurements under actual driving conditions using measurement instrumentation installed on the vehicle, "Portable Emissions Measurement System" (PEMS), are now being conducted not only to estimate emission factors, but also to test the attainment of regulatory standards. As a matter of fact, such measurements led to the discovery in 2014 of the large exceedance of regulatory standards by some Volkswagen diesel vehicles (Thompson et al., 2014). A measurement program was conducted in 2016 in France on diesel vehicles driving on a closed circuit according to an official driving cycle, called the "new European driving cycle" (NEDC), used to test fuel consumption and air pollutant emissions. This program showed that among 45 vehicles corresponding to the Euro 6 standard (see Table 2.4), only 3 were in attainment of the standard (80 mg NO_x km^{-1}) and 18 exceeded the standard by more than a factor of 5 (i.e., >400 mg NO_x km^{-1}) (MEEM, 2016). These results suggest that the emission control systems of some vehicles are over-optimized for the regulatory standard tests.

Reduction of the Pollutant Content in the Fuels

The reduction of vehicle emissions does not only pertain to the air pollutants listed in the emission standards such as the "Euro" standards, but must also address the emissions of two regulated pollutants potentially present in the fuel: sulfur and lead. Lead was used in the form of tetraethyl lead as an antiknock agent to minimize engine knocking, which results from the heterogeneity of the combustion process within the engine combustion chamber. However, lead may cause adverse health effects, including a mental development delay in children, as well as other adverse health effects in adults (see Chapter 12). As a result, lead was eliminated from fuels in the United States in 1995 and in Europe starting in 2000 (some delays were granted to a few countries, including Italy, Spain, Greece, and Portugal). Substitution products such as toluene and ethanol are now used as antiknock agents. Thus, the reduction of lead emissions from on-road traffic was performed by replacing lead by other compounds with lesser health impacts (note, however, that VOC such as toluene and ethanol are ozone precursors and that toluene is a precursor of secondary PM). Although tetraethyl lead is now forbidden in fuels in Europe and North America, it is still used in some countries of Latin America and Africa. In addition, lead is present in some engine oils.

Sulfur is present in fossil fuels, such as coal and oil. It is emitted after fuel combustion in the form of sulfur oxides, mostly as sulfur dioxide (SO_2) and a small fraction as sulfuric acid (H_2SO_4). SO_2 is a regulated pollutant, which is also a precursor of acid rain and secondary PM. The reduction of sulfur emissions from on-road traffic was obtained by limiting the sulfur content of diesel and gasoline fuels. For example, this limit is 15 ppm (15 parts per million, i.e., 15 mg of S per kg of fuel) for diesel since 2006 and 10 ppm for gasoline since 2017 in the United States; it is 10 ppm in France since 2009. Therefore, emission control is obtained for sulfur via a pretreatment used to reach the fuel content standard. The usual pretreatment method is called hydrodesulfurization. It is an industrial process in which molecular hydrogen (H_2) is added and sulfur is removed as hydrogen sulfide (H_2S). This process takes place at high temperature (300 to 400 °C) and pressure (10 to

100 atm), in the presence of catalysts such as molybdenum (Mo), cobalt (Co), and nickel (Ni) (Song, 2003).

Emission Control of Gaseous Pollutant Exhaust

Carbon monoxide (CO) was the first gaseous air pollutant from on-road traffic to be regulated. It was regulated in the 1970s in the United States and in France (CAA, 1970; Hugrel and Joumard, 2006). The emission control process involves the oxidation of CO on a catalyst to convert it to carbon dioxide (CO_2). The CO emission control is typically greater than 80 %. The catalyst, platinum or palladium, favors the reaction of CO with oxygen to form CO_2:

$$2\ CO\ +\ O_2 \rightarrow 2\ CO_2 \tag{R2.9}$$

This type of catalyst is also used to oxidize hydrocarbons (HC) into CO_2:

$$2\ C_nH_{2n+2}\ +\ (3n+1)\ O_2 \rightarrow 2n\ CO_2 + 2(n+1)\ H_2O \tag{R2.10}$$

where here the hydrocarbon C_nH_{2n+2} is an alkane. The emission control system is called a two-way catalytic converter, because there are two oxidation "ways," one for CO and one for HC. The first two-way catalytic converter was developed in 1956 by Eugène Houdry, a French engineer who had moved to the United States (U.S. patent N° 2,742,437).

It is also possible to partially reduce NO_x to molecular nitrogen, N_2, with a catalyst, which is generally rhodium:

$$2\ NO\ +\ 2\ CO \rightarrow N_2 + 2\ CO_2 \tag{R2.11}$$

A system that allows the oxidation of CO and HC and the reduction of NO_x is called a three-way catalytic converter (since there are three chemical reaction "ways," two for oxidation and one for reduction). The catalysts may be located in different parts of the converter or in the same area. The catalytic converter consists of a ceramic or metallic substrate, which is covered by oxidized compounds such as alumina (aluminum oxide) or cerine (cerium oxide), where the catalysts are located. The oxygen content and the temperature affect the efficiency of the oxidation and reduction reactions. The proper performance of a three-way catalytic converter requires that the mixture injected into the engine be in a range that is neither too rich in fuel, nor in oxygen, i.e., close to a stoichiometric fuel/air ratio (1:1). If the mixture is too rich in oxygen, then NO_x reduction will not take place. If the mixture is too rich in fuel, then oxygen will react more completely with the fuel and it will not be present in sufficient amount to oxidize CO and HC efficiently. A three-way catalytic converter may be used on gasoline cars because it is possible to maintain the mixture in the desired range of the fuel/air ratio. However, only two-way catalytic converters are used for diesel engines because the mixture of diesel engines is oxygen rich and, therefore, prevents the efficient use of a three-way system.

The diesel oxidation catalytic converters (DOC) generally use platinum and palladium for the oxidation of CO and HC. The efficiency of DOC is on the order of 90 % for a temperature greater than 400 °C. A DOC may also partially oxidize the organic fraction of diesel particles, which helps decrease slightly the PM emissions (this organic fraction is sometimes called the soluble organic fraction, SOF, because this fraction is measured in the laboratory using an organic solvent extraction). However, SO_2 may also be oxidized in the process, thereby leading to the formation of particulate sulfate. Therefore, it is desirable to use a very low sulfur content fuel. A DOC does not lead to any NO_x reduction.

Another method, besides the three-way catalytic converter, to control NO_x emissions consists of recirculating a fraction of the exhaust gases into the combustion chamber. The addition of exhaust gases leads to a lower temperature and lower oxygen content in the combustion chamber, which leads to less NO_x formation. However, the decrease in the oxygen content tends to increase the formation of particles (since the combustion is not as complete) and also may decrease engine performance. This method, which is called "exhaust gas recirculation " (EGR), has a lower efficiency in terms of NO_x emission control compared to the three-way catalytic converters. It was used when the regulations were less stringent, for example with the Euro 2 standards (see Table 2.4). It may also be used in combination with a three-way catalytic converter to improve the overall NO_x emission control in gasoline vehicles, and it is used for some diesel engines since DOC do not address NO_x emission control.

Increasingly stringent standards for NO_x emissions (for example, 60 mg km^{-1} for gasoline vehicles and 80 mg km^{-1} for diesel passenger cars in Europe, see Table 2.4) have required more efficient emission control technologies. The two main technologies are the NO_x trap and selective catalytic reduction (SCR). These technologies may be used in combination with EGR.

The NO_x trap is a sequential system that adsorbs NO_x on the surface of the trap. This trap contains several metals, such as platinum (to convert NO into NO_2), barium (to trap NO_2 as baryum nitrate), and rhodium (to reduce NO_2 to N_2). When the trap becomes saturated with NO_x (after about ten minutes or ten kilometers), it must be regenerated for a few seconds. The regeneration is performed by adding some fuel (diesel); the fuel-rich exhaust leads to a reducing gas containing molecular hydrogen (H_2), which then converts NO_x to NH_3, with a subsequent reaction between NH_3 and NO_x to form N_2. However, this system depends on the efficiency of the EGR, which regulates the NO_x concentration. As the EGR efficiency decreases significantly outside a range of ambient temperature of about 17 to 35 °C, the NO_x trap efficiency varies greatly depending on the ambient conditions. Although the efficiency may be as high as 70 % under favorable ambient conditions, it may be as low as 30 % under low or high ambient temperatures.

SCR is a system that is conceptually similar to that used for coal-fired power plants (see Section 2.3.1). Instead of using ammonia, urea, $(NH_2)_2C=O$, which is less toxic than ammonia, is used in vehicles. Urea is transformed into ammonia (NH_3) in the exhaust before reacting with NO_x. To ensure that N_2 is produced by the reaction between the oxidized form of nitrogen (NO_x) and its reduced form (NH_3), a catalyst is used (vanadium oxide, for example). The SCR efficiency is on the order of 90 %. A SCR is, therefore, much more efficient than a NO_x trap. However, its cost is much higher than that of a NO_x trap.

In the testing program conducted in France in 2016 (see the section on regulatory standards), diesel vehicles equipped with a SCR appeared to be performing better than those equipped with a NO_x trap, since only about a quarter (27 %) of the former exceeded the emission standard by a factor of 5, whereas almost half (47 %) of the latter exceeded it by the same factor (MEEM, 2016).

Emission Control of Particulate Exhaust

Particles emitted in the vehicle exhaust are mostly ultrafine particles (i.e., particles with a diameter <0.1 µm), which grow rapidly once released in the atmosphere via condensation and coagulation to become fine particles (i.e., with diameters between 0.1 and 2.5 µm, see Chapter 9). Diesel vehicles without a diesel particle filter (DPF) are the biggest emitters of particles among on-road sources. These particles consist of a black-carbon core on which sulfate and semi-volatile organic compounds have condensed (Morawska et al., 2008; Seigneur, 2009). These particles also contain metals originating from the fuel, the oil, and engine wear. The control of particle emissions is achieved with a DPF. Such a filter captures most of the particles before their emission to the atmosphere. In Europe, DPF are required to comply with the regulations for trucks and buses (since Euro IV) and for passenger cars (since Euro 5). Deposition of particles on the filter leads after a while to a pressure drop in the exhaust flow, which may induce some poorer engine performance. Therefore, the filter needs to be regenerated from time to time. This regeneration process is performed by oxidizing the particulate carbon deposited on the filter into gaseous CO_2, which is then released with the exhaust. Since sulfate is not combustible, it is important to avoid particulate sulfate deposition on the filter and to use diesel fuel with very low sulfur content. Several methods are available to regenerate a DPF. The two main approaches are active regeneration and passive regeneration. The oxidation of particulate carbon to CO_2 occurs at very high temperatures (i.e., via incineration). Two options are available: (1) to increase the temperature of the exhaust to reach values suitable for this oxidation to take place (this is active regeneration) or (2) allowing oxidation to take place at lower temperatures using a catalyst and a gaseous oxidant (this is passive regeneration). To increase the temperature, diesel fuel is injected into the exhaust (post-injection); then soot particles are incinerated at temperatures in the range of 550 to 600 °C. Passive regeneration may take place continuously if the exhaust temperature is high enough and if the temperature needed for the incineration process to take place can be reduced.

The continuous regeneration trap (CRT) allows passive regeneration to occur. In a CRT, a catalyst oxidizes NO into NO_2 upstream of the DPF and uses NO_2 to facilitate the oxidation of particulate carbon to CO_2 within the filter. This is the catalyzed DPF. The range of temperatures suitable for incineration to take place is then lowered by about 100 °C (i.e., <500 °C). The regeneration of the filter may occur under certain driving conditions (for example, on the freeway) when the exhaust temperature is sufficiently high. However, passive regeneration is generally not sufficient and some active regeneration must be applied from time to time. Therefore, a post-injection of diesel fuel is performed when a pressure drop due to the accumulation of particulate matter on the filter is detected downstream of the DPF. The disadvantage of this method is the increase of the fraction of NO_2 (a regulated pollutant) in the NO_x emissions; it may increase from a few percent (~5 % typically) to about 50 % (e.g., Kousoulidou et al., 2008).

Another method consists of incorporating the catalyst within the fuel (an additive is used), which implies that it is present in the carbon particles deposited on the filter. This is called the additivated DPF. The contact between the particles and the catalyst is then optimal and particulate matter is oxidized by O_2, rather than by NO_2. The temperature needed for this oxidation to take place is lowered by about 100 °C, as with a catalyzed DPF. A post-injection is also used on a regular basis for the active regeneration of the DPF. The main advantage of this method compared to the catalyzed DPF is that NO_2 is not involved in the oxidation of the carbonaceous particulate matter and that there is no need to increase the NO_2 fraction of the NO_x emissions. Then, a NO_x emission control system may be added upstream of an additivated DPF.

2.4 Numerical Modeling of Air Pollutant Emissions

As mentioned, the equations used to represent the emissions of air pollutants from a source given the source activity and the corresponding pollutant emission factor are generally simple. Nevertheless, the sheer number of sources and pollutants and the spatial and temporal resolution of the emission inventory needed for air pollution modeling make the development of an emission inventory a challenging task. Therefore, numerical models have been developed to make the development of spatially distributed and temporally resolved emission inventories more accessible to practitioners of air pollution modeling.

One may distinguish emissions modeling systems that focus on one aspect of the emission inventory (e.g., biogenic emissions, on-road mobile source emissions) and emissions modeling systems that are comprehensive and cover all source categories. The latter may include some of the former as sub-models.

Two main approaches have been used when developing a numerical emissions model: (1) those that are based on a relational database management system (RDMS) suitable for managing large amounts of data (e.g., Access, SQL, PostgreSQL) and (2) those that are based on programming languages suitable for fast numerical computations (e.g., Fortran, C++, Python). There are advantages and shortcomings to both approaches. Emissions models based on an RDMS offer the user easy access to the intermediate steps carried out during the data processing from the input data to the final emission inventory. However, the calculation of a new emission inventory using an RDMS-based model may be time-consuming. Emissions models based on a numerical computing programming language are typically computationally efficient. However, they are not conducive to providing information on the intermediate calculation steps. Some emissions modeling systems may combine both approaches, for example using an overall RDMS-based model in combination with some sub-models coded with a numerical computing programming language. A large number of emissions modeling systems are available and only a few examples are provided next.

For modeling emissions from mobile sources, data are needed on traffic flow, fleet composition, and emission factors. In the U.S., the MOVES emissions model is available from the U.S. EPA (www.epa.gov/moves). It treats both on-road and non-road mobile sources and has replaced the MOBILE5 and NONROAD models. MOVES uses a combination of RDMS (MySQL and SQL scripts for data input and output) and numerical computing (Java) programming languages. The EMFAC model is available from the California Air Resources Board for on-road mobile sources (www.arb.ca.gov/msei/categories.htm). In Europe, the Copert 4 system provides the necessary data to calculate emission factors for various vehicle types and driving conditions (www.eea.europa.eu/themes/air/links/guidance-and-tools/copert4-road-transport-emissions-model). Some emissions models based on the Copert 4 emission factor algorithms have been developed to facilitate the emission calculations for applications to a modeling domain and period (e.g., the Pollemission model of Cerea and Inria written in Python; https://github.com/pollemission).

For modeling biogenic emissions, the Model of Emissions of Gases and Aerosols from Nature (MEGAN) is widely used (Guenther et al., 2012). The U.S. EPA has developed the Biogenic Emission Inventory System (BEIS) (www.epa.gov/air-emissions-modeling/biogenic-emission-inventory-system-beis). In Europe, the biogenic emissions model of Simpson et al. (1999) has also been used. MEGAN and BEIS3 are written in Fortran. All biogenic emissions models require information on land use and land cover as well as meteorological information (solar radiation and temperature).

Among the overall emissions modeling systems, one may distinguish the Sparse-Matrix Open Kernel Emission (SMOKE) model of the U.S. EPA (www.cmascenter.org/smoke/), which is written in Fortran, and most other emissions modeling systems, which use an RDMS overall structure. Among those latter modeling systems, one may mention the French *Inventaire national spatialisé* (INS; http://emissions-air.developpement-durable.gouv.fr). SMOKE has been adapted for application to Europe (Bieser et al., 2011).

Problems

Problem 2.1 Emissions of air pollutants and greenhouse gases from on-road vehicles
Two vehicles travel 10,000 km annually each. The gasoline vehicle has a fuel consumption of 7.4 liters per 100 km and the diesel vehicle has a fuel consumption of 5.7 liters per 100 km. The fuel densities are assumed to be 0.755 g cm^{-3} for gasoline and 0.845 g cm^{-3} for diesel. These two types of fuel contain a large number of hydrocarbons. The following average molecular formulas are used to represent these fuels: heptane (C_7H_{16}) for gasoline and an average theoretical formula, $C_{16}H_{29}$, for diesel.

a. Assuming that combustion is complete (therefore, all hydrocarbon molecules are converted to CO_2), what are the annual emissions of CO_2 of each vehicle?
b. Assuming that the fuels of both vehicles are in attainment of the European sulfur content regulation of 10 ppm by weight (mg kg^{-1}) and that the sulfur exhaust emissions occur as SO_2, what are the annual SO_2 emissions of each vehicle?

Problem 2.2 Control of on-road traffic emissions

The car fleet is assumed to consist of 60 % diesel and 40 % gasoline vehicles. Furthermore, in terms of emission standards, half of these vehicles (both gasoline and diesel) are Euro 4 vehicles (bought between 2006 and 2010 in Europe) and half are Euro 5 (bought in 2011 or later). The emission factors for Euro 4, Euro 5, and Euro 6 mid-size passenger cars are as follows:

	NO_x Emission factor (g km^{-1})	
Speed (km h^{-1})	70	80
Gasoline Euro 4	0.030	0.025
Gasoline Euro 5	0.019	0.017
Gasoline Euro 6	0.021	0.018
Diesel Euro 4	0.421	0.441
Diesel Euro 5	0.481	0.481
Diesel Euro 6	0.162	0.161

Note that although the emission standard may not change between two Euro categories, the emission factors may differ, due to the variability of emissions among the used vehicles that are tested to develop those emission factors. Here, this is the case for Euro 5 and Euro 6 gasoline vehicles. On the other hand, the emission standard became more stringent from Euro 4 to Euro 5 for diesel vehicles, but the emission factors do not reflect this change. However, the more stringent emission standards for Euro 6 diesel vehicles are reflected in the decrease of the emission factors.

To reduce NO_2 concentrations during air pollution episodes in Paris, it is assumed that a decrease in NO_x emissions from on-road traffic on the Paris ring road (*boulevard périphérique*) is targeted. The speed limit on the ring road was 80 km h^{-1} until 2013. Traffic on the ring road is about 7 million vehicle-km per day. Which one of the two following options would be the most efficient to reduce NO_x emissions?

- Reduce the speed limit from 80 km h^{-1} to 70 km h^{-1} (i.e., the speed limit since January 2014).
- Do not allow Euro 4 vehicles (both gasoline and diesel) on the ring road, while maintaining the speed limit at 80 km h^{-1} and the same number of vehicles-km per day. Assume that the car fleet will then be renewed and that the Euro 4 vehicles will be replaced by Euro 6 vehicles.

References

André, M., R. Joumard, R. Vidon, P. Tassel, and P. Perret, 2006. Real-world European driving cycles, for measuring pollutant emissions from high- and low-powered cars, *Atmos. Environ.*, **40**, 5944–5953.

Andreae, M.O. and P. Merlet, 2001. Emissions of trace gases and aerosols from biomass burning, *Global Biogeochem. Cycles*, **15**, 955–966.

References

Bates, T.S., B.K. Lamb, A. Guenther, J. Dignon, and R.E. Stoiber, 1992. Sulfur emissions to the atmosphere from natural sources, *J. Atmos. Chem.*, **14**, 315–337.

Bieser, J., A. Aulinger, V. Matthias, M. Quante, and P. Builjtes, 2011. SMOKE for Europe – adaptation, modification and evaluation of a comprehensive emission model for Europe, *Geosci. Model Dev.*, **4**, 47–68.

Bouwman, A.F., D.S. Lee, W.A.H. Asman, F.J. Dentener, K.W. van der Hock, and J.G.J. Olivier, 1997. A global high-resolution emission inventory for ammonia, *Global Biogeochem. Cycles*, **11**, 561–587.

CAA, 1970. Clean Air Act Extension of 1970, 42 U.S. Code, § 7401 et seq.

CFR, 2016. U.S. Code of Federal Regulations, *Title 42 – The Public Health and Welfare*, Chapter 85 – Air Pollution Prevention and Control, Subchapter 111 – General Provisions, § 7602 – Definitions.

Chen, R., V. Aguiléra, V. Mallet, F. Cohn, D. Poulet, and F. Brocheton, 2017. A sensitivity study of road transportation emissions at metropolitan scale, *J. Earth Sci. Geotech. Eng.*, **7**, 151–173.

Code de l'Environnement, 2016. *Titre II – Air et atmosphère*, Article L220-2. Available at: https://Legifrance.gouv.fr.

Copert, 2006. COPERT 4: Estimating emissions from road transport, European Environmental Agency (EEA), Copenhagen, Denmark, www.eea.europa.eu/publications/copert-4-2014-estimating-emissions.

EC, 2016. Emissions Database for Global Air Research (EDGAR) version 4.3.1, European Commission, Joint Research Centre, Ispra, Italy, http://edgar.jrc.ec.europa.eu.

Flagan, R.C. and J.H. Seinfeld, 1988. *Fundamentals of Air Pollution Engineering*, Prentice Hall, Englewood Cliffs, NJ.

Franco, V., M. Kousoulidou, M. Muntean, L. Ntziachristos, S. Hausberger, and P. Dilara, 2013. Road vehicle emission factors development: A review, *Atmos. Environ.*, **70**, 84–97

Gong, S.L., L.A. Barrie, and M. Lazare, 2002. Canadian Aerosol Module (CAM): A size-segregated simulation of atmospheric aerosol processes for climate and air quality models. 2. Global sea-salt aerosol and its budgets, *J. Geophys. Res.*, **107**, D24, 4779.

Guenther, A.B., X. Jiang, C.L. Heald, T. Sakuyanontvittaya, T. Duhl, L.K. Emmons, and X. Wang, 2012. The Model of Emissions of Gases and Aerosols from Nature version 2.1 (MEGAN2.1): An extended and updated framework for modeling biogenic emissions, *Geosci. Model Dev.*, **5**, 1471–1492.

Hugrel, C. and R. Joumard, 2006. *Directives et facteurs agrégés d'émission des véhicules routiers en France de 1970 à 2025*, Technical report Inrets/LTE n° 0611, Institut français des sciences et technologies des transports, de l'aménagement et des réseaux (Ifsttar), Bron, France.

Khalil, M.A.K. and R.A. Rasmussen, 1990. The global cycle of carbon monoxide, *Chemosphere*, **20**, 227–242.

Kopacz, M., D.J. Jacob, J.A. Fisher, J.A. Logan, L. Zhang, I.A. Megretskaia, R.M. Yantosca, K. Singh, D.K. Henze, J.P. Burrows, M. Buchwitz, I. Khlystova, W.W. McMillan, J.C. Gille, D.P. Edwards, A. Eldering, V. Thouret, and P. Nedelec, 2010. Global estimates of CO sources with high resolution by adjoint inversion of multiple satellite datasets (MOPITT, AIRS, SCIAMACHY, TES), *Atmos. Chem. Phys.*, **10**, 855–876.

Kousoulidou, M., L. Ntziachristos, G. Mellios, and Z. Samaras, 2008. Road-transport emission projections to 2020 in European urban environments, *Atmos. Environ.*, **42**, 7465–7475.

Lee, C., R.V. Martin, A. van Donkelaar, H. Lee, R.R. Dickerson, J.C. Hains, N. Krotkov, A. Richter, K. Vinnikov, and J.J. Schwab, 2011. SO_2 emissions and lifetimes: Estimates from inverse modeling using in situ and global, space-based (SCIAMACHY and OMI) observations, *J. Geophys. Res.*, **116**, D06304.

Logan, J.A., 1983. Nitrogen oxides in the troposphere: Global and regional budgets, *J. Geophys. Res.*, **88**, 10785–10807.

MEEM, 2016. *Rapport final de la commission indépendante mise en place par la Ministre Ségolène Royal après la révélation de l'affaire Volkswagen – Contrôle des émissions de polluants atmosphériques et de CO_2 mené sur 86 véhicules*, Ministère de l'Environnement, de l'Énergie et de la Mer, Paris, France.

Mieville, A., C. Granier, C. Liousse, B. Guillaume, F. Mouillot, J.-F. Lamarque, J.-M. Grégoire, and G. Pétron, 2010. Emission of gases and particles from biomass burning during the 20th century using satellite data and an historical reconstruction, *Atmos. Environ.*, **44**, 1469–1477.

Morawska, L., Z. Ristovski, E.R. Jayaratne, D.U. Keogh, and X. Ling, 2008. Ambient nano and ultrafine particles from motor vehicle emissions: Characteristics, ambient processing and implications on human exposure, *Atmos. Environ.*, **42**, 8113–8138.

Seigneur, C., 2009. Current understanding of ultra fine particulate matter emitted from mobile sources, *J. Air Waste Manage. Assoc.*, **59**, 3–17.

Simpson, D., W. Winiwarter, G. Börjesson, S. Cinderby, A. Ferreiro, A. Guenther, C. Hewitt, R. Janson, M. Khalil, S. Owen, T. Pierce, H. Puxbaum, M. Shearer, U. Skiba, R. Steinbrecher, L. Tarrason, and M. Oquist, 1999. Inventorying emissions from nature in Europe. *J. Geophys. Res.*, **104**, 8113–8152.

Song, C., 2003. An overview of new approaches to deep desulfurization for ultra-clean gasoline, diesel fuel and jet fuel, *Catalysis Today*, **86**, 211–263.

Srivastava, R.K. and W. Josewicz, 2001. Flue gas desulfurization: The state of the art, *J. Air Waste Manage. Assoc.*, **51**, 1676–1688.

Srivastava, R.K., R.E. Hall, S. Khan, K. Culligan, and B.W. Lanu, 2005. Nitrogen oxides emission control options for coal-fired electric utility boilers, *J. Air Waste Manage. Assoc.*, **55**, 1367–1388.

Thompson, G.J., D.K. Carder, M.C. Besch, A. Thiruvengadam, and H.K. Kappanna, 2014. *In-use emissions testing of light-duty diesel vehicles in the United States*, Final report, Centre for Alternative Fuels, Engines & Emissions, West Virginia University, Morgantown, WV.

Waked, A., C. Afif, and C. Seigneur, 2012. An atmospheric emission inventory of anthropogenic and biogenic sources for Lebanon, *Atmos. Environ.*, **50**, 88–96.

Wang, L.K., N.C. Pereira, and Y.-T. Hung, eds., 2005. *Advanced Air and Noise Pollution Control*, Humana Press, Totowa, NJ.

Zender, C.S., H. Bian, and D. Newman, 2003. Mineral dust entrainment and deposition (DEAD) model: Description and 1990s dust climatology, *J. Geophys. Res.*, **108**, D14, 4416.

3 Meteorology: General Circulation

Air pollution is directly affected by various aspects of meteorology, such as winds, which transport pollutants (in some cases over long distances); turbulence, which disperses air pollutants; solar radiation, which initiates photochemical reactions leading to the formation of ozone, fine particles, and acid rain; high pressure systems, which are conducive to air pollution episodes because of their calm and sunny conditions; and precipitations, which scavenge air pollutants and transfer them to other media (e.g., acid rain). Therefore, it is essential to understand general meteorological features before addressing in detail the processes that are specific to air pollution. This chapter presents first some general considerations on the atmosphere (chemical composition, pressure, and temperature). Next, the main aspects of the general atmospheric circulation are described.

3.1 General Considerations on the Atmosphere

3.1.1 Atmospheric Chemical Composition

The Earth's atmosphere is the gas layer that surrounds the Earth. It is composed mostly of molecular nitrogen (N_2) and molecular oxygen (O_2) in proportions of 78 % and 21 %, respectively. Argon is present at <1 %. Carbon dioxide (CO_2) is present at about 400 ppm (parts per million), i.e., about 0.04 %. Water vapor is on average about 1 %, but it displays great spatio-temporal variability.

3.1.2 Atmospheric Pressure

The atmosphere is a gas that can be considered to be ideal at the pressures observed in the Earth's atmosphere. Therefore, the ideal gas law applies:

$$PV = nRT \qquad (3.1)$$

where P is atmospheric pressure (atm), V is the air volume of interest (m^3), T is temperature (K), n is the number of moles of air in the volume of interest, and R is the ideal gas law constant ($R = 8.206 \times 10^{-5}$ atm m^3 $mole^{-1}$ K^{-1}; if P is expressed in pascal (Pa), then $R = 8.314$ J mol^{-1} K^{-1}).

Atmospheric pressure results from the gravitational force of the planet Earth on the atmospheric gases. It decreases with height. The hydrostatic equation can be written as follows (where $z = 0$ at the Earth's surface):

$$dP(z) = -\rho_a(z)\, g\, dz \qquad (3.2)$$

where z is altitude (m), P is pressure (here in Pa, i.e., kg m^{-1} s^{-2}), ρ_a is the density of the air (kg m^{-3}), and g is the gravitational constant (9.81 m s^{-2}). For an incompressible fluid: $\rho_a(z) =$ constant; therefore, the relationship between pressure and altitude is linear. For example, pressure increases almost linearly with depth in the oceans (with small variations due to changes in temperature and salt-content and a slight compressibility at very high pressure). The atmosphere is a gas and, therefore, a compressible fluid, implying that its density increases with pressure. The density of the air may be calculated according to the ideal gas law as follows:

$$\rho_a(z) = MM_{air}\, \frac{n}{V} = \frac{MM_{air}\, P(z)}{R\, T(z)} \qquad (3.3)$$

where MM_{air} is the average molar mass of the air, i.e., a weighted average of the molar masses of the various constituents of the air (0.029 kg mole^{-1} for dry air). Therefore, the dry air density is 1.184 kg m^{-3} at 1 atm and 25 °C; it is 1.225 kg m^{-3} at 1 atm and 15 °C. Replacing $\rho_a(z)$ in the hydrostatic equation leads to:

$$\frac{dP(z)}{P(z)} = -\frac{MM_{air}\, g\, dz}{R\, T(z)} \qquad (3.4)$$

Thus:

$$P(z) = P(0)\, \exp\left(-\int_0^z \frac{MM_{air}\, g\, dz}{R\, T(z)}\right) \qquad (3.5)$$

This equation may be written as a function of a characteristic height, $h_a(z)$:

$$h_a(z) = \frac{R\, T(z)}{MM_{air}\, g} \qquad (3.6)$$

$$P(z) = P(0)\, \exp\left(-\int_0^z \frac{dz}{h_a(z)}\right) \qquad (3.7)$$

This characteristic height varies as a function of altitude since temperature varies as a function of altitude. However, the variation of atmospheric temperature (expressed here in K) as a function of altitude is significantly less than that of pressure. If one assumes that temperature is constant, atmospheric pressure may then be written as a decreasing exponential function:

$$P(z) = P(0)\, \exp\left(-\frac{z}{h_a}\right) \qquad (3.8)$$

where $h_a \approx 8$ km for a temperature of 273 K and $R = 8.314$ J mol^{-1} K^{-1}. Thus, atmospheric pressure is about 0.88 atm at an altitude of 1 km and 0.29 atm at an altitude of 10 km. Figure 3.1 depicts this variation of atmospheric pressure with altitude.

The assumption of a quasi-constant temperature is in part justified by the fact that absolute temperature does not vary as much as pressure if one assumes that atmospheric

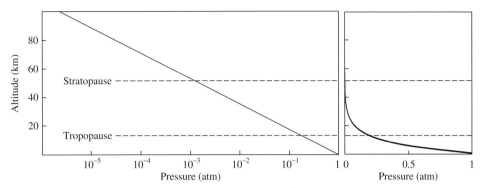

Figure 3.1. Vertical profile of atmospheric pressure. Atmospheric pressure is shown as a function of altitude with a logarithmic scale (left figure) and a linear scale (right figure). The tropopause and stratopause altitudes are approximate and given only as qualitative information.

processes are adiabatic (see Section 3.1.3). For example, the relationship between temperature and pressure results in a change of only 3 % in temperature for a change of 10 % in pressure.

The concentration of air molecules may be calculated according to the ideal gas law. Therefore, for an atmosphere with a pressure of 1 atm and a temperature of 25 °C (i.e., 298 K), the molar concentration is as follows:

$$\frac{n}{V} = \frac{P}{RT} = \frac{1}{8.206 \times 10^{-5} \times 298} = 40.9 \text{ moles m}^{-3} \qquad (3.9)$$

The air density ($\rho_a = MM_{air}\, n / V$) is proportional to its molar concentration; it increases proportionately with pressure and inversely proportionately with temperature (K). Therefore, warm air is less dense than cold air.

The unit of pressure "atmosphere" (atm) corresponds to the average atmospheric pressure observed at sea level at the latitude of Paris, France ($\approx$ 49 °N). It varies at the Earth's surface depending on altitude and temperature. For a given location, atmospheric pressure varies slightly, by a few % (see the discussion of high and low pressure systems as part of the general circulation description in Section 3.2.3). On average, it is 0.977 atm at the Earth's surface because of altitude. Atmospheric pressure may be expressed with different units. For a reference pressure of 1 atm: $P = 1$ atm $= 1.013 \times 10^5$ Pa $= 1,013$ mbar (millibar) $= 1,013$ hPa (hectopascal) $= 760$ Torr (or mm of mercury).

The total mass of the Earth's atmosphere, M_A, may be calculated according to the definition of pressure being the force exerted by the atmosphere (i.e., the mass of the atmosphere multiplied by the acceleration of gravity) per unit surface area. The Earth's surface area is 5.1×10^8 km^2, i.e., 5.1×10^{14} m^2. Therefore, the corresponding pressure expressed in pascal (i.e., N m^{-2} or kg m^{-1} s^{-2}) is:

$$P = \frac{M_A \times 9.81}{5.1 \times 10^{14}} = 0.977 \text{ atm} = 9.9 \times 10^4 \text{ Pa} \qquad (3.10)$$

Thus: $M_A = 5.15 \times 10^{18}$ kg.

3.1.3 Vertical Structure of the Atmosphere

The temperature vertical profile may be derived from the first principle of thermodynamics and from the ideal gas law, with the assumption that the atmosphere is adiabatic. The first principle of thermodynamics may be written as follows:

$$dU = dQ + dW \qquad (3.11)$$

where the first term, dU, represents the change in the internal energy of the system (expressed here per mass, J kg^{-1}), dQ (J kg^{-1}) represents the amount of heat added to the system, and dW (J kg^{-1}) represents the amount of work done by the system (or on the system if $dW < 0$). For an adiabatic process, $dQ = 0$. Furthermore, for a parcel of air moving vertically, the expression for an ideal gas leads to:

$$dU = c_v\, dT \qquad (3.12)$$

where c_v is the specific heat capacity of the air at constant volume (J kg^{-1} K^{-1}):

$$dW = -P\, dV' \qquad (3.13)$$

where V' is the specific volume (i.e., the volume of an air parcel of 1 kg), which is equivalent to the inverse of the air density ($V' = V / (n\, MM_{air})$), where MM_{air} is expressed here in kg mole^{-1}. Therefore:

$$c_v\, dT = -P\, dV' \qquad (3.14)$$

It is more convenient to use only T and P as state variables and V' may be replaced as a function of T and P. The ideal gas law may be written as a function of the specific volume:

$$PV' = \frac{R}{MM_{air}}\, T \qquad (3.15)$$

By derivation:

$$dV' = \frac{R}{MM_{air}} \left(\frac{dT}{P} - \frac{T\, dP}{P^2} \right) \qquad (3.16)$$

Thus:

$$c_v\, dT = \frac{R\, T\, dP}{MM_{air}\, P} - \frac{R}{MM_{air}}\, dT \qquad (3.17)$$

To obtain the vertical profile of temperature, one changes dP to dz according to the hydrostatic relationship (see Equation 3.2):

$$c_v\, dT + \frac{R}{MM_{air}}\, dT = -\frac{R\, T\, \rho_a\, g\, dz}{MM_{air}\, P} \qquad (3.18)$$

Substituting ρ_a as a function of P, T, and MM_{air} (see Equation 3.3):

$$dT \left(c_v + \frac{R}{MM_{air}} \right) = -g\, dz \qquad (3.19)$$

Given that $c_p = (c_v + (R/MM_{air}))$, where c_p is the specific heat capacity at constant pressure:

$$dT = -\frac{g}{c_p} dz \qquad (3.20)$$

Thus, for a given altitude difference:

$$\frac{\Delta T(z)}{\Delta z} = -\frac{g}{c_p} \qquad (3.21)$$

Therefore, the adiabatic vertical gradient of the atmospheric temperature is linear. Given $c_p \approx 1{,}000$ J kg^{-1} K^{-1}, the temperature gradient is -9.8 K km^{-1}. This gradient corresponds to a dry atmosphere, i.e., without any water vapor. An adiabatic change with water vapor will affect temperature differently (see Chapter 4.2). As a result, the temperature adiabatic gradient for a wet atmosphere is less than that of a dry atmosphere. It is on average -6.5 K km^{-1}; it is less in regions with high humidity (for example, in the tropics) and it tends toward the dry adiabatic gradient in regions with low humidity (for example, in deserts).

Near the Earth's surface, the atmosphere is influenced by heat transfer between the surface and the atmosphere, as well as by turbulence generated by obstacles (e.g., vegetation, buildings, complex terrain). Therefore, the temperature vertical profile is not necessarily adiabatic (see Chapter 4.2). This part of the troposphere that is affected by the Earth's surface is called the planetary boundary layer (PBL). The part of the troposphere located above the PBL is called the free troposphere. The troposphere corresponds to about ¾ of the total mass of the atmosphere. In the stratosphere, i.e., the atmospheric layer just above the troposphere, the strong absorption of ultraviolet radiation by oxygen and ozone (see Chapter 7) leads to an increase in temperature and the temperature vertical profile increases with altitude. The boundary between the troposphere and the stratosphere is called the tropopause. It may be defined in several ways:

– Thermal definition: the altitude at which the temperature vertical profile starts to increase
– Chemical definition: the altitude at which the ozone concentration starts to increase
– Dynamic definition: the altitude at which the dynamic vertical motions (characterized by the value of the potential vorticity, a variable that combines a representation of the dynamics of the flow and of the thermal stratification) decrease

These three processes (thermal, chemical, and dynamic) are related: the presence of oxygen leads via the absorption of solar radiation to the formation of ozone and an increase in temperature, and this temperature increase reduces vertical air motions (see Chapter 4). However, these definitions lead to slightly different values of the altitude of the tropopause. The definition of the tropopause by the World Meteorological Organization (WMO) is the lowest altitude at which the temperature decreases by less than 2 °C per km (and does not exceed this value on average for the next 2 km). An appropriate chemical definition is a value of 2.5×10^{-7} kg of ozone per kg of air, i.e., about 0.15 ppm for dry air. An appropriate dynamic definition is a value of the potential vorticity of 1.5 to 2 PVU ("Potential vorticity unit," 1 PVU = 10^{-6} K m^2 kg^{-1} s^{-1}).

The atmosphere may be seen as comprising several layers of different properties. The troposphere ranges from the Earth's surface to the tropopause. The altitude of the tropopause varies in space and time. It is greater at the equator (up to 17 km) and lower at the poles (about 7 km). Long-haul commercial flights take place at about 10 km altitude, i.e., in the upper troposphere. Supersonic commercial aircraft, such as Concorde, used to cruise in the stratosphere.

The stratosphere is generally defined as the atmospheric layer where temperature increases with height. Continuing upward, eventually the oxygen and ozone concentrations become too low to lead to any significant warming of the air, and temperature starts to decrease with altitude. This, the stratopause, occurs at about 50 km in altitude. The next layer is called the mesosphere.

The mesosphere ranges up to an altitude of about 85 km. The atmospheric layer above it is called the thermosphere (it includes the ionosphere where solar radiation leads to ionizing reactions producing gaseous ions). The thermosphere ranges up to about 500 to 1,000 km altitude. Beyond the thermosphere are the exosphere and the magnetosphere. The boundary between the atmosphere and outer space is generally considered to be at about 100 km altitude, at what is called the Kármán line. Here, the density of the atmosphere has become very low; it no longer behaves as a fluid and is not suitable for aeronautics. Beyond this boundary, gas molecules have few interactions among each other and those molecules that have enough kinetic energy may escape the Earth's gravitational field and move into outer space.

Figure 3.2 depicts the average temperature vertical profile in the atmosphere, except for the planetary boundary layer. Near the Earth's surface, the air temperature is affected by the surface and its vertical profile is not necessarily adiabatic (see Chapter 4).

The mean radius of the Earth is 6,371 km. However, the Earth is not a perfect sphere as it is slightly flatter at the poles. Its radius varies from 6,378 km at the equator to 6,357 km at the poles. The atmosphere represents a fine layer surrounding this sphere.

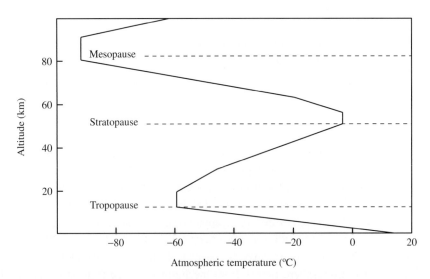

Figure 3.2. Typical vertical profile of the atmospheric temperature.

It corresponds to a thickness less than 2 % of the radius of the Earth if one considers the Kármán line and less than 1 % if one only includes the troposphere and stratosphere. The Earth's surface is located entirely within the troposphere since the highest peaks of the Himalayas reach altitudes only slightly above 8,800 m. In terms of the atmospheric environment, i.e., the processes that affect the Earth's surface and the impact of human activities on the atmospheric environment, it is sufficient to focus on the troposphere and the stratosphere.

3.1.4 Scales of Atmospheric Phenomena

Atmospheric phenomena cover a wide range of spatial and temporal scales, and it is useful to present briefly the concept of atmospheric scales and the corresponding phenomena.

The planetary scale corresponds to phenomena that cover important parts of the Earth and have temporal characteristics that are either permanent (or semi-permanent) or periodic. This category includes phenomena that constitute the general circulation (described in Section 3.2), as well as seasonal phenomena that recur every year, such as weather regimes (described in Section 3.3) and the monsoon.

The synoptic scale covers areas ranging from a few hundreds to several thousands of kilometers and time periods of several days. Major synoptic phenomena include low-pressure systems and high-pressure systems.

The mesoscale covers regional phenomena, which are affected by the regional physiographic configuration (coastal areas, mountainous areas, etc.). It ranges from a few kilometers up to the synoptic scale (i.e., several hundreds of kilometers). The corresponding time scales range from a few minutes up to one or two days. One may distinguish several subcategories of the mesoscale, which are, starting from the synoptic scale, the alpha mesoscale (approximately from 200 to 2000 km and from 6 h to 2 days), which covers part of the synoptic scale, the beta mesoscale (approximately from 20 to 200 km and from 30 min to 6 h), and the gamma mesoscale (approximately from 2 to 20 km and from 2 to 30 min). Examples of mesoscale meteorological phenomena include tropical storms at the alpha mesoscale, land/sea breezes and mountain/valley breezes at the beta mesoscale, and thunderstorms at the gamma mesoscale.

The aerologic scale (sometimes called the misoscale) covers areas ranging from a few hundred meters to several kilometers, with time scales on the order of a few minutes to an hour. Characteristic phenomena include tornadoes and isolated thunderstorms.

Finally, the microscale (which may include the misoscale) includes phenomena at the scale of a cloud or an eddy. Therefore, the corresponding areas may range from a few tens of centimeters to several hundreds of meters and the timescales range from a few seconds to several tens of minutes.

Note that the phenomena included in those different categories are related and that these scales are, therefore, interrelated. Furthermore, the spatial and temporal scales mentioned earlier in this section are approximate, because there is a large variability among the spatio-temporal characteristics of meteorological phenomena.

3.2 General Atmospheric Circulation

3.2.1 General Considerations

The general atmospheric circulation is governed by four major factors:

- Solar radiation
- The presence of water
- The presence of oxygen
- The rotation of the Earth

Solar radiation is more important at the equator than at the poles. Therefore, the amount of incoming energy absorbed by the Earth and its atmosphere is greater at the equator than at the poles. Furthermore, a lesser fraction is sent back into space from the tropics than from higher latitudes because of water vapor that is present, on average, in larger amount in the tropical atmosphere than at other latitudes. In addition, snow and ice, which are ubiquitous in the polar regions, lead to more radiation being reflected by the Earth's surface and, therefore, less absorption of radiative energy at the poles and in boreal regions than in the tropics. These phenomena, which are related mostly to the water cycle (evaporation in the tropics with the associated cloud formation and latent heat, presence of ice and snow in the polar regions) lead to a thermal energy difference between the equator (or more generally the tropics) and the polar regions. On average, the Earth is at equilibrium and a transfer of heat from the equator toward the poles must occur to prevent an accumulation of heat at the equator and a depletion at the poles (see Chapter 5).

The presence of oxygen in the Earth's atmosphere leads to the absorption of ultraviolet radiation by oxygen and ozone molecules in the stratosphere. These phenomena are described in Chapter 7. This absorption of radiative energy leads to a temperature increase in the stratosphere and, consequently, an inversion of the vertical temperature gradient above the tropopause. Such a temperature inversion leads to a quasi suppression of vertical motions of air parcels through the tropopause region (see Chapter 4).

The rotation of the Earth leads to a deviation of the atmospheric flow toward the right in the northern hemisphere and toward the left in the southern hemisphere. These deviations result from the Coriolis force, which is explained next.

3.2.2 Coriolis Force

The Earth's rotation leads to the fact that an air parcel moving along one direction in a space-fixed coordinate system will move along a different trajectory seen from an Earth-based rotating coordinate system. A simple demonstration consists in following the trajectory of an air parcel in the northern hemisphere that is moving southward. Let us assume that for an observer in a space-fixed coordinate system, the initial motion of this air parcel follows a straight-line trajectory, i.e., along a meridian (meridional trajectory, which has a constant longitude) and moves for a time step Δt from a high

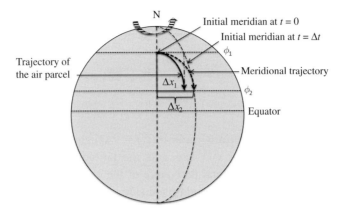

Figure 3.3. Schematic representation of the Coriolis force. The effect of the Earth's rotation is shown for a southward meridional atmospheric flow in the northern hemisphere.

latitude, ϕ_1, toward a lower latitude, ϕ_2. If the Earth's rotation (angular velocity) is Ω_T, at Δt the meridian is located east of its initial location in a space-fixed coordinate system at a zonal distance (i.e., along its new latitude ϕ_2): $\Delta x_2 = \Omega_T r_T \cos(\phi_2) \Delta t$, where r_T is the mean radius of the Earth. On the other hand, the air parcel has moved along a trajectory that corresponds to its initial velocity, $\Omega_T r_T \cos(\phi_1)$, and the zonal distance between the location of the air parcel and its initial location is $\Delta x_1 = \Omega_T r_T \cos(\phi_1) \Delta t$. Since the air parcel is moving southward toward the equator: $\phi_1 > \phi_2$, therefore, $\cos(\phi_1) < \cos(\phi_2)$. Thus, the air parcel is slightly behind the meridian corresponding to its initial location after the time step Δt ($\Delta x_1 < \Delta x_2$). Figure 3.3 illustrates this example.

Conversely, if the air parcel trajectory moves northward in the northern hemisphere, $\phi_1 < \phi_2$, and therefore, $\cos(\phi_1) > \cos(\phi_2)$. Then, the air parcel is slightly ahead of the initial meridian after a time step Δt ($\Delta x_1 > \Delta x_2$). In both cases, the air parcel trajectory has been deviated toward the right. The same demonstration can be carried out for the southern hemisphere, where the air parcel trajectory will be deviated to the left.

This conceptual description of the effect of the Earth's rotation on the direction of the atmospheric flow is actually approximate. Indeed, according to this demonstration, the motion of an air parcel along a latitude (zonal trajectory), for example eastward, would not lead to any deviation. Nevertheless, the Earth's rotation affects also air parcels with zonal trajectories and the associated deviations are also toward the right in the northern hemisphere and toward the left in the southern hemisphere.

The deviations of the atmospheric flow toward the right in the northern hemisphere and toward the left in the southern hemisphere result from a virtual force called the Coriolis force after Gustave-Gaspard Coriolis, who formulated its mathematical representation in 1835 (it had been identified as early as in the 17th century concerning the trajectories of cannon balls by Riccioli and Grimaldi). The Coriolis force is zero at the equator and is maximum at the poles.

The acceleration of the flow velocity may be defined simply according to the Navier-Stokes equations of fluid mechanics (see Chapter 4 for a more complete description of these equations). Here, we ignore the vertical wind velocity, which is typically negligible compared to the horizontal components:

$$\begin{aligned} \frac{\partial u}{\partial t} + u\frac{\partial u}{\partial x} + v\frac{\partial u}{\partial y} &= f\,v - \frac{\partial P}{\rho_a\,\partial x} \\ \frac{\partial v}{\partial t} + u\frac{\partial v}{\partial x} + v\frac{\partial v}{\partial y} &= -f\,u - \frac{\partial P}{\rho_a\,\partial y} \end{aligned} \qquad (3.22)$$

where f is the Coriolis parameter: $f = 2\,\Omega_T \sin(\phi)$, where Ω_T is the angular velocity of the Earth (radian s^{-1}) and ϕ is the latitude (°). The terms on the left-hand side correspond to the total derivative of the components of the horizontal wind velocity. The first terms on the right-hand side correspond to the Coriolis force and the second terms correspond to the pressure gradient force. At equilibrium (i.e., without any acceleration):

$$\frac{du}{dt} = \frac{dv}{dt} = 0 \qquad (3.23)$$

Changing total derivatives into partial derivatives:

$$\frac{\partial u}{\partial t} + u\frac{\partial u}{\partial x} + v\frac{\partial u}{\partial y} = \frac{\partial v}{\partial t} + u\frac{\partial v}{\partial x} + v\frac{\partial v}{\partial y} = 0 \qquad (3.24)$$

Therefore, the atmospheric flow is given by the following equations:

$$\begin{aligned} f\,v - \frac{\partial P}{\rho_a\,\partial x} &= 0 \\ -f\,u - \frac{\partial P}{\rho_a\,\partial y} &= 0 \end{aligned} \qquad (3.25)$$

These equations correspond to the geostrophic wind, i.e., the case where there is equilibrium between the pressure gradient force and the Coriolis force. Therefore:

$$\begin{aligned} u &= -\frac{1}{f\,\rho_a}\frac{\partial P}{\partial y} \\ v &= \frac{1}{f\,\rho_a}\frac{\partial P}{\partial x} \end{aligned} \qquad (3.26)$$

In the very simple case of the general circulation described in Section 3.2.3, the atmospheric pressure gradients are along the meridians, therefore: $\partial P / \partial x = 0$. Then:

$$u = -\frac{1}{f\,\rho_a}\frac{\partial P}{\partial y} \qquad (3.27)$$

The pressure gradient between a latitude of 30 °N (high-pressure region) and a latitude of 60 °N (low-pressure region) is negative (see Section 3.2.3). Therefore, u is positive, which corresponds to mid-latitude winds blowing eastward (i.e., westerly winds). This simple calculation does not account for the complexity of atmospheric motions, which (1) are not at equilibrium and (2) include a vertical component. Furthermore, the Coriolis parameter tends toward 0 when the latitude tends toward zero (at the equator); therefore, this calculation is not strictly applicable in the tropics.

3.2.3 Conceptual Representation of the Atmospheric General Circulation

The heat excess at the equator and the heat deficiency at the poles lead to a circulation transferring this heat from the tropics toward the polar regions (see Chapter 5). At the equator, the heat excess leads to a less dense atmosphere (according to the ideal gas law) and, therefore, an upward air motion toward the tropopause. This upward motion leads to a low pressure at the surface. The upward motion stops at the tropopause since the temperature inversion creates a stable layer, which tends to suppress vertical air motions. An air parcel moving upward must then move away from the equator toward the tropics. The Coriolis force deviates this air parcel eastward. Once its motion toward the poles has slowed down sufficiently (at about 30 °N and 30 °S), it cannot remain at the same altitude, because the space east of its position is already occupied by another air parcel; therefore, it must subside. This subsidence leads to high-pressure regions in the subtropical regions. The cycle is completed when air from the high-pressure regions moves toward the low-pressure region at the equator. This cycle was proposed by George Hadley, who was an attorney and amateur meteorologist. Accordingly, these atmospheric circulations are called the Hadley cells. The high-pressure regions, which are at about 30 °, are characterized by little rain and as a result many of the world deserts are located at those latitudes: for example, the Sahara desert in Africa, the Sonora desert in North America, the Arabian desert in Asia, and the Victoria desert in Australia.

A cycle similar to that of the Hadley cells occurs in the polar regions. These polar cells result from a subsidence of cold and dry air over the poles. At the surface, this descending air must move toward lower latitudes. At latitudes around 60 °, these air parcels run into air parcels that move toward higher latitudes from the high-pressure subtropical regions. Thus, an ascending motion occurs around 60 °. At these latitudes, the ascending air parcels lead to water vapor condensation and, therefore, heavy precipitation. These regions are associated with polar fronts. The upward air parcels meet the tropopause, but at altitudes lower than at the equator. These air parcels move then toward the poles where they subside, or toward lower latitudes where around 30 ° they meet air parcels originating from the equator.

The Hadley cells do not transfer energy from the equator to the poles because the air parcels subside and return toward the equator in the subtropical regions. The transfer continues in the temperate regions, i.e., the mid-latitudes, with a cycle that ranges from about 30 ° to 60 °. Those are the Ferrel cells. The transfer is completed with the polar cells, which move air parcels toward the poles via the ascending air motions at about 60 °.

Schematically, there is a low-pressure region at the equator and, in both hemispheres, a high-pressure region at about 30 °, a low-pressure region at about 60 °, and high-pressure regions at the poles. In the northern hemisphere, these high- and low-pressure regions are mostly limited to the oceans, because the presence of land masses affects these flow cycles significantly. Semi-permanent high-pressure regions around 30 ° include the Bermuda High in the Atlantic Ocean and the Pacific High in the Pacific Ocean. Semi-permanent low-pressure regions around 60 ° include the Aleutian Low in the Pacific Ocean and the Icelandic Low in the Atlantic Ocean. On the other hand, in the southern hemisphere, where land masses are few, there are three major high-pressure regions in the Atlantic, Indian, and Pacific Oceans, which are better defined than those of the northern hemisphere. There is also a low-pressure belt at about 60 °. On land, low- and high-pressure regions occur due to phenomena related to energy balance and atmospheric flows specific to land masses (for example, the Siberian High). The geographical locations of these low- and high-pressure regions vary according to season, since the energy balance depends on solar radiation. For example, the Pacific and Bermuda Highs are present at lower latitudes in winter than in summer.

The atmospheric flow occurs from high-pressure regions toward low-pressure regions, i.e., from latitudes of about 30 ° toward the equator (subtropical and tropical regions) and toward higher latitudes of about 60 ° (temperate regions) and from the poles toward the 60 ° latitudes (polar and subpolar regions). The Coriolis force affects these air flows, which initially are meridional flows (i.e., along a meridian).

In the subtropical and tropical regions, the air flowing toward the equator will be deviated toward the right in the northern hemisphere and toward the left in the southern hemisphere, i.e., westward. Thus, the winds in those regions are mostly easterly (i.e., originating from the east; more precisely northeasterly winds in the northern hemisphere and southeasterly winds in the southern hemisphere). They are typically called the trade winds, because they favored trade with ships sailing between Europe and the American continents. In the temperate regions, the deviation of winds toward the right in the northern hemisphere and toward the left in the southern hemisphere leads to prevailing winds originating mostly from the southwest in the northern hemisphere and from the northwest in the southern

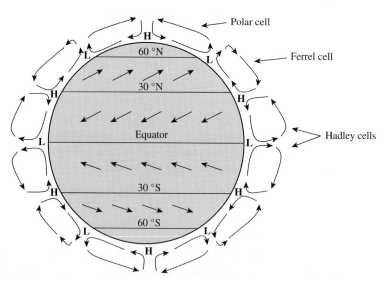

Figure 3.4. Atmospheric general circulation. This simplified representation depicts schematically the Hadley cells in the subtropical and tropical regions, the Ferrel cells in the temperate regions, the polar cells, the prevailing westerly winds in the mid-latitudes, and the easterly trade winds in the tropical regions. The semi-permanent high-pressure regions (H) are mostly at about 30 ° latitude and at the poles; the semi-permanent low-pressure regions (L) are mostly at 60 ° latitude.

hemisphere. In the polar regions, the air flows originating from the poles are deviated westward. Figure 3.4 summarizes schematically these main aspects of the atmospheric general circulation. More detailed descriptions are available in the scientific literature (for example, James, 1994).

3.3 Meteorological Regimes

3.3.1 Main Regimes

The processes described in Section 3.2 represent a very simple overview of the general atmospheric circulation. As mentioned, the main characteristics of this general circulation occur over the oceans and little over the continents. Indeed, the presence of complex terrain modifies the heat exchanges and atmospheric flows. Furthermore, the chaotic nature of meteorology implies that a small difference in an initial state of the atmosphere can lead to significant differences a few days later. Therefore, within the conceptual representation presented in Section 3.2, several scenarios may develop in relation with permanent structures (e.g., the Hadley cells) and semi-permanent ones (e.g., the Bermuda High and the Icelandic Low in the case of Europe; the Pacific High and the Aleutian Low in the case of North America). These meteorological scenarios are nevertheless limited in terms of their main characteristics, and they can be grouped in several regimes. For example, a representation using four meteorological regimes per season (i.e., 16 weather regimes in total) can be used for Europe. All days of a meteorological regime are not identical; however, they share similar characteristics at the synoptic scale. Figure 3.5 summarizes such a classification for the northern Atlantic Ocean and Europe for winter (December to February) and summer (June to August). These maps represent pressure anomalies compared to means calculated from meteorological data over the period 1950–2008. The frequencies of each regime are shown. Transition days lead to the change from one regime to the next. However, all regimes do not necessarily evolve into all other regimes. Spring and fall can also be represented with such regimes, which are intermediate between those of winter and summer.

The meteorological regimes called "North Atlantic Oscillation" (NAO) correspond to variations in the atmospheric pressure between the Bermuda High (as measured in Lisbon) and the Icelandic Low (as measured in Reykjavik). The NAO+ regime corresponds to periods when the pressure difference between these two locations is greater than the mean and the NAO- regime corresponds to the opposite situation. The Atlantic Ridge regime corresponds to a high-pressure region (barometric ridge) that originates from the Bermuda High and spreads northward over the Atlantic Ocean. Similarly, the Atlantic Talweg regime corresponds to a low-pressure region (barometric valley) that originates from the Icelandic Low and spreads southward over the northern Atlantic Ocean. The blocking regimes correspond to high-pressure and low-pressure regions that are stationary with respect to one another and, therefore, lead to meteorological conditions over the corresponding regions that vary little during the period of the blocking regime.

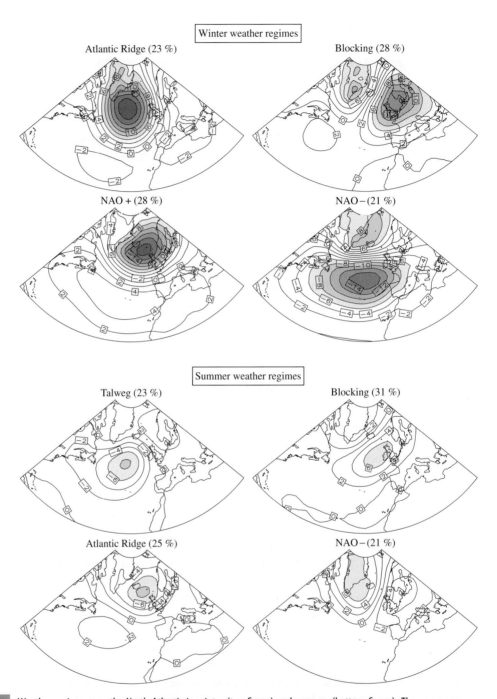

Figure 3.5. Weather regimes over the North Atlantic in winter (top figure) and summer (bottom figure). The occurrence frequencies for the period 1950–2008 are shown in terms of anomalies in the atmospheric pressure at sea level (PSL, hPa) with respect to the seasonal mean. Winter corresponds to December-January-February and summer corresponds to June-July-August. Source: Lecœur (2013).

Weather regimes have similarly been developed for other world regions. For example, Robertson and Ghil (1999) used six regimes to characterize meteorology during winter over the western United States.

3.3.2 Fronts and Blocks

A front is a virtual surface between two air masses of different properties (in terms of temperature, humidity, and pressure). One typically distinguishes cold fronts and warm fronts.

In a cold front, a cold air mass moves toward a warm air mass. Since the cold air is denser than the warm air, the cold air mass moves underneath the warm air mass, thereby lifting the warm air mass. The humid warm air encounters lower temperature aloft, which leads to the formation of clouds ahead of the cold front.

In a warm front, a warm air mass moves toward a cold air mass. The warm air is less dense than the cold air; therefore, the warm air mass moves on top of the cold air mass. Then, clouds form over the cold air mass.

One also considers stationary and occluded fronts. A stationary front corresponds to a virtual surface separating two air masses, one cold and the other warm, that are nearly stationary with respect to each other (for example, in the case where wind directions are parallel to the front). An occluded front corresponds to the case where a cold front catches up with a warm front. Then, the warm air mass is lifted above both cold air masses surrounding it.

Fronts are associated with precipitations, which result from the cloud formation in the frontal zone. These precipitations scavenge atmospheric pollutants (see Chapter 11) and a frontal situation is generally associated with an improvement in air quality.

A block results from the interruption of the zonal circulation in the mid-latitudes. This interruption is due to the presence of a high-pressure system (or that of a low-pressure system), which deviates the prevailing winds toward the north (or the south) in the northern hemisphere. In Europe, the blocking regimes correspond to the presence of a high-pressure system located over western Europe. Such a blocking regime may last several days and leads in the high-pressure region to meteorological conditions characterized by low wind speeds and clear skies. These conditions are conducive to an increase of the air pollution because of a reduced dispersion of the pollutants and the increased formation of secondary pollutants due to photochemical reactions enhanced by the increased solar radiation (see Chapters 6, 8, and 9).

Examples: Air pollution episodes in the Paris region

The two selected air pollution episodes occurred in the Paris region in December 2013 (winter conditions) and March 2014 (spring conditions). A meteorological analysis was conducted by Météo France, the French national weather service, for these episodes (see Figure 3.6, Tonnelier et al., 2015).

The episode occurring from December 8 to 12, 2013 corresponds to air pollution dominated by primary particulate matter (PM), i.e., those atmospheric particles that are

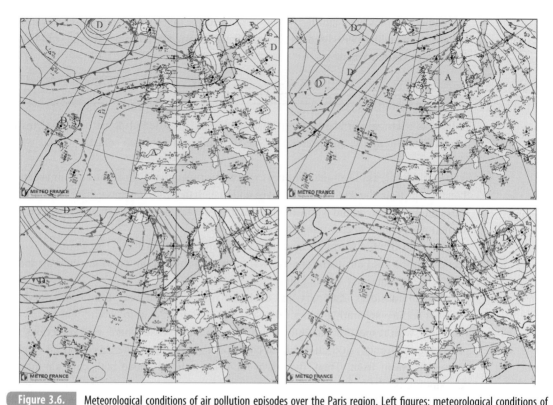

Figure 3.6. Meteorological conditions of air pollution episodes over the Paris region. Left figures: meteorological conditions of a winter air pollution episode dominated by primary PM on December 8, 2013 (top figure) and on December 13, 2013 (bottom figure, end of the episode); right figures: meteorological conditions of a spring air pollution episode dominated by secondary PM on March 11, 2014 (top figure) and on March 15, 2014 (bottom figure, end of the episode). "A" stands for "anticyclone" (i.e., high-pressure system); "D" for "depression" (i.e., low-pressure system). Source: Tonnelier et al., *Pollution Atmosphérique*, Special issue, March 2015, reproduced with permission of the authors.

directly emitted into the atmosphere (see Chapter 9). The PM_{10} hourly concentrations exceeded 100 µg m^{-3} and the black-carbon hourly concentrations (mostly emitted by diesel vehicles and biomass combustion) reached 12 µg m^{-3}. On December 8, a high-pressure system covered western Europe (1025 to 1035 hPa) and low-pressure regions were present over the northern Atlantic. These conditions correspond to a blocking regime (see Figure 3.6). The prevailing winds over the Paris region were from the southeast with low wind speeds. A strong temperature inversion was present in the planetary boundary layer, not only at night, but also during the day because of the low wintertime solar radiation. The temperature varied between −2 and 8 °C. This episode ended on December 13, with the end of the blocking regime and a transition toward a regime with westerly winds, which brought maritime air masses. This new NAO+ regime was then characterized by low-pressure systems coming from the northern Atlantic and a high-pressure region centered over the Atlantic south of Spain.

The episode occurring from March 11 to 14, 2014 corresponds to air pollution with PM_{10} hourly concentrations exceeding 160 µg m^{-3}. Unlike the winter episode mentioned, this spring episode was dominated by the secondary fraction of PM, which results from the condensation of semi-volatile compounds formed in the atmosphere via photochemical reactions of gaseous pollutants (see Chapter 9). These reactions were enhanced by a strong springtime solar radiation and clear skies. The meteorological conditions correspond to a blocking regime with a high-pressure system over western Europe (1025 to 1030 hPa) and low-pressure systems over the northern Atlantic. In the Paris region, low wind speeds followed the moderate wind speeds of the previous days with prevailing northeastern wind directions. The temperature inversion that was present at night remained in the morning, but disappeared during the day due to strong solar radiation. The temperature varied between 8 and 20 °C. The moderate temperature during the day favored chemical kinetics leading to the formation of nitric acid and semi-volatile organic compounds (SVOC) via oxidation of gaseous precursors. On the other hand, the low nighttime and morning temperature favored the condensation of these semi-volatile compounds on existing particles and, therefore, the formation of a significant secondary PM fraction (ammonium nitrate and secondary organic aerosols). This episode ended in a manner similar to that of the other one mentioned, i.e., via the arrival of maritime air masses. However, the transition was less pronounced than for the winter episode, because the high-pressure system located over the Atlantic Ocean was still present in the vicinity of France, i.e., at a higher latitude than in winter.

Problems

Problem 3.1 Calculation of atmospheric pressure
Calculate the atmospheric pressure and temperature at an altitude of 5 km, given that atmospheric pressure and temperature at sea level are 1 atm and 20 °C, respectively, and that atmospheric conditions are dry.

Problem 3.2 Coriolis force
a. Calculate the Coriolis parameter at a latitude of 45 °.
b. Compare at that latitude the accelerations due to a pressure gradient of 8 mbar over a distance of 400 km and the Coriolis force for a wind speed of 10 m s^{-1}.
c. Calculate the geostrophic wind speed at that latitude and for this pressure gradient.

References

Coriolis, G.-G., 1835. Sur les équations du mouvement relatif des systèmes de corps, *J. École royale polytechnique*, **15**, 144–154.

James, I.N., 1994. *Introduction to Circulating Atmospheres*, Cambridge University Press, Cambridge, UK.

Lecœur, È., 2013. *Influence de l'évolution climatique sur la qualité de l'air en Europe*, Ph.D. thesis, Université Paris-Est, France.

Robertson, A.W. and M. Ghil, 1999. Large-scale weather regimes and local climate over the western United States, *J. Climate*, **12**, 1796–1813.

Tonnelier, J.-P., S. Guidotti, B. Lossec, and F. Baraer, 2015. La météorologie et les pics de pollution. Étude de cas de décembre 2013 et mars 2014, Pollution Atmosphérique, *Special issue*, March 2015, 78–91.

4 Air Pollution Meteorology

Population exposure to air pollution occurs mostly near the Earth's surface. Furthermore, most air pollution sources are located near the Earth's surface (some exceptions include tall stacks, aircraft emissions, and volcanic eruptions). Therefore, the meteorological phenomena of the lower layers of the atmosphere are the most relevant to understand and analyze air pollution. The part of the atmosphere that is in contact with the Earth's surface and is affected by it is called the atmospheric planetary boundary layer (PBL). This chapter describes the dynamic processes that take place within the PBL. Those include, in particular, turbulent atmospheric flows and heat transfer processes, which affect air pollution near the surface. Those processes are often referred to as "air pollution meteorology," because they are the most relevant to air pollution. The major equations governing these processes are presented. A more detailed description of the PBL is available in books such as those by Stull (1988) and Arya (2001).

4.1 The Atmospheric Planetary Boundary Layer

The PBL can be defined according to its dynamic or thermal processes. From a dynamic viewpoint, the PBL is the atmospheric layer that is affected by the Earth's surface, either because of mechanical effects due to terrain, buildings, vegetation, etc., or indirectly because of heat transfer processes related to the evaporation of water or the anthropogenic production of heat (for example, the urban heat island effect). From a thermal viewpoint, the PBL is the atmospheric layer that is affected by the temporal variation of solar radiation, for example via its ambient temperature.

The PBL height varies as a function of terrain and solar radiation. It is on the order of a kilometer, but it is lower in winter when there is little solar radiation (on the order of a few hundred meters) and greater in summer when solar radiation is maximum (on the order of 2,000 to 3,000 m).

Within the PBL, one distinguishes the surface layer, which is located near the surface. Within the surface layer, vertical fluxes of momentum, heat, and mass transfer are nearly constant ($\sim \pm 10$ %). The surface layer has a thickness that is less than 10 % of that of the PBL. One should note, however, that within a forest or an urban canopy, the atmospheric flows are generally complex and the vertical fluxes are most likely not constant.

Above the PBL, the effect of the Earth's surface is negligible: this is the free atmosphere. When one is interested in air pollution at local or urban scales, it is generally appropriate to take into account only the PBL. On the other hand, when one is interested in air pollution at regional, continental, or global scales, the free troposphere must be taken into account, because air pollutants can be transported over long distances in the free troposphere and precipitation events, which scavenge air pollutants, generally include a significant fraction of the free troposphere.

4.2 Vertical Motions of Air Parcels

4.2.1 Potential Temperature

The change of pressure and temperature as a function of altitude is obtained via the hydrostatic and ideal gas laws (see Chapter 3). Furthermore, a relationship between pressure and temperature may be obtained by assuming that air parcels undergo adiabatic changes. Therefore, the law of Laplace applies:

$$PV^\gamma = \text{constant} \tag{4.1}$$

where γ is the adiabatic index of the gas (here, air), also known as the Laplace coefficient ($\gamma = c_p / c_v$, where c_p and c_v are the specific heat capacities at constant pressure and volume, respectively). Applying the ideal gas law (see Chapter 3), the following relationship between pressure and temperature is obtained under adiabatic conditions:

$$PT^{\gamma/(1-\gamma)} = \text{constant} \tag{4.2}$$

For a diatomic gas (which is the case of the atmosphere, since molecular nitrogen and oxygen account for 99 % of the air), $\gamma = 7/5 = 1.4$. Thus:

$$PT^{-3.5} = \text{constant} \tag{4.3}$$

Therefore:

$$TP^{-0.29} = \text{constant} \tag{4.4}$$

The potential temperature, θ, is defined as the temperature of an air parcel brought back to sea level at a standard pressure (i.e., $P_0 = 1$ atm) under adiabatic conditions. Therefore:

$$TP^{-0.29} = \theta P_0^{-0.29} \tag{4.5}$$

Thus:

$$\theta = T \left(\frac{P_0}{P}\right)^{0.29} \tag{4.6}$$

4.2.2 Neutral Atmosphere

A constant vertical profile of the potential temperature is defined as neutral. Let us consider an air parcel with uniform properties (pressure, temperature, and density). Under neutral conditions, this air parcel will remain in equilibrium with its environment as it moves upward or downward, since its temperature will change adiabatically and will, therefore, remain consistent with the temperature of the surrounding air.

4.2.3 Unstable Atmosphere

The atmosphere is unstable if the vertical profile of the potential temperature is negative, i.e., if the actual temperature decreases faster with height than the adiabatic gradient. An air parcel moving upward becomes warmer than that of its surroundings, because its adiabatic change in temperature is less than the decrease of the surrounding atmospheric temperature. Therefore, according to the ideal gas law, the density of this air parcel becomes less than that of the surrounding air. Thus, this air parcel keeps moving upward. Similarly, an air parcel moving downward undergoes an adiabatic change and its temperature change is less than that of the gradient of the surrounding atmosphere. Therefore, the surrounding temperature becomes greater than that of the air parcel. Thus, this air parcel is denser than the surrounding air and it continues to move downward. Vertical motions in an unstable atmosphere are enhanced because a small perturbation in the vertical direction leads to continuous motion in the direction of the initial perturbation (see Figure 4.1). Such unstable air parcels lead to vertical mixing of air pollution, at least up to an altitude where the atmosphere becomes neutral or stable (see Section 4.2.4). An unstable atmospheric layer is called a mixing layer. It occurs generally within the PBL during the day, because solar radiation warms up the Earth's surface, but does not affect the temperature of the

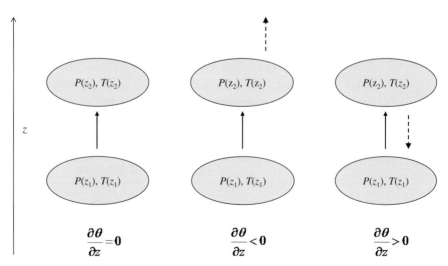

Figure 4.1. Schematic representation of vertical motions of an air parcel for neutral conditions (left figure), unstable conditions (middle figure), and stable conditions (right figure). The solid arrows indicate the initial perturbation of the air parcel, which undergoes an adiabatic change, and the dashed arrows indicate the following motion of the air parcel, which is due to the difference between the temperature of the air parcel and that of the surrounding air (i.e., density difference).

lower atmosphere (since the air absorbs little ultraviolet radiation within the troposphere, see Chapter 5). As a result, the vertical potential temperature gradient becomes negative.

4.2.4 Stable Atmosphere

The atmosphere is stable if the vertical profile of the potential temperature becomes positive. As an air parcel moves upward, its temperature follows an adiabatic change and it becomes lower than that of the surrounding air. Therefore, the air parcel becomes denser than the surrounding air. As a result, it tends to move back downward. Similarly, an air parcel moving downward undergoes an adiabatic change and its temperature becomes greater than that of the surrounding air. Thus, this air parcel, being less dense than the surrounding air, tends to move back upward. Therefore, a stable atmosphere tends to suppress vertical motions of air parcels. Thus, such an atmospheric layer may become stratified, because there are no, or few, transfers between the different layers (strata) of the atmosphere (see Figure 4.1). A stable atmospheric layer may be present aloft, for example above a mixing layer, or at the surface. This latter case occurs at night when the infrared radiation emitted by the Earth's surface toward the atmosphere leads to a decrease of the surface temperature and, accordingly, of the surface layer temperature. Therefore, a temperature inversion develops: the gradient of the actual temperature becomes positive and the gradient of the potential temperature becomes also positive. (Note that an atmospheric layer may be stable without a temperature inversion, as long as the actual temperature gradient is less than the adiabatic gradient.) Then, the atmospheric layer near the surface is stable at night. This stable layer will gradually become thicker as the surface temperature keeps decreasing. The PBL layer located above this stable layer is called the residual layer. Pollutants located within this residual layer are then isolated from the surface. However, during daytime, as solar radiation warms the surface, thereby leading to the gradual disappearance of the surface-based stable layer, the previously stable and residual layers merge to form a new mixing layer. Pollutants that were formerly present within the residual layer are mixed vertically and may then interact with the surface. The evolution of the atmospheric layer near the Earth's surface is depicted schematically in Figure 4.2. The stratosphere shows a positive vertical profile of the potential temperature; therefore, it is a stable atmospheric layer with few vertical air motions.

4.2.5 Effect of Moisture

This description of temperature gradients applies to dry atmospheres. If the atmosphere is moist, its vertical temperature gradient under adiabatic conditions will be less than that of a dry atmosphere. The water content, q (i.e., the specific humidity expressed in g of water per g of moist air), corresponds to a relative humidity (%), which is greater when temperature decreases, because the saturation vapor pressure of water decreases with temperature. In other words, for a given quantity (mass) of water per quantity of air (mass or volume), relative humidity increases when temperature decreases. Accordingly, for a given relative humidity, the water content of the atmosphere decreases when temperature decreases.

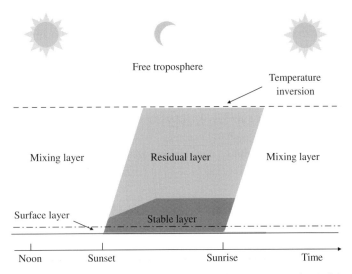

Figure 4.2. Schematic representation of the daily cycle of the planetary boundary layer. Source: After Stull (1988).

Humidity

Relative humidity (%) may be converted to absolute humidity using the method proposed by McRae (1980). According to Dalton's law, the water vapor concentration expressed in parts per million (ppm; 1 ppm = 1 molecule per million molecules of air) is:

$$[H_2O(ppm)] = 10^4 RH(\%) \frac{P_{s,H_2O}(T)}{P} \quad (4.7)$$

where T is the ambient temperature, $P_{s,H_2O}(T)$ is the saturation vapor pressure of water, and P is the atmospheric pressure. The saturation vapor pressure of water depends on temperature according to the following polynomial formula:

$$P_{s,H_2O}(T) = P \exp(13.3185\,\omega_T - 1.9760\,\omega_T^2 - 0.6445\,\omega_T^3 - 0.1299\,\omega_T^4) \quad (4.8)$$

where $\omega_T = 1 - (373.15/T)$, where T is expressed in K. For example, at $P = 1$ atm and $T = 25$ °C, $P_{s,H2O}$ (298 K) = 0.031 atm; at $T = 5$ °C, $P_{s,H2O}$ (278 K) = 0.0086 atm. At a relative humidity of 50 % and a temperature of 5 °C, there are about 4300 ppm of water vapor.

Converting % to (g m^{-3}) is performed as follows:

$$[H_2O(g\ m^{-3})] = 0.18\ RH(\%) \frac{P_{s,H_2O}(T)}{R\,T} \quad (4.9)$$

Thus, a relative humidity of 50 % at 1 atm and 5 °C corresponds to 3.4 g of water vapor per m^3 of air. The water vapor mass per volume of air is called the volumetric absolute humidity. The water vapor mass concentration at saturation ($RH = 100$ %) increases with temperature. For example, at 1 atm, there are 9.4, 17, and 30 g of water vapor per m^3 at 10, 20, and 30 °C, respectively.

Absolute humidity may also be defined as the mass of water vapor per mass of air. It can be defined per mass of dry air or per mass of moist air. The absolute humidity per mass of dry air is usually called the mixing ratio (by weight) and the absolute humidity per mass of moist air is usually called the specific humidity. The molar mass of dry air is 29 g mole^{-1}; therefore, according to the ideal gas law, the mass in g of one m^3 of dry air is: 29 $P_{dry\ air}$ /(RT). The mixing ratio is thus related to relative humidity as follows:

$$[H_2O(g/g\ dry\ air)] = 0.18 RH(\%) \frac{P_{s,H_2O}(T)}{R\ T} \frac{R\ T}{29\ P_{dry\ air}} = 0.62 \frac{RH(\%)}{100} \frac{P_{s,H_2O}(T)}{P_{dry\ air}} \quad (4.10)$$

According to Dalton's law, the total pressure is the sum of the partial pressures of dry air and water vapor:

$$P = P_{dry\ air} + \frac{RH(\%)}{100} P_{s,H_2O}(T) \quad (4.11)$$

Therefore:

$$[H_2O(g/g\ dry\ air)] = 0.62 \frac{RH(\%)}{100} \frac{P_{s,H_2O}(T)}{(P - \frac{RH(\%)}{100} P_{s,H_2O}(T))} \quad (4.12)$$

The specific humidity, q, is related to relative humidity as follows:

$$q = \frac{1}{(1.61 \frac{P}{P_{s,H_2O}(T)} \frac{100}{RH(\%)}) - 0.61} \quad (4.13)$$

Given that water vapor is a small fraction of the total mass of an air parcel, the mixing ratio and the specific humidity generally have very close values. For example, at T = 25 °C and P = 1 atm, $P_{s,H2O}$ = 0.031 atm (see previous equation). For a relative humidity of 100 %, the mixing ratio is calculated to be 0.0198 g H$_2$O/g dry air and the specific humidity, q, is 0.0195 g H$_2$O/g moist air, i.e., a difference of less than 2 %.

Virtual Temperature

The virtual temperature of a moist air parcel, T_v, is defined as the temperature that dry air would have for the same values of pressure and density as moist air. The relationship between the actual temperature, T, and the virtual temperature, T_v, (in K) is a function of the specific humidity, q (g of water/g of moist air), of the air parcel. It is given approximately by the following formula (see, for example, Wallace and Hobbs, 2006, for the exact relationship and the derivation of this simplified relationship):

$$T_v = T(1 + 0.61q) \quad (4.14)$$

The ideal gas law applies to a moist air parcel if T_v is used instead of T. Since water vapor has a molar mass (18 g mole^{-1}) that is less than that of dry air (29 g mole^{-1}), unsaturated moist air is less dense than dry air. Therefore, according to the ideal gas law, the virtual temperature is always greater than the actual temperature (by a few degrees at most), as shown by Equation 4.14. A virtual potential temperature, θ_v, may be defined as the potential temperature that dry air would have for the same values of pressure and density:

$$\theta_v = \theta \left(1 + 0.61q\right) \qquad (4.15)$$

Vertical Temperature Gradients

The adiabatic gradient of saturated moist air is, in absolute value, less than that of dry air. This gradient is due to the fact that at saturation, a moist air parcel moving upward will cool down and, therefore, will undergo partial condensation of water. Since condensation produces heat, the temperature of the air parcel decreases less than in the case of dry air. The atmospheric conditions can be categorized as follows in terms of the vertical temperature gradients for dry air and air saturated with water (the gradients are expressed here in absolute values and for cases where temperature decreases with altitude):

- Temperature gradient > dry adiabatic gradient: unstable atmosphere
- Temperature gradient = dry adiabatic gradient: neutral atmosphere if dry
- Saturated adiabatic gradient < temperature gradient < dry adiabatic gradient: stable, neutral, or unstable atmosphere depending on humidity
- Temperature gradient < saturated adiabatic gradient: stable atmosphere

4.3 Winds and Turbulence

4.3.1 Wind Direction in the PBL

As discussed in Chapter 3, the wind direction in the free troposphere results from the balance between two forces: the pressure gradient force and the Coriolis force (which is due to the Earth's rotation). In the PBL, another force comes into play: the frictional force, which represents the effect of the friction of the flow on the Earth's surface. Therefore, the wind direction at the surface results from the balance among these three forces: the pressure gradient force, the Coriolis force, and the frictional force. The frictional force is opposite to the wind direction. Furthermore, the Coriolis force is proportional to the wind speed. Since the wind speed decreases near the surface (tending toward zero at or near the surface, see Section 4.3.2), the Coriolis force also decreases near the surface. Thus, the influence of the pressure gradient force on the wind direction is greater near the surface. The theoretical solution for the northern hemisphere leads to a wind direction near the surface that is 45 ° to the left of the geostrophic wind direction. Thus, the wind direction evolves from that direction near the surface toward the geostrophic wind direction at the height within the PBL where the frictional force becomes negligible. The change in wind direction as a function of height is called the Ekman spiral. The depth of this Ekman layer, where the change in wind direction occurs, depends on the effect of the surface. It is greater for a situation with complex terrain (urban canopy, forest ...) than for a smooth surface (sea, snow, ice ...). The equations that govern the vertical profile of the wind speed and direction are described next.

4.3.2 Equations of Atmospheric Flow and Turbulence

The Navier-Stokes equations are used to calculate the atmospheric flow. They are expressed in Cartesian coordinates as follows:

$$\frac{\partial u}{\partial t} + u\frac{\partial u}{\partial x} + v\frac{\partial u}{\partial y} + w\frac{\partial u}{\partial z} - f\,v = -\frac{1}{\rho_a}\frac{\partial P}{\partial x} + \frac{1}{\rho_a}\left(\frac{\partial \tau_{xx}}{\partial x} + \frac{\partial \tau_{yx}}{\partial y} + \frac{\partial \tau_{zx}}{\partial z}\right)$$

$$\frac{\partial v}{\partial t} + u\frac{\partial v}{\partial x} + v\frac{\partial v}{\partial y} + w\frac{\partial v}{\partial z} + f\,u = -\frac{1}{\rho_a}\frac{\partial P}{\partial y} + \frac{1}{\rho_a}\left(\frac{\partial \tau_{xy}}{\partial x} + \frac{\partial \tau_{yy}}{\partial y} + \frac{\partial \tau_{zy}}{\partial z}\right) \quad (4.16)$$

$$\frac{\partial w}{\partial t} + u\frac{\partial w}{\partial x} + v\frac{\partial w}{\partial y} + w\frac{\partial w}{\partial z} = -\frac{1}{\rho_a}\frac{\partial P}{\partial z} + \frac{1}{\rho_a}\left(\frac{\partial \tau_{xz}}{\partial x} + \frac{\partial \tau_{yz}}{\partial y} + \frac{\partial \tau_{zz}}{\partial z}\right) - g$$

where the terms τ_{ij} represent the shear stresses (kg m^{-1} s^{-2}), which are due to the fluid viscosity. For a newtonian fluid, Newton's law of viscosity applies. It is written for fluid motion in a single direction as follows:

$$\tau_{yx} = \mu_{v,a}\frac{\partial u}{\partial y} \quad (4.17)$$

where $\mu_{v,a}$ is the dynamic viscosity of the air (kg m^{-1} s^{-1}). This equation may be seen as the flux in the y direction of the momentum in the x direction. This law, defined for a fluid moving in a given direction (here, x), has been generalized by Navier, Stokes, and Poisson for a fluid moving in three dimensions. Then, the shear stresses for momentum in the x direction are as follows:

$$\tau_{xx} = 2\mu_{v,a}\frac{\partial u}{\partial x} + \left(\mu_D - \frac{2}{3}\mu_{v,a}\right)\left(\frac{\partial u}{\partial x} + \frac{\partial v}{\partial y} + \frac{\partial w}{\partial z}\right)$$

$$\tau_{yx} = \mu_{v,a}\left(\frac{\partial u}{\partial y} + \frac{\partial v}{\partial x}\right) \quad (4.18)$$

$$\tau_{zx} = \mu_{v,a}\left(\frac{\partial u}{\partial z} + \frac{\partial w}{\partial x}\right)$$

The other terms for the y and z directions, τ_{xy}, τ_{yy}, τ_{zy}, τ_{xz}, τ_{yz}, and τ_{zz}, are written similarly. Clearly: $\tau_{xy} = \tau_{yx}$, $\tau_{yz} = \tau_{zy}$, and $\tau_{xz} = \tau_{zx}$. In these equations, $\mu_{v,a}$ is the dynamic viscosity of the fluid and μ_D is the dilatational viscosity of the fluid ($\mu_D = 0$ for a monoatomic gas and $\mu_D \approx 0$ for air).

Boussinesq introduced several assumptions that apply to the PBL:

- The kinematic viscosity and the molecular thermal conductivity are constant (i.e., their dependence on pressure and temperature can be neglected).
- The heat produced by the viscous forces is negligible compared to solar radiation in the heat transfer equation.
- The atmosphere is considered to be incompressible, since it is an open system (except at the Earth's surface).

- The fluctuations of the state variables (pressure, temperature, and density) are small compared to their mean values.
- The ratio of the pressure fluctuation and the mean pressure is small compared to the ratio of the temperature fluctuation and the mean temperature and compared to the ratio of the density fluctuation and the mean density.
- The fluctuations of the density are important only when the density is multiplied by the gravitational constant; indeed, the Earth's surface introduces a constraint on the system in the vertical direction.

The last two assumptions are commonly called the Boussinesq approximation or hypothesis: they imply that the change in the air density can be neglected except in the term ($\rho_a\, g\, dz$) of the hydrostatic equation (see Equation 3.2).

Using the Boussinesq hypothesis, the continuity equation for momentum is as follows:

$$\frac{\partial u}{\partial x} + \frac{\partial v}{\partial y} + \frac{\partial w}{\partial z} = 0 \qquad (4.19)$$

For u, the friction term may be written as follows (assuming that $\mu_{v,a}$ is spatially uniform):

$$\frac{1}{\rho_a}\left(\frac{\partial \tau_{xx}}{\partial x} + \frac{\partial \tau_{yx}}{\partial y} + \frac{\partial \tau_{zx}}{\partial z}\right) = \frac{\mu_{v,a}}{\rho_a}\left(\frac{\partial}{\partial x}\left(2\frac{\partial u}{\partial x} - \frac{2}{3}\left(\frac{\partial u}{\partial x} + \frac{\partial v}{\partial y} + \frac{\partial w}{\partial z}\right)\right)\right.$$
$$\left. + \frac{\partial}{\partial y}\left(\frac{\partial u}{\partial y} + \frac{\partial v}{\partial x}\right) + \frac{\partial}{\partial z}\left(\frac{\partial u}{\partial z} + \frac{\partial w}{\partial x}\right)\right) \qquad (4.20)$$

Changing the order of the derivations:

$$\frac{1}{\rho_a}\left(\frac{\partial \tau_{xx}}{\partial x} + \frac{\partial \tau_{yx}}{\partial y} + \frac{\partial \tau_{zx}}{\partial z}\right) = \frac{\mu_{v,a}}{\rho_a}\left(\left(\frac{\partial^2 u}{\partial x^2} + \frac{\partial^2 u}{\partial y^2} + \frac{\partial^2 u}{\partial z^2}\right) - \frac{2}{3}\frac{\partial}{\partial x}\left(\frac{\partial u}{\partial x} + \frac{\partial v}{\partial y} + \frac{\partial w}{\partial z}\right)\right.$$
$$\left. + \frac{\partial}{\partial x}\left(\frac{\partial u}{\partial x} + \frac{\partial v}{\partial y} + \frac{\partial w}{\partial z}\right)\right) \qquad (4.21)$$

According to the continuity equation, the last two terms are zero. Since the kinematic viscosity, $v_{v,a}$(m^2 s^{-1}), is equal to the ratio of the dynamic viscosity and the fluid density, $\mu_{v,a}/\rho_a$:

$$\frac{1}{\rho_a}\left(\frac{\partial \tau_{xx}}{\partial x} + \frac{\partial \tau_{yx}}{\partial y} + \frac{\partial \tau_{zx}}{\partial z}\right) = v_{v,a}\left(\frac{\partial^2 u}{\partial x^2} + \frac{\partial^2 u}{\partial y^2} + \frac{\partial^2 u}{\partial z^2}\right) \qquad (4.22)$$

Thus, the Navier-Stokes equations may be written as follows:

$$\frac{\partial u}{\partial t} + u\frac{\partial u}{\partial x} + v\frac{\partial u}{\partial y} + w\frac{\partial u}{\partial z} - fv = -\frac{1}{\rho_a}\frac{\partial P}{\partial x} + v_{v,a}\nabla^2 u$$
$$\frac{\partial v}{\partial t} + u\frac{\partial v}{\partial x} + v\frac{\partial v}{\partial y} + w\frac{\partial v}{\partial z} + fu = -\frac{1}{\rho_a}\frac{\partial P}{\partial y} + v_{v,a}\nabla^2 v \qquad (4.23)$$
$$\frac{\partial w}{\partial t} + u\frac{\partial w}{\partial x} + v\frac{\partial w}{\partial y} + w\frac{\partial w}{\partial z} = -\frac{1}{\rho_a}\frac{\partial P}{\partial z} + v_{v,a}\nabla^2 w - g$$

where u, v, and w are the wind velocities (m s^{-1}) in the x, y, and z directions, respectively, f is the Coriolis parameter (s^{-1}), ρ_a is the air density (kg m^{-3}), P is the atmospheric pressure (Pa), $v_{v,a}$ is the kinematic viscosity of the air (m^2 s^{-1}), and g is the acceleration due to gravity (m s^{-2}).

The Reynolds number, Re, may be used to estimate whether a flow is turbulent or laminar. It represents the ratio of inertial forces and viscous forces, and it is defined as follows:

$$\mathrm{Re} = \frac{u l_c}{v_{v,a}} \quad (4.24)$$

where u is the velocity of the flow, l_c is a characteristic length of the flow, and $v_{v,a}$ is the kinematic viscosity ($v_{v,a} = 1.55 \times 10^{-5}$ m^2 s^{-1} at 1 atm and 25 °C). For a wind velocity of 10 m s^{-1} and a characteristic length corresponding to the PBL height, say 1,000 m:

$$\mathrm{Re} = 10 \times 1000/(1.55 \times 10^{-5}) = 7 \times 10^8$$

A flow is considered to be turbulent when Re > 3,000. Therefore, the atmospheric flow within the PBL is without any doubt turbulent.

In a turbulent flow, the velocity varies randomly around a mean value. Therefore, it may be written as the sum of a mean value and a term that represents the fluctuation around that mean value (according to the statistical representation of Reynolds):

$$u = \bar{u} + u' \quad (4.25)$$

where $\bar{u}$ represents the mean value of the velocity and u' represents the random fluctuation around that mean value. The Navier-Stokes equations may then be written as a function of these mean and random terms. Applying the continuity equation for momentum, the Navier-Stokes equations may be rewritten as follows:

$$\begin{aligned}
\frac{\partial u}{\partial t} + \frac{\partial u^2}{\partial x} + \frac{\partial uv}{\partial y} + \frac{\partial uw}{\partial z} - fv &= -\frac{1}{\rho_a}\frac{\partial P}{\partial x} + v_{v,a} \nabla^2 u \\
\frac{\partial v}{\partial t} + \frac{\partial vu}{\partial x} + \frac{\partial v^2}{\partial y} + \frac{\partial vw}{\partial z} + fu &= -\frac{1}{\rho_a}\frac{\partial P}{\partial y} + v_{v,a} \nabla^2 v \\
\frac{\partial w}{\partial t} + \frac{\partial wu}{\partial x} + \frac{\partial wv}{\partial y} + \frac{\partial w^2}{\partial z} &= -\frac{1}{\rho_a}\frac{\partial P}{\partial z} + v_{v,a} \nabla^2 w - g
\end{aligned} \quad (4.26)$$

By decomposing the velocities as the sum of a mean value and a fluctuation around that mean value (Reynolds decomposition), we obtain for u (similar equations may be written for v and w):

$$\frac{\partial(\bar{u}+u')}{\partial t} + \frac{\partial(\bar{u}+u')^2}{\partial x} + \frac{\partial(\bar{u}+u')(\bar{v}+v')}{\partial y} + \frac{\partial(\bar{u}+u')(\bar{w}+w')}{\partial z} - f(\bar{v}+v')$$
$$= -\frac{1}{\rho_a}\frac{\partial P}{\partial x} + v_{v,a} \nabla^2 (\bar{u}+u') \quad (4.27)$$

Averaging over time leads to the Navier-Stokes equation averaged according to the Reynolds decomposition, i.e., RANS for "Reynolds-averaged Navier-Stokes." Since the mean of the fluctuation term around the mean is by definition zero, the following relationships apply:

$$\bar{\bar{u}} = \bar{u}$$
$$\overline{u'} = 0 \qquad (4.28)$$

The RANS equation for u is then as follows:

$$\frac{\partial \bar{u}}{\partial t} + \frac{\partial \bar{u}^2}{\partial x} + \frac{\partial \overline{u'^2}}{\partial x} + \frac{\partial \overline{u}\overline{v}}{\partial y} + \frac{\partial \overline{u'v'}}{\partial y} + \frac{\partial \overline{u}\overline{w}}{\partial z} + \frac{\partial \overline{u'w'}}{\partial z} - f\,\bar{v} = -\frac{1}{\rho_a}\frac{\partial P}{\partial x} + \nu_{v,a}\,\nabla^2 \bar{u} \qquad (4.29)$$

The terms that include the random variables u', v', and w' represent turbulence, and their products are called the Reynolds stresses:

$$R_{xx} = \overline{u'^2} \; ; \; R_{xy} = \overline{u'v'} \; ; \; R_{xz} = \overline{u'w'} \qquad (4.30)$$

Similar terms may be obtained for v and w. Note that these Reynolds stresses, as defined here, have units of m^2 s^{-2} and not the units of a standard stress, which are kg m^{-1} s^{-2}; the unit of a standard stress would be obtained if the terms on the right-hand sides were multiplied by the fluid density.

There is a set of nine Reynolds stresses, which are called the Reynolds tensor. Because of symmetry between R_{xy} and R_{yx}, R_{xz} and R_{zx}, and R_{yz} and R_{zy}, there are six unknowns in the Navier-Stokes equations, which result from the turbulent nature of the atmospheric flow. New equations must, therefore, be introduced to "close" this system. Several approaches have been proposed to close the RANS equations with an explicit representation of the turbulent terms.

4.3.3 Closure of the Equations of Turbulent Flows

The Boussinesq hypothesis allows one to introduce a relationship between the momentum turbulent term and the mean value of the flow velocity. One introduces the turbulent kinematic viscosity, K_M (m^2 s^{-1}), which is defined as follows (here for the x and z directions) by analogy with the shear stress ($\tau_{zx} = \mu_{v,a} \frac{\partial u}{\partial z}$):

$$-\overline{u'w'} = K_M \frac{\partial \bar{u}}{\partial z} \qquad (4.31)$$

However, note that this analogy is approximate, because the dynamic viscosity, $\mu_{v,a}$, is a property of the fluid (here, the air), whereas the turbulent kinematic viscosity is a characteristic of the flow. K_M must be defined. Prandtl introduced the concept of a characteristic mixing length, l_m, which characterizes the length over which a turbulent eddy will retain its properties before mixing with the surrounding fluid. A dimensional analysis implies that the turbulent kinematic viscosity may be expressed as the product of a characteristic velocity, u_* (called the friction velocity), and this characteristic mixing length, l_m:

$$K_M = u_* \, l_m \qquad (4.32)$$

The friction velocity is defined as the square root of the product of the fluctuations of the horizontal and vertical velocities:

$$-\overline{u'w'} = u_*^2 \qquad (4.33)$$

The friction velocity represents the amplitude of the vertical turbulent flux of the horizontal momentum within the surface layer. Equations 4.31 to 4.33 lead to the following results:

$$u_*^2 = u_* \, l_m \frac{\partial \overline{u}}{\partial z}; \quad \text{i.e.,} \quad u_* = l_m \frac{\partial \overline{u}}{\partial z} \qquad (4.34)$$

The mixing length is not constant, but depends on the distance from the wall (here, the Earth's surface). This relationship uses the von Kármán constant, κ, taken here to be equal to 0.4:

$$l_m = \kappa \, z \qquad (4.35)$$

Therefore, the turbulent kinematic viscosity is expressed as follows:

$$K_M = \kappa \, u_* \, z \qquad (4.36)$$

The friction velocity, u_*, is on the order of 0.1 to 0.5 m s^{-1} for standard atmospheric conditions.

4.3.4 Kinetic Energy of the Atmospheric Flow

The specific kinetic energy of the flow (i.e., kinetic energy per unit mass; m^2 s^{-2}) is defined as follows:

$$E_k = \frac{1}{2}(u^2 + v^2 + w^2) = \frac{1}{2}((\overline{u}+u')^2 + (\overline{v}+v')^2 + (\overline{w}+w')^2) \qquad (4.37)$$

The mean specific kinetic energy, E_k, is defined as follows:

$$\begin{aligned} \overline{E_k} &= \frac{1}{2}(\overline{u}^2 + \overline{v}^2 + \overline{w}^2) + \frac{1}{2}(\overline{u'^2} + \overline{v'^2} + \overline{w'^2}) \\ k_{MKE} &= \frac{1}{2}(\overline{u}^2 + \overline{v}^2 + \overline{w}^2) \\ k_{TKE} &= \frac{1}{2}(\overline{u'^2} + \overline{v'^2} + \overline{w'^2}) \\ \overline{E_k} &= k_{MKE} + k_{TKE} \end{aligned} \qquad (4.38)$$

where k_{MKE} is the specific kinetic energy of the mean flow and k_{TKE} is the mean specific kinetic energy of the turbulent component of the flow (called TKE for "turbulent kinetic energy"). An equation may be derived for the change in TKE. This equation contains a term, ε, which represents the TKE mean dissipation rate (m^2 s^{-3}). Several computational fluid dynamics (CFD) models use the TKE equation to close the RANS equations. Among the models that are the most widely used, one may mention the k-l_m, k-ε, and k-ω models (ω is the TKE specific mean dissipation rate). For a more in-depth description of turbulent flow modeling, see, for example, Nieuwstadt et al. (2016).

4.3.5 Vertical Profile of Wind Speed in the Surface Layer

Wind speed changes with height. The vertical profile of wind speed follows a logarithmic law under adiabatic conditions, which results from the assumptions introduced with the first-order closure of turbulence and the use of the Prandtl mixing length (according to Equations 4.34 and 4.35):

$$\frac{\partial u}{\partial z} = \frac{u_*}{l_m} = \frac{u_*}{\kappa z} \qquad (4.39)$$

In the surface layer, the wind speed in a given horizontal direction is given by a differential equation (assuming here for simplicity that the wind direction is in the x direction; i.e., $v = 0$):

$$du(z) = \frac{u_*}{\kappa} \frac{dz}{z} \qquad (4.40)$$

For a viscous fluid, the no-slip assumption is generally applied, i.e., the fluid velocity is zero at the surface. However, such a boundary condition does not apply here since the logarithm of zero is not defined and one must introduce a height at which the wind velocity is zero. This height, z_0, is called the roughness length. Thus, the wind velocity follows the following profile:

$$u(z) = \frac{u_*}{\kappa} \ln(\frac{z}{z_0}) \quad \text{for } z \geq z_0 \qquad (4.41)$$

The height at which the wind speed becomes zero can be estimated experimentally by extrapolating the wind speed in the vertical direction. Thus, the roughness length can be estimated. This roughness length represents the effect of terrain and obstacles (buildings, vegetation, etc.) on the wind speed vertical profile. As a first approximation, it can be estimated to be in the range of 1/30 to 1/10 of the height of the terrain elements. For example, over grassland with grass about 3 cm high, the roughness length would be in the range of 1 to 3 mm; in an urban area with buildings about 10 m high, the roughness length would be in the range of 30 cm to 1 m.

This logarithmic representation of the vertical wind speed profile is idealized, because no momentum is taken into account below z_0. Furthermore, in the case where obstacles affect the atmospheric flow significantly (for example, in a forest or in an urban canopy), a displacement height, d_b, is introduced, which corresponds to a vertical translation of the surface toward the top of the canopy. For example, for a built area, one may define the displacement height empirically as a function of the fraction of the area that is built, λ_b, and of the mean building height, h_b (MacDonald et al., 1998):

$$d_b = h_b \left[1 - \frac{1 - \lambda_b}{4^{\lambda_b}} \right] \qquad (4.42)$$

For an urban area with a built fraction of, say, 50 % ($\lambda_b = 0.5$), $d_b = 0.75\ h_b$. In the limit, d_b tends toward 0 as λ_b tends toward 0 and d_b tends toward h_b as λ_b tends toward 1.

The vertical profile of the wind speed is then defined above a height $(d_b + z_0)$, at which the wind speed becomes zero. If one wants to represent the flow below that height, one must introduce another model of the atmospheric flow within the canopy. For example, in an urban canopy, experimental data have led to the assumption that the vertical profile of the wind speed may be represented by an exponential function (MacDonald, 2000):

$$u(z) = \xi_b \, u(h_b) \, \exp(\beta_b(\frac{z}{h_b} - 1)) \tag{4.43}$$

where ξ_b and β_b are empirical parameters, which depend on the built area configuration, $u(h_b)$ is the wind speed at the top of the canopy (which can be taken to be the wind speed above the canopy multiplied by the cosine of the angle of the wind direction and the street-canyon axis), and h_b is the average building height. For example, the following parameterizations may be used (Masson, 2000):

$$\begin{aligned} \beta_b &= \frac{h_b}{2W_s} \\ \xi_b &= 1 \text{ for } \frac{h_b}{W_s} \leq \frac{1}{3} \\ \xi_b &= 1 + 3\left(\frac{2}{\pi} - 1\right)\left(\frac{h_b}{W_s} - \frac{1}{3}\right) \text{ for } \frac{1}{3} \leq \frac{h_b}{W_s} \leq \frac{2}{3} \\ \xi_b &= \frac{2}{\pi} \text{ for } \frac{2}{3} \leq \frac{h_b}{W_s} \end{aligned} \tag{4.44}$$

where W_s is the mean street width. The term h_b / W_s is sometimes called the aspect ratio of the street canyon (Landsberg, 1981). Near the street surface, the no-slip condition cannot be satisfied by an exponential profile and a logarithmic profile is used. The vertical profile of the wind speed in an urban area may, therefore, be seen as an ensemble of three connected profiles, which are, starting at the surface:

– a logarithmic profile near the surface
– an exponential profile within the urban canopy
– a logarithmic profile above the urban canopy

These three profiles must connect at heights that are defined by the canopy configuration and atmospheric flow in order to obtain a continuous vertical wind profile (e.g., MacDonald, 2000).

For non-adiabatic conditions, vertical motions are affected by the temperature vertical profile. For neutral adiabatic conditions, one may define the non-dimensional wind shear number, Φ_M, as follows:

$$\frac{du}{dz} \frac{\kappa \, z}{u_*} = \Phi_M \tag{4.45}$$

where by definition, $\Phi_M = 1$, according to Equation 4.39. This approach, which involves the use of dimensionless variables (sometimes called "numbers"), has been widely used to define the characteristics of the PBL. Once the standard variables (vertical fluxes, etc.) have been represented using dimensionless functions, the results are applicable regardless of the absolute dimensions. This approach is, therefore, called "similarity theory." For the PBL, one refers to the similarity theory of Monin-Obukhov (after the names of the two Russian scientists, Andrei Monin and Alexander Obukhov, who proposed it in 1954). This approach has been useful to estimate non-dimensional variables using experimental data for a wide range of atmospheric conditions. Thus, Businger and co-workers (1971) and Dyer (1974) have independently developed corrected vertical profiles for unstable and stable conditions. These profiles use a dimensionless height, ζ. For stable conditions:

$$\Phi_M = 1 + 4.7\zeta \tag{4.46}$$

For unstable conditions:

$$\Phi_M = (1 - 15\zeta)^{-1/4} \tag{4.47}$$

These functions were determined experimentally with the assumption that the von Kármán constant was $\kappa = 0.35$. For a value $\kappa = 0.4$, the following relationships should be used (Jacobson, 2005):

$$\begin{aligned} &\text{Stable conditions:} & \Phi_M &= 1 + 6\zeta \\ &\text{Neutral conditions:} & \Phi_M &= 1 \\ &\text{Unstable conditions:} & \Phi_M &= (1 - 19.3\zeta)^{-1/4} \end{aligned} \tag{4.48}$$

In these equations, the dimensionless height, ζ, is defined as (z/L_{MO}), where L_{MO} is the Monin-Obukhov length (as discussed in Section 4.4.3). During daytime, the absolute value of L_{MO} corresponds to the height above which turbulence is generated more by buoyancy (i.e., thermal processes) than by wind shear (i.e., mechanical processes). Its formulation is provided in Equation 4.59. L_{MO} is negative for unstable conditions and is positive for stable conditions. For neutral conditions, L_{MO} is infinite and, therefore, $\zeta = 0$.

The vertical profiles of the wind speed are then obtained by integration:

$$\int_0^u du = \int_{z_0}^z \Phi_M(z) \frac{u_*}{\kappa} \frac{dz}{z} \tag{4.49}$$

Analytical solutions may be obtained. For neutral conditions, the solution is straightforward and was provided in Equation 4.41. For stable conditions:

$$u(z) = \frac{u_*}{\kappa} \left[\ln\left(\frac{z}{z_0}\right) + (\Phi_M - \Phi_{M_0}) \right] \tag{4.50}$$

where Φ_{M0} corresponds to the value of Φ_M at $z = z_0$. For unstable conditions:

$$u(z) = \frac{u_*}{\kappa} \left[\ln\left(\frac{1 - \Phi_M}{1 + \Phi_M}\right) - \ln\left(\frac{(1 - \Phi_{M_0})}{(1 + \Phi_{M_0})}\right) + 2\left(\arctan(\Phi_M^{-1}) - \arctan(\Phi_{M_0}^{-1})\right) \right] \tag{4.51}$$

For neutral conditions, $\zeta = 0$ and $\Phi_M = 1$; therefore, the two profiles represented by Equations 4.50 and 4.51 then become equivalent to the logarithmic profile given by Equation 4.41.

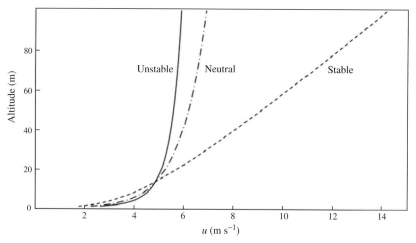

Figure 4.3. Vertical profiles of the horizontal wind speed in the surface layer for stable, neutral, and unstable conditions above 1 m. The assumptions are a roughness length (z_0) of 0.1 m, a friction velocity (u_*) of 0.3, 0.4, and 0.5 m s^{-1}, respectively, and a Monin-Obukhov length (L_{MO}) of 50 m, infinite, and −20 m, respectively.

Figure 4.3 depicts vertical profiles of the horizontal wind speed in a surface layer of 100 m depth for stable conditions (Equation 4.50), neutral conditions (Equation 4.41), and unstable conditions (Equation 4.51). The assumptions regarding the roughness length, the friction velocity, and the Monin-Obukhov length are provided in the figure caption. The vertical gradient of the wind speed is much more pronounced when the atmosphere is stratified, and it becomes rapidly negligible above the surface when the conditions are unstable.

4.4 Heat Transfer

4.4.1 Heat Transfer Equation

The heat transfer equation may be written as follows, using the virtual potential temperature, θ_v:

$$\frac{\partial \theta_v}{\partial t} + u\frac{\partial \theta_v}{\partial x} + v\frac{\partial \theta_v}{\partial y} + w\frac{\partial \theta_v}{\partial z} = \frac{\theta_v}{c_{p,\,dry}\,T_v}\frac{dQ}{dt} \tag{4.52}$$

where T_v is the virtual temperature (K), Q is the amount of heat exchanged by the system (J kg^{-1}), $c_{p,dry}$ is the specific heat capacity of dry air (J kg^{-1} K^{-1}), and the term on the right-hand side represents the sources and sinks of heat, i.e., the rates of heat transfer due to changes in water phases (condensation, evaporation, freezing, and melting), solar radiation (absorption of solar radiation in the ultraviolet, visible, and infrared ranges and infrared radiation emission by the Earth's surface and by greenhouse gases), and human activities (heating, transportation, industrial activities, etc.). This equation may be rewritten as follows:

$$\frac{\partial \theta_v}{\partial t} + \frac{\partial u\theta_v}{\partial x} + \frac{\partial v\theta_v}{\partial y} + \frac{\partial w\theta_v}{\partial z} - \theta_v \left(\frac{\partial u}{\partial x} + \frac{\partial v}{\partial y} + \frac{\partial w}{\partial z} \right) = \frac{\theta_v}{c_{p,\,dry}\, T_v} \frac{dQ}{dt} \quad (4.53)$$

The Boussinesq assumption implies (see Equation 4.19) that the fifth term is zero.

4.4.2 Sensible Heat Flux

Turbulence affects the atmospheric flow, as well as any associated variable. Thus, temperature is a turbulent variable, which can be decomposed following the Reynolds approach using a mean value and a random component around that mean value. The heat transfer equation in its Reynolds average form is written as follows:

$$\theta_v = \overline{\theta_v} + \theta_v'$$

$$\frac{\partial \overline{\theta_v}}{\partial t} + \bar{u}\frac{\partial \overline{\theta_v}}{\partial x} + \bar{v}\frac{\partial \overline{\theta_v}}{\partial y} + \bar{w}\frac{\partial \overline{\theta_v}}{\partial z} + \frac{\partial \overline{u'\theta_v'}}{\partial x} + \frac{\partial \overline{v'\theta_v'}}{\partial y} + \frac{\partial \overline{w'\theta_v'}}{\partial z} = \frac{\overline{\theta_v}}{c_{p,\,dry}\, T_v} \frac{dQ}{dt} \quad (4.54)$$

A first-order closure may be performed using the turbulent heat transfer coefficient, K_H (m² s⁻¹):

$$\frac{\partial \overline{u'\theta_v'}}{\partial x} = -\frac{\partial}{\partial x}\left(K_H \frac{\partial \overline{\theta_v}}{\partial x} \right)$$

$$\frac{\partial \overline{v'\theta_v'}}{\partial y} = -\frac{\partial}{\partial y}\left(K_H \frac{\partial \overline{\theta_v}}{\partial y} \right) \quad (4.55)$$

$$\frac{\partial \overline{w'\theta_v'}}{\partial z} = -\frac{\partial}{\partial z}\left(K_H \frac{\partial \overline{\theta_v}}{\partial z} \right)$$

The last term represents heat transfer via the vertical turbulent flux, which transfers heat from the Earth's surface toward the atmosphere (see Chapter 5). It is related to the sensible heat flux, which is defined as follows (J m⁻² s⁻¹):

$$F_{Qs} = \rho_a\, c_{p,\,dry}\, \overline{w'\theta_v'} \quad (4.56)$$

Then, the heat transfer equation is written as follows (using for simplicity the notation θ_v for $\overline{\theta_v}$ and u, v, and w for $\bar{u}$, $\bar{v}$, and $\bar{w}$):

$$\frac{\partial \theta_v}{\partial t} + u\frac{\partial \theta_v}{\partial x} + v\frac{\partial \theta_v}{\partial y} + w\frac{\partial \theta_v}{\partial z} - \nabla K_H \nabla \theta_v = \frac{\theta_v}{c_{p,\,dry}\, T_v} \frac{dQ}{dt} \quad (4.57)$$

4.4.3 Variables Characterizing the PBL Stability

In the surface layer, a temperature scale, θ_*, may be introduced, which is related to the sensible heat flux at the surface and which characterizes the production of turbulence via heat transfer:

$$\theta_* = -\frac{\overline{(w'\theta')}_0}{u_*} \tag{4.58}$$

Using the friction velocity, which characterizes the production of turbulence via mechanical processes (flow affected by terrain and/or obstacles), and the temperature scale, a length scale, L_{MO}, may be defined, which is called the Monin-Obukhov length:

$$L_{MO} = \frac{u_*^2\,\theta_v}{\kappa\,g\,\theta_*} \tag{4.59}$$

Thus, the Monin-Obukhov length (Monin and Obukhov, 1954) represents during daytime the height above which the production of turbulence via buoyancy dominates the production of turbulence via mechanical processes (i.e., wind shear). A negative value of L_{MO} implies that the atmosphere is unstable because of heat transfer. A positive value of L_{MO} corresponds to a stable atmosphere. Under neutral conditions, L_{MO} tends to infinity, since the heat flux tends toward zero and, therefore, θ_* tends toward zero. The following ranges of L_{MO} are typical of atmospheric conditions: very unstable conditions when L_{MO} ranges between 0 and –10 m, unstable conditions when L_{MO} is less than –10 m, neutral conditions when L_{MO} has very high values (>100 m in absolute value), stable conditions when L_{MO} is greater than 30 m, and very stable conditions when L_{MO} ranges between 0 and 30 m. However, these ranges are rough estimates.

The non-dimensional Richardson numbers are also used to characterize the atmospheric stability (they are named after the English mathematician and physicist Lewis Fry Richardson). The flux Richardson number is expressed as follows:

$$\mathrm{Ri}_f = \frac{g}{T_v}\frac{\overline{w'\theta'}}{\overline{u'w'}\frac{\partial u}{\partial z}} \tag{4.60}$$

The gradient Richardson number is expressed as follows:

$$\mathrm{Ri}_g = \frac{g}{T_v}\frac{\frac{\partial \theta}{\partial z}}{\left(\frac{\partial u}{\partial z}\right)^2} \tag{4.61}$$

where T_v is in K. The bulk Richardson number, Ri_b, is derived from Ri_g by using discretized vertical gradients of temperature and horizontal wind speed, obtained from meteorological measurements at distinct heights:

$$\mathrm{Ri}_b = \frac{g}{T_v}\frac{\frac{\Delta \theta}{\Delta z}}{\left(\frac{\Delta u}{\Delta z}\right)^2} \tag{4.62}$$

A positive Richardson number corresponds to a stable atmosphere and a negative Richardson number corresponds to an unstable atmosphere. For a neutral atmosphere, $\mathrm{Ri} = 0$. The Richardson numbers depend on the height at which they are estimated, since the vertical gradients of wind speed and temperature are functions of z. At a given height (or within a given atmospheric layer in the case of Ri_g and Ri_b), they represent the ratio of the heat-generated turbulence (represented by the sensible heat flux or the vertical temperature gradient) and mechanically generated turbulence (represented by the friction velocity and/

or the vertical gradient of the horizontal wind speed). Therefore, the Richardson number is qualitatively the inverse of the Monin-Obukhov length (in a dimensionless form) and there is a theoretical relationship between Ri_g (or Ri_b) and the dimensionless form of the Monin-Obukhov length:

$$\frac{z}{L_{MO}} = \varphi \, Ri_g \tag{4.63}$$

where z is the height at which Ri_g (or Ri_b) is estimated and φ is a dimensionless parameter, which depends on the meteorological conditions; φ is a function of (z / L_{MO}) in the relationship proposed by Monin and Obukhov (1954). For example, according to experiments conducted over the sea, φ may vary between 4 and 21 depending on the temperature gradient and the wind speed (Hsu, 1989). Other relationships between L_{MO} and Ri_b have been developed based on theoretical considerations and experimental data, in order to estimate L_{MO} from the same measurements as those used to estimate Ri_b.

The convective velocity, w_*, is sometimes used to quantify the intensity of vertical motions within the PBL during unstable conditions:

$$w_* = \left(\frac{g \, z_i \overline{(w'\theta_v')}_0}{\theta_v} \right)^{1/3} \tag{4.64}$$

where g is the acceleration of gravity, z_i is the PBL height, θ_v is the virtual potential temperature, and $\overline{(w'\theta_v')}_0$ represents the sensible heat source at the surface. The convective velocity is on the order of 1 to 2 m s^{-1} for daytime conditions with strong solar radiation.

For calm atmospheric conditions ($u \approx 0$), the friction velocity tends toward zero and, therefore, the Monin-Obukhov length also tends toward zero. Therefore, it is not appropriate as a measure of atmospheric stability if heat-generated convection is important. Under such conditions, one may use the convective temperature scale, θ_{*C}, which is defined with respect to w_* (instead of θ_*, which is defined with respect to u_*, as shown in Equation 4.58), to characterize the importance of the local thermal convection:

$$\theta_{*C} = -\frac{\overline{(w'\theta')}_0}{w_*} \tag{4.65}$$

4.4.4 Vertical Temperature Profiles

The vertical heat flux can be expressed in a dimensionless manner, introducing the variable Φ_H, similarly to the vertical momentum flux:

$$\frac{d\theta_v}{dz} \frac{\kappa \, z}{\theta_*} = \Phi_H \tag{4.66}$$

where θ_v is the virtual potential temperature and θ_* is the potential temperature scale defined previously. The virtual potential temperature profile is obtained by integrating in the vertical direction:

$$\int_{\theta_{v0}}^{\theta_v} d\theta_v = \int_{z_0}^{z} \Phi_H(z) \frac{\theta_*}{\kappa} \frac{dz}{z} \tag{4.67}$$

For a neutral atmosphere (i.e., under adiabatic conditions), $\Phi_H = Pr_t$, where Pr_t is the turbulent Prandtl number. This dimensionless number is defined as the ratio of the turbulent diffusion coefficients for momentum and heat. Businger and coworkers estimated $Pr_t = 0.74$ assuming that the von Kármán constant was $\kappa = 0.35$. Assuming that $\kappa = 0.4$ (i.e., the value most commonly used), then using $Pr_t = 0.95$ leads to the same results as those obtained by Businger et al. Similarly to the approach used to obtain the vertical profiles of the horizontal wind speed, empirical values have been developed for Φ_H to obtain vertical profiles of the virtual potential temperature under stable and unstable conditions. These profiles are expressed as a function of a dimensionless height, $\zeta = z/L_{MO}$. The functions of Φ_H developed by Businger and coworkers used the assumption that $\kappa = 0.35$. The functions presented next correspond to $\kappa = 0.4$ (Jacobson, 2005).

Under stable conditions:

$$\Phi_H = 0.95 + 7.8\zeta \tag{4.68}$$

Under unstable conditions:

$$\Phi_H = 0.95 \left(1 - 11.6\zeta\right)^{-1/2} \tag{4.69}$$

Then, the vertical profiles of the virtual potential temperature are as follows. Under neutral conditions:

$$\theta_v(z) = \theta_v(z_0) + \frac{\theta_*}{\kappa} 0.95 \ln(\frac{z}{z_0}) \tag{4.70}$$

Since there is no sensible heat flux: $\overline{w'\theta'} = 0$, therefore, $\theta_* = 0$ and $\theta_v(z) = \theta_v(z_0)$. Thus, the potential temperature is constant with height under neutral conditions.

Under stable conditions:

$$\theta_v(z) = \theta_v(z_0) + \frac{\theta_*}{\kappa} \left[0.95 \ln(\frac{z}{z_0}) + \frac{7.8}{L_{MO}}(z - z_0)\right] \tag{4.71}$$

Under unstable conditions:

$$\theta_v(z) = \theta_v(z_0) + \frac{\theta_*}{\kappa} 0.95 \left[\ln\left[\frac{0.95 - \Phi_H}{0.95 + \Phi_H}\right] - \ln\left[\frac{0.95 - \Phi_{H_0}}{0.95 + \Phi_{H_0}}\right]\right] \tag{4.72}$$

where Φ_{H0} corresponds to the value of Φ_H at $z = z_0$. The vertical profiles of the virtual potential temperature given by Equations 4.71 and 4.72 tend toward a constant value when the atmospheric conditions become neutral, since the Monin-Obukhov length then tends toward infinity and θ_* tends toward zero.

4.4.5 Latent Heat Flux

The latent heat flux corresponds to the water vapor flux associated with evapotranspiration of vegetation and evaporation of water from the surface. This vertical transport of water vapor from the Earth's surface to the atmosphere results from the heat added to the surface by solar radiation. When the water vapor condenses later in the atmosphere (leading to cloud formation), the condensation process leads to heat production. Therefore, the water vapor flux from the surface to the atmosphere corresponds to a hidden heat flux (the Latin "latens" means hidden).

4.5 Local Circulations: Land-Sea Breezes and Mountain-Valley Winds

The phenomena of land-sea breezes in coastal regions and mountain-valley winds in mountainous regions result from temperature differences at the surface, that lead to differences in air density and, therefore, pressure gradients.

In a coastal region, the land surface has a greater temperature change during the diurnal cycle than the sea surface. This difference is due to the fact that land has a smaller heat capacity (1,350 kJ m^{-3} K^{-1} for dry soil) than water (4,200 kJ m^{-3} K^{-1}); i.e., soil requires less heat than water to increase its temperature by one degree; conversely, a given heat flux will lead to a greater temperature increase on land than at sea. Furthermore, heat transfer occurs in soil only by conduction, whereas heat transfer occurs both by conduction and convection (vertical mixing) in water. Therefore, during the day, solar radiation leads to a greater increase in temperature over land than at sea. As a result, the air layer near the land surface becomes warmer than the air layer near the sea surface. Since warm air is less dense than cold air, the air over land will rise, leading to a decrease in pressure. The pressure gradient corresponds to greater pressure over the sea than over land and it leads to a wind coming from the sea toward the land, i.e., a sea breeze. This air circulation at the surface is then completed aloft by a returning flow from the land toward the sea.

At night, the land surface releases the heat absorbed during the day back to the atmosphere via infrared radiation (see Chapter 5). Thus, the land surface temperature decreases faster than that of the sea surface. Then, a land breeze sets up at night, which is in the opposite direction of the daytime sea breeze. The temperature difference is generally less at night than during the day. Therefore, a land breeze is generally weaker than the corresponding sea breeze. Figure 4.4 depicts these land-sea breeze processes.

In a mountainous region, the hills and snowless mountain tops warm up more during the day than the valleys, because the former have better exposure to solar radiation. As a result, a phenomenon similar to that of the land-sea breeze takes place. During the day, the air near the mountain top becomes warmer than that of the valleys. There

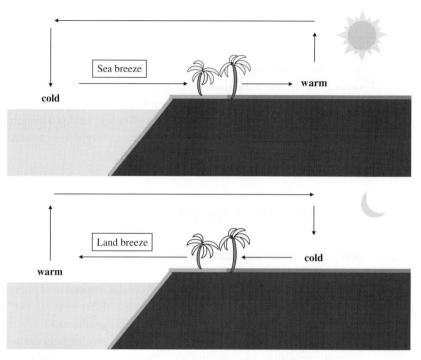

Figure 4.4. Schematic representation of sea and land breezes. Top figure: daytime sea breeze; bottom figure: nighttime land breeze. The terms "warm" and "cold" are relative to one another and characterize the temperature gradient.

is, therefore, a slight decrease in the atmospheric pressure near the mountain top compared to the valley and the atmospheric flow at the surface will be from the valley toward the mountain top. This is the valley breeze. The atmospheric local circulation is completed by a reverse atmospheric flow aloft. At night, the infrared radiation from the surface is more efficient at the mountain top, which cools down faster than the valley. Therefore, a mountain breeze sets up from the mountain top toward the valley. If the flow occurs along the axis of a valley, a similar phenomenon will occur between the upper and lower parts of the valley. Figure 4.5 depicts these mountain-valley breeze processes.

4.6 Meteorological Modeling

Numerical modeling of the atmospheric flows is performed using meteorological models. The equations that must be solved numerically include the Reynolds-averaged Navier-Stokes (RANS) equations, the heat transfer equation, radiative transfer through the atmosphere (see Chapter 5), and mass conservation for water. The equations for the atmospheric flows are by nature chaotic, i.e., the solution is very sensitive to the initial conditions (Lorenz, 1963). This chaotic behavior is the main reason why meteorological

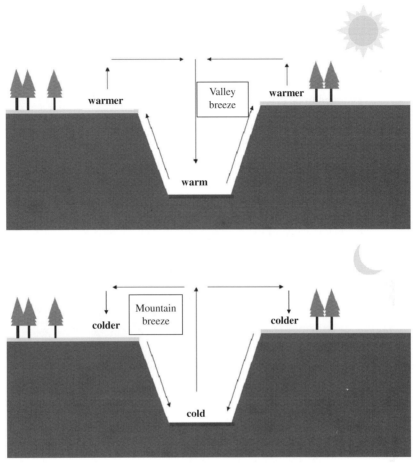

Figure 4.5. Schematic representation of the mountain and valley breezes. Top figure: daytime valley breeze; bottom figure: nighttime mountain breeze. The terms "warm" and "cold" reflect the daytime and nighttime conditions, respectively.

forecasts degrade after a few days. Currently, the Weather Research & Forecasting model (WRF; www.wrf-model.org/) is the main numerical model being used to generate three-dimensional (3D) fields of pressure, temperature, humidity, winds, and turbulence, as input data for air quality models. For air quality studies pertaining to past periods, it is possible to obtain 3D meteorological fields from organizations such as the United States National Centers for Environmental Prediction (NCEP; www.ncep.noaa.gov/) and the European Centre for Medium-Range Weather Forecasts (ECMWF; www.ecmwf.int/). These fields are re-analyses, i.e., meteorological data have been assimilated into the numerical model to obtain an optimal representation of the meteorological fields (Kalnay, 2003). Descriptions of the processes and equations of meteorological models are available, for example, in the books by Pielke (2013) and Jacobson (2005).

Problems

Problem 4.1 Potential temperature and atmospheric stability
The temperature is 20 °C at sea level and 16 °C at an altitude of 500 m. The atmosphere is assumed to be dry.

a. Calculate the potential temperature at 500 m.
b. Is the atmosphere stable, neutral, or unstable?

Problem 4.2 Reynolds number and atmospheric turbulence
a. Calculate the Reynolds number in the atmospheric planetary boundary layer (PBL) under the following conditions: a low wind speed of 2 m s^{-1}, a wintertime PBL height of 500 m, and a kinematic viscosity of the air of 1.55×10^{-5} m^2 s^{-1}.
b. Is this value of the Reynolds number characteristic of a turbulent or laminar regime?

Problem 4.3 Friction velocity and roughness length
The following measurements of wind speed as a function of height are available:

Altitude (m)	Wind speed (m s^{-1})
0.25	3.76
0.5	4.62
1	5.31
2	6.11
4	6.75
8	7.72
16	8.59

Calculate the roughness length, z_0, and the friction velocity, u_*, corresponding to these experimental data, assuming that the atmosphere is neutral.

Problem 4.4 Land-sea breezes
In a coastal region where land-sea breezes occur, a freeway is located along the seashore. A residential area is located inland a few hundred meters from the freeway. Will this residential area be affected by the freeway traffic emissions during the day or at night?

References

Arya, P.S., 2001. *Introduction to Micrometeorology, International Geophysics Series, 79,* Academic Press, New York.
Businger, J.A., J.C. Wyngaard, Y. Izumi, and E.F. Bradley, 1971. Flux profile relationships in the atmospheric surface layer, *J. Atmos. Sci.,* **28,** 181–189.
Dyer, A.J., 1974. A review of flux-profile relations, *Bound.-Lay. Meteor.,* **1,** 363–372.
Hsu, S.A., 1989. The relationship between the Monin-Obukhov stability parameter and the bulk Richardson number at sea, *J. Geophys. Res.,* **94,** 8053–8054.

References

Jacobson, M.Z., 2005. *Fundamentals of Atmospheric Modeling*, Cambridge University Press, Cambridge, UK.

Kalnay, E., 2003. *Atmospheric modeling, data assimilation and predictability*, Cambridge University Press, Cambridge, UK.

Landsberg, H.E., 1981. *The Urban Climate, International Geophysics Series, 28*, Academic Press, New York.

Lorenz, E.N., 1963. Deterministic nonperiodic flow, *J. Atmos. Sci.*, **20**, 130–141.

MacDonald, R.W., 2000. Modelling the mean velocity profile in the urban canopy layer, *Bound.-Lay. Meteor.*, **97**, 25–45.

MacDonald, R.W., R.F. Griffiths, and D.J. Hall, 1998. An improved method for the estimation of surface roughness of obstacle arrays, *Atmos. Environ.*, **32**, 1857–1864.

Masson, V., 2000. A physically-based scheme for the urban energy budget in atmospheric models, *Bound.-Lay. Meteor.*, **94**, 357–397.

McRae, G.J., 1980. A simple procedure for calculating atmospheric water vapor concentration, *J. Air Pollut. Control Assoc.*, **30**, 394.

Monin, A.S. and A.M. Obukhov, 1954. Basic laws of turbulent mixing in the surface layer of the atmosphere, *Tr. Akad. Nauk. SSSR Geophiz. Inst.*, **24**, 163–187; available in English at: www2.mmm.ucar.edu/wrf/users/phys_refs/SURFACE_LAYER/eta_part1.pdf.

Nieuwstadt, F.T.M., B.J. Boersma, and J. Westerweel, 2016. *Turbulence – Introduction to theory and applications of turbulent flows*, Springer, Switzerland.

Pielke, R.A., Sr., 2013. *Mesoscale Meteorological Modeling*, International Geophysics Series, 98, Academic Press, New York.

Stull, R.B., 1988. *An Introduction to Boundary Layer Meteorology*, Kluwer Academic Publishers, Dordrecht, The Netherlands.

Wallace, J.M. and P.V. Hobbs, 2006. *Atmospheric Science – An Introductory Survey*, Academic Press, Amsterdam.

5 Atmospheric Radiative Transfer and Visibility

Solar radiation is essential to life on Earth and is one of the major factors governing the atmospheric general circulation. Furthermore, solar radiation plays a major role in air pollution, since it leads to photochemical reactions when its radiative energy breaks apart some molecules. Then, these photochemical reactions initiate chemical and physico-chemical transformations that contribute to various forms of air pollution, including ozone, fine particles, and acid rain. In addition, the Earth emits radiation, which may be partially absorbed by anthropogenic greenhouse gases and some fraction of particulate matter, thereby leading to climate change. Finally, understanding how radiation is transferred through the atmosphere is useful to estimate the effect of air pollution on atmospheric visibility. This chapter describes first the radiative transfer processes in the atmosphere, i.e., solar radiation, its absorption by oxygen and ozone in the stratosphere, and its scattering by gases and particles. It also describes the emission of infrared radiation by the Earth and its absorption by greenhouse gases. Next, the effect of air pollution on atmospheric visibility is presented via the calculation of visual range and a brief discussion of the colors resulting from air pollution.

5.1 General Considerations on Atmospheric Radiative Transfer

5.1.1 Solar Radiation

The Sun consists of about three-quarters of hydrogen and one quarter of helium. Helium is formed by a fusion reaction between hydrogen atoms, which generates energy. This energy escapes from the Sun as electromagnetic radiation. This solar radiation comprises photons that carry energy, which is a function of wavelength. The radiation frequency, $v(s^{-1})$, associated with a photon is related to the wavelength of the radiation, λ (m), via the speed of light, c (m s^{-1}):

$$v = \frac{c}{\lambda} \tag{5.1}$$

The energy of a photon, E_p (J), of wavelength λ is:

$$E_p = h\,v = \frac{h\,c}{\lambda} \tag{5.2}$$

where h is the Planck constant ($h = 6.626 \times 10^{-34}$ J s). Therefore, the energy of a photon increases as its wavelength decreases.

The electromagnetic spectrum covers a wide range of wavelengths. Visible light ranges from about 400 to 700 nm, with violet light at about 400 nm, blue light at about 450 nm, green light at about 550 nm, and red light at about 650 nm. Below 400 nm is ultraviolet (UV) radiation. At wavelengths far below UV radiation are X rays (from about 0.001 to 10 nm) and gamma rays (from about 0.1 to 1 pm). Above 700 nm is infrared (IR) radiation. At greater wavelengths are radar wavelengths (from about 1 mm to 1 m), as well as television and radio wavelengths. Since the energy of radiation decreases with increasing wavelength, UV radiation has more energy than IR radiation.

Solar radiation covers a range of wavelengths. The Sun emits radiation, which corresponds approximately to that of a black body at a temperature of about 5,800 K. In comparison, IR radiation reemitted by the Earth corresponds to that of a black body at about 300 K. (The term "black body" corresponds to the assumption that this body absorbs all radiation completely and does not reflect it; however, the temperature of this body leads to radiation that is maximum at a wavelength that depends on the temperature of this body and, therefore, it is not actually black.)

Solar radiation is partially scattered and absorbed by gases present in the atmosphere and, consequently, some of the radiation does not reach the Earth's surface. Oxygen and nitrogen molecules absorb solar radiation below 100 nm. Between 100 and 200 nm, molecular oxygen strongly absorbs solar radiation. Between 200 and 280 nm, ozone (a product of the oxygen photolysis, see Chapter 7) also absorbs radiation very effectively. Therefore, solar radiation reaching the Earth's surface corresponds mostly to wavelengths greater than 280 nm. Figure 5.1 shows the solar radiation spectrum at the top of the atmosphere, as well as at

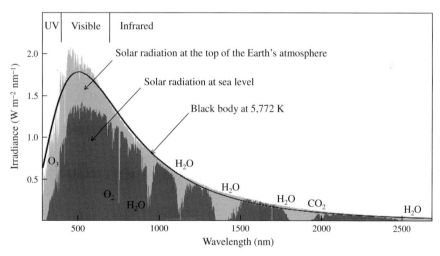

Figure 5.1. Solar radiation spectrum reaching the Earth. The solid black line corresponds to the theoretical irradiance of a black body at 5,772 K located at the distance of the Sun and calculated at the top of the Earth's atmosphere. The light gray shaded area corresponds to solar radiation at the top of the Earth's atmosphere and the dark gray shaded area corresponds to solar radiation reaching the Earth's surface after scattering and absorption by atmospheric gases and reflection by clouds. Gases absorbing radiation are indicated at their major absorption wavelengths. Source of the data: NREL (2017); the black body irradiance was calculated using Planck's law.

sea level, i.e., after scattering and absorption by atmospheric gases. Scattering of solar radiation by cloud droplets is also taken into account in this figure.

Triatomic gases such as water vapor (H_2O), carbon dioxide (CO_2), and ozone (O_3) form dipoles, which interact with the electromagnetic radiation at IR wavelengths. These molecules are, therefore, greenhouse gases, because they do not absorb much solar radiation, which is mostly in the UV and visible light range (except ozone, which absorbs some UV light, see Chapter 7), but they absorb IR light, which is emitted by the Earth. For example, CO_2 shows its maximum absorption at about 15 μm, which corresponds to the wavelength range near the maximum IR emission by the Earth's surface.

5.1.2 Radiance and Irradiance

Radiance (or intensity), I, represents the radiative energy per unit time (W or J s^{-1}) per unit surface area (m^{-2}), and per unit solid angle (sr^{-1}). It can be integrated over all wavelengths (W m^{-2} sr^{-1}) or expressed monochromatically (spectral radiance) per unit wavelength (W m^{-2} sr^{-1} nm^{-1}).

Irradiance, E_e, represents the radiative energy flux going through a surface. Therefore, it corresponds to the component of the radiance that is perpendicular to the surface and integrated over all solid angles of the hemisphere corresponding to the side of the surface where the source is located. If the angle between the direction of the radiation and the direction perpendicular to the surface is θ_r:

$$E_e(\lambda) = \int_\Omega I(\lambda) \cos(\theta_r) \, d\Omega \tag{5.3}$$

where Ω is the direction of the incoming radiation. Irradiance may be integrated over all wavelengths (W m^{-2}) or expressed monochromatically per unit wavelength (monochromatic or spectral irradiance in W m^{-2} nm^{-1}).

5.1.3 Radiative Budget of the Earth's Atmosphere

The spectral radiance of a black body is governed by Planck's law:

$$I = \frac{2 h c^2 \lambda^{-5}}{\left(\exp\left(\frac{hc}{k_B \lambda T}\right) - 1\right)} \tag{5.4}$$

where k_B is the Boltzmann constant (1.381×10^{-23} J K^{-1}) and T is the temperature in K. Therefore, radiance increases as the temperature of the black body increases, for all wavelengths. The wavelength corresponding to the maximum radiance, λ_{max}, is approximately given by Wien's law:

$$\lambda_{max} = \frac{h c}{(5 k_B T)} = \frac{2.9 \times 10^6}{T} \tag{5.5}$$

where λ is in nm and T in K. Therefore, the greater the temperature, the smaller the wavelength corresponding to the maximum radiance. The estimation of the effective temperature of the Sun's surface (photosphere) is on the order of 5,770 to 5,780 K. For a temperature of 5,772 K (NASA, 2017), the wavelength corresponding to the maximum radiance of the Sun is calculated as follows:

$$\lambda_{max} = \frac{2.9 \times 10^6}{5772} \approx 500 \text{ nm}$$

This wavelength corresponds to a blue-green light. However, blue light is preferentially scattered by the molecules of the Earth's atmosphere and the Sun's appearance is yellow or even red when the Sun is near the horizon (see the discussion of atmospheric visibility in Section 5.2). Outside the atmosphere, the Sun's appearance is white because its radiation spectrum covers all visible wavelengths. The full spectrum of the colors of solar radiation is clearly visible in a rainbow, because the water droplets decompose this radiation spectrum due to refraction of the incoming radiation, which modifies the direction of the photon trajectories depending on their wavelength.

The Stefan-Boltzmann law relates the irradiance of a black body to its temperature. It is derived from Planck's law by integrating over all wavelengths:

$$E_e = \sigma_{SB} T^4 \tag{5.6}$$

where σ_{SB} is the Stefan-Boltzmann constant (5.67×10^{-8} W m^{-2} K^{-4}).

Thus, the irradiance of the Sun at its source is:

$$E_{e,S} = 5.67 \times 10^{-8} \times (5772)^4 = 6.29 \times 10^7 \text{W m}^{-2}$$

At the distance of the Earth's orbit, this irradiance is smaller, since it has been dispersed over a greater sphere, which has a radius equal to the distance between the Sun and the Earth. This irradiance of solar radiation at the Earth's orbit is called the solar constant, E_S:

$$E_S = 6.29 \times 10^7 \left(\frac{\text{radius of the Sun}}{\text{distance Earth--Sun}}\right)^2 \tag{5.7}$$

Thus:

$$E_S = 6.29 \times 10^7 \times \left(6.96 \times 10^5 / (1.496 \times 10^8)\right)^2 = 1{,}361 \text{W m}^{-2}$$

where the distances are expressed in km. This value corresponds to that obtained by irradiance measurements performed via the Solar Radiation and Climate Experiment (SORCE) NASA satellite (Kopp and Lean, 2011). The radiative energy intercepted by the Earth corresponds to the radiative energy flux (irradiance) integrated over the Earth's cross-section, i.e., $E_s \pi r_T^2$, where r_T is the Earth's radius (6,371 km). Therefore, the average radiative flux reaching the top of the Earth's atmosphere is equal to this value divided by the Earth's surface area, $\pi r_T^2 E_S / (4 \pi r_T^2)$, i.e., $E_S / 4$, which corresponds to about 340 W m^{-2}.

However, part of the solar radiation is reflected toward space by clouds, atmospheric particles, and the Earth's surface. This fraction, called albedo, is on the order of 30 %. Therefore, the solar radiative flux absorbed by the Earth and its atmosphere, $E_{e,E}$, is estimated to be on average about 235 W m^{-2} and about 105 W m^{-2} are reflected toward space.

If the Earth is considered to be a black body and if one assumes that the radiative flux received by the Earth (235 W m^{-2}) is at equilibrium with that reemitted by the Earth, then the temperature at the Earth's surface is estimated from the Stefan-Boltzmann law:

$$T = \left(\frac{E_{e,E}}{\sigma_{SB}}\right)^{\frac{1}{4}} = \left(\frac{235}{5.67 \times 10^{-8}}\right)^{\frac{1}{4}} = 254 \text{ K} = -19 \text{ °C} \qquad (5.8)$$

The average temperature at the Earth's surface is actually greater, since it is on average 288 K (15 °C). The difference between these two temperatures is due mostly to the absorption of part of the radiative energy reemitted by the Earth by some gases present in the Earth's atmosphere. These gases are called greenhouse gases (GHG). The radiative budget of the Earth must, therefore, take into account the effect of these GHG, as well as some other important processes.

First, the radiative energy reflected to space may be decomposed into a fraction reflected by the atmosphere (clouds and particles), which is about 75 W m^{-2}, and a fraction reflected by the Earth's surface (in particular by snow and ice, which have a strong albedo), which is about 30 W m^{-2}. The fraction absorbed may be decomposed into a fraction absorbed by the atmosphere (mostly in the UV), which is 67 W m^{-2}, and a fraction absorbed by the Earth's surface, which is 168 W m^{-2}.

Radiation emitted by the Earth is located in the IR range. Therefore, this radiation is partially absorbed by GHG (see Figure 5.2). Natural GHG are water vapor (H_2O) and

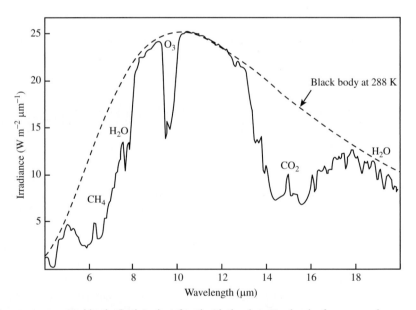

Figure 5.2. Radiation spectrum emitted by the Earth in the infrared with the absorption bands of some greenhouse gases (CO_2, H_2O, CH_4, and O_3). The solid line corresponds to a simulation with the MODTRAN radiative transfer model (Spectral Sciences, Inc. and U.S. Air Force Research Laboratory; http://modtran.spectral.com) for a standard U.S. atmosphere at an altitude of 50 km. The dashed line corresponds to the radiation flux of a black body at 288 K at the same altitude. The atmospheric windows through which the Earth's radiation flux escapes to space appear clearly between 8 and 9 μm and between 11 and 13 μm.

CO_2. In addition, anthropogenic activities lead to greater CO_2 atmospheric concentrations, since it is a product of combustion. Other GHG that are due to anthropogenic activities include methane (CH_4), nitrous oxide (N_2O), and ozone (O_3). The relative contributions of these GHG to the absorption of IR radiation are presented in Chapter 14. All GHG absorb about 350 W m^{-2}. This amount corresponds to about 90 % of the IR radiative flux emitted by the Earth's surface, which is 390 W m^{-2}; therefore, 40 W m^{-2} are directly emitted into space through a window of the IR spectrum where there is little absorption by GHG.

In addition, the Earth's surface emits heat (1) via water evaporation, which releases heat during its subsequent condensation in the atmosphere leading to cloud and fog formation (latent heat flux), and (2) via heat fluxes due to atmospheric turbulence (sensible heat flux); see Chapter 4 for the description of the physical processes corresponding to these heat fluxes. These two fluxes correspond to about 102 W m^{-2} (78 and 24 W m^{-2}, respectively).

Therefore, the total amount of energy absorbed by the atmosphere is 519 W m^{-2}, which includes 67 W m^{-2} via absorption in the UV, 350 W m^{-2} via absorption in the IR, and 102 W m^{-2} via the latent and sensible heat fluxes. At equilibrium, the atmosphere reemits this energy: 62 % (324 W m^{-2}) toward the Earth's surface and 38 % (195 W m^{-2}) toward space. The fact that the most important fraction is emitted toward the Earth's surface results from the vertical distribution of the atmospheric density, which is greater near the Earth's surface.

In summary, the radiative flux at the top of the atmosphere is 340 W m^{-2}, the atmosphere absorbs and reemits 519 W m^{-2}, and the Earth receives and reemits 522 W m^{-2}, of which 30 W m^{-2} correspond to the Earth's albedo. Table 5.1 summarizes the different components of these energy fluxes in terms of their sources and associated processes.

Table 5.1. Summary of the energy balance of the Earth and its atmosphere (W m^{-2}).

Earth	Absorbed energy (source)	Emitted energy (process)
	168 (Sun; absorption)	390 (IR)
	324 (GHG)	102 (heat)
Budget for the Earth*	492	492
Atmosphere	Absorbed energy (source)	Emitted energy (process)
	67 (Sun)	324 (IR toward the Earth)
	350 (Earth; IR)	195 (IR toward space)
	102 (Earth; heat)	
Budget for the atmosphere	519	519
Top of the atmosphere	Incoming energy (source)	Outgoing energy (source)
	340 (Sun)	75 (cloud albedo)
		30 (Earth's albedo)
		195 (IR from GHG)
		40 (IR form the Earth)
Budget at the top of the atmosphere	340	340

* In addition, 30 W m^{-2} are directly reflected by the Earth (albedo).

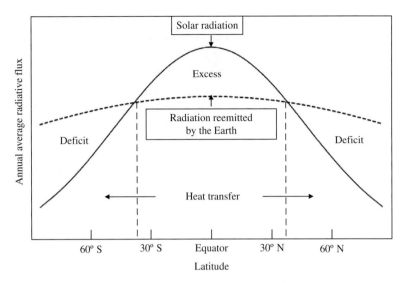

Figure 5.3. Schematic representation of the meridional profile of the Earth's energy budget.

The radiative energy reemitted by the Earth's surface as IR radiation is 390 W m^{-2}. Applying the black body formula for this irradiance (see Equation 5.8) leads to a temperature of 288 K (i.e., 15 °C), which is consistent with the mean temperature at the Earth's surface. The wavelength corresponding to the maximum radiation emitted by the Earth may be calculated by Wien's law (Equation 5.5): it is 10 μm, which corresponds to the infrared range.

More solar radiation is absorbed at the equator than at the poles, because of the zenith angle (i.e., the angle between the direction of the direct solar radiation and the direction perpendicular to the surface). The Earth reemits also more IR radiation at the low latitudes (tropical regions) than at the high latitudes (polar regions). However, the gradient from the equator to the poles is less pronounced for reemission than for absorption. This difference is due (1) to a greater reflectance (i.e., greater albedo) of the Earth's surface at the poles than in the tropical regions and (2) to the presence of more water vapor (greater humidity) in the tropical regions and the associated clouds and latent heat release aloft. The energy budget leads, therefore, to a net gain in the tropical regions and a net loss in the polar regions (see Figure 5.3). These energy differences must, therefore, lead to a transfer of energy from the equator to the poles, in order to maintain global equilibrium for the Earth. This energy transfer corresponds to the Hadley cells in the tropical and subtropical regions and the polar cells near the poles. In addition, the ocean currents transfer also some heat from the equator toward regions at higher latitudes. Finally, the water vapor emitted in the tropical regions is transported toward mid-latitude regions, which corresponds to a latent heat flux from the low toward the higher latitudes.

5.2 Atmospheric Visibility

Here, the absorption and scattering of atmospheric radiation by molecules and particles are briefly described. Next, the calculation of the visual range (i.e., the distance at which the human eye can see an object in the atmosphere) is presented. Finally, some examples of the effect of some air pollutants in terms of discoloration of the atmosphere are provided.

5.2.1 Absorption of Atmospheric Radiation

Some gases and particles absorb atmospheric radiation. The amount of light absorbed is proportional to the absorption efficiency of the absorbing species, k_a (m^2 g^{-1}), its concentration, C (g m^{-3}), and the distance traveled by the radiation, dx. The absorption coefficient (m^{-1}) is defined as the product of the absorption efficiency and the concentration:

$$b_a = k_a C \tag{5.9}$$

Assuming a uniform concentration of the absorbing species, the change in the radiance may be calculated as a function of the distance traveled by the radiation according to Beer-Lambert's law:

$$dI = -b_a I \, dx \tag{5.10}$$

where dI is the amount of radiance absorbed over the distance dx. Integrating from 0 to x:

$$I = I_0 \exp(-b_a x) \tag{5.11}$$

where I_0 is the initial radiance at $x = 0$. In the case of an object, it is called the intrinsic radiance.

5.2.2 Scattering of Atmospheric Radiation

Atmospheric radiation is scattered by gas molecules, particles, and droplets (for example, cloud and fog droplets). Mie theory describes the scattering of radiation by molecules and particles. For gas molecules and ultrafine particles (<100 nm), the wavelengths of the visible range are greater than the molecule or particle size. This range is called the Rayleigh scattering regime, where radiation is scattered as a function of wavelength following a λ^{-4} function. Therefore, molecules and ultrafine particles scatter preferentially blue light (between 400 and 450 nm) compared to the other colors of the visible spectrum and scatter red light the least. A clear sky during daytime, looking away from the Sun, represents this scattered radiation and, therefore, appears blue. At sunset or sunrise, solar radiation along the line of sight of an observer looking at the horizon travels through a maximum distance of the atmosphere (almost tangential to the Earth's surface) and red light, which is the least affected by scattering, dominates and gives the red appearance of sunsets and sunrises.

Atmospheric radiation is not scattered in an isotropic manner, but is scattered differently in various directions. The anisotropy of scattered radiation is represented mathematically by the phase function, $p(\Omega' \rightarrow \Omega)$, where Ω' and Ω are the directions of the incoming radiation and scattered radiation, respectively. In the Rayleigh regime, scattering occurs mostly along the incoming radiation direction, in equal amounts forward and backward. The amount of radiation scattered in the other directions is less and it is minimum in the direction perpendicular to the incoming radiation (half the amount scattered in the direction of the incoming radiation).

For particles, Mie theory predicts phase functions that correspond to forward scattering dominating for coarse particles (i.e., those particles greater in diameter than 2.5 µm). Actually, forward scattering is already dominant for particles with a diameter greater than 0.5 µm. For example, in presence of fog, the intensity of a light is preferentially scattered forward along the direction of the incoming light.

Beer-Lambert's law may be applied to scattering to calculate the fraction of radiation that is not scattered out of the line of sight. The scattering coefficient, b_s, is defined as the product of the scattering efficiency, k_s, and the concentration of the scattering species (molecule or particle):

$$dI = -b_s I \, dx \qquad (5.12)$$

where dI is the amount of radiation scattered along the distance dx. Integrating from 0 to x:

$$I = I_0 \exp(-b_s x) \qquad (5.13)$$

In the general case where both absorption and scattering take place, an extinction coefficient, b_e, may be defined, which corresponds to the sum of both phenomena:

$$b_e = b_a + b_s \qquad (5.14)$$

Beer-Lambert's law is then written as follows:

$$I = I_0 \exp(-b_e x) \qquad (5.15)$$

5.2.3 Optical Thickness

Optical thickness (also called optical depth), τ, is defined as the integral value of the extinction coefficient along a given distance. For example, in the case of a measurement performed by a satellite-borne instrument, it is the value integrated over the atmospheric height between the Earth's surface and the satellite:

$$\tau = \int_0^z b_e(z) \, dz \qquad (5.16)$$

The optical thickness is dimensionless. The extinction coefficient, b_e, is placed within the integral because it may vary with distance (here altitude), for example, if the concentrations of the absorbing and/or scattering species vary with distance.

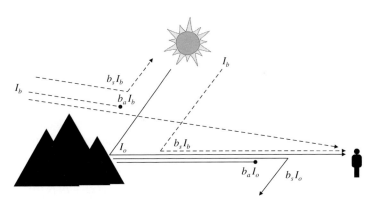

Figure 5.4. Schematic representation of visible atmospheric radiation between an object and an observer. See text for the definitions of the various terms.

5.2.4 Visual Range

The visual range is the utmost distance at which an observer can perceive a black object (see Figure 5.4). The criterion of a 2 % contrast between the object and the background is typically used as the perception threshold for the human eye. Contrast, Co, is defined as the relative difference between the background radiance, I_b, and that of the object, I_o:

$$Co = \frac{(I_b - I_o)}{I_b} \tag{5.17}$$

For a black object, $I_o(x=0) = 0$ (since it does not reflect any light) and its intrinsic contrast, i.e., at $x = 0$, is 1.

Over a distance dx along the line of sight between the object and the observer, the radiance increases due to scattering of background light into the line of sight and the radiance of the object decreases due to light extinction along the line of sight. According to the radiative transfer equation (Chandrasekhar, 1960; Goody et Yung, 1989), the radiance from an object, $I_o(\Omega, 0)$, reaching the observer at a distance x and in a direction Ω, is calculated as follows:

$$I_o(\Omega, x) = I_o(\Omega, 0)\exp(-\tau_x) + \int_0^{\tau_x} \frac{\omega_a(\tau')}{4\pi} \int_{\Omega'=4\pi} I_b(\Omega', \tau')\, p(\Omega' \to \Omega, \tau')\, d\Omega' \exp(-\tau')d\tau' \tag{5.18}$$

where τ_x is the optical thickness, ω_a is the albedo, and $p(\Omega' \to \Omega, \tau')$ is the phase function. The first term on the right-hand side of the equation represents the decrease in the radiance due to particles and gases that absorb and scatter light, and the second term represents the radiance added by scattering of the background light into the line of sight of the observer. At $x = 0$, the solution corresponds to the intrinsic radiance of the object, $I_o(\Omega, 0)$, and, when $x \to \infty$, the solution tends toward the background radiance, I_b, (i.e., the object is no longer perceptible). This integral equation may be solved numerically in the general case. For the simple case where the scattering and absorption coefficients are uniform between the object and the observer, this equation may be written as follows:

$$I_o(\Omega, x) = I_o(\Omega, 0) \exp(-\tau_x) + \int_0^{\tau_x} I_b \exp(-\tau') \, d\tau' \qquad (5.19)$$

To calculate the contrast, the radiance originating from the object and the background radiance (i.e., at the horizon near the object) must be calculated. The change in the radiance coming from the object due to light scattering is obtained from Equation 5.19. By derivation, the solution for an incremental distance dx is as follows:

$$dI_o(x) = \left(-b_s I_o(x) + b_s I_b\right) dx \qquad (5.20)$$

This equation corresponds to the difference between the radiance added and that subtracted along the line of sight. By definition, the background radiance is spatially uniform, i.e., $I_b(x) = I_b$, a constant value. The change in the contrast may be calculated along the distance dx:

$$\frac{dCo(x)}{dx} = \frac{d\left(1 - \frac{I_o(x)}{I_b}\right)}{dx} = -\frac{1}{I_b} \frac{dI_o(x)}{dx} \qquad (5.21)$$

According to the relationship for $dI_o(x)$, i.e., Equation 5.20:

$$\frac{dCo(x)}{dx} = -\frac{(b_s I_b - b_s I_o(x))}{I_b} \qquad (5.22)$$

The ratio on the right-hand side is equal to $b_s \, Co(x)$. Therefore, the contrast follows Beer-Lambert's law:

$$dCo = -b_s \, Co \, dx \qquad (5.23)$$

Integrating from 0 to x:

$$Co = Co_0 \exp(-b_s \, x) \qquad (5.24)$$

where Co_0 is the intrinsic contrast of the object, often chosen to be 1 by default. If the perception threshold criterion is chosen to be 2 % ($Co / Co_0 = 0.02$) at 550 nm (i.e., the wavelength at which perception is maximum for the human eye), the visual range, VR, which is the maximum distance at which an object can be perceived, is given by the following equation:

$$0.02 = \exp(-b_s \, VR) \qquad (5.25)$$

Solving for VR leads to the Koschmieder relationship (1924):

$$VR = \frac{-\ln(0.02)}{b_s} = \frac{3.9}{b_s} \qquad (5.26)$$

Several assumptions are implicit in this equation:

– The atmosphere is spatially uniform in terms of light scattering.
– The background radiance is the same at the object and at the observer location.
– The observed object is black and thus has an intrinsic contrast of 1 (i.e., at $x = 0$).

- The minimum contrast that is perceptible is 2 % (a value up to 5 % is sometimes used).
- The Earth's curvature is neglected (i.e., this relationship cannot be applied to atmospheres approaching Rayleigh scattering, see example).
- The extinction coefficient represents only scattering and does not include absorption (see discussion later in this section).

The relationship for the change in contrast as a function of distance is different for light absorption. In the case of absorption, both the background radiance and that of the object decrease similarly with distance in the case where absorption occurs between the object and the observer:

$$dI_o(x) = \left(-b_a I_o(x)\right) dx \qquad (5.27)$$

and:

$$dI_b(x) = \left(-b_a I_b(x)\right) dx \qquad (5.28)$$

Thus, the difference between the two radiances decreases according to Beer-Lambert's law. Therefore, the contrast does not follow Beer-Lambert's law, but remains constant in the case of absorption. If one considers both scattering and absorption, the change of the radiances with distance may be written as follows:

$$dI_o(x) = \left(b_s I_b(x) - (b_s + b_a) I_o(x)\right) dx \qquad (5.29)$$

and:

$$dI_b(x) = \left(-b_a I_b(x)\right) dx \qquad (5.30)$$

The solutions to these differential equations are as follows:

$$I_o(x) = I_o(0) \exp\left(-(b_s + b_a) x\right) + I_b(0) \exp(-b_a x)\left(1 - \exp(-b_s x)\right) \qquad (5.31)$$

$$I_b(x) = I_b(0) \exp(-b_a x) \qquad (5.32)$$

Thus:

$$Co = Co_0 \exp(-b_s x) \qquad (5.33)$$

The contrast of the object with respect to the background is not affected by the absorption of light between the object and the observer. On the other hand, the perception of the object becomes more difficult because the transfer of its radiance decreases due to absorption. In addition, if the air pollution extends beyond the distance at which the object is located, the background radiance will be affected more by absorption than it was in the previous calculation, where it was assumed that absorption occurs only between the object and the observer. Absorption will then decrease the background radiance and make it darker. The effect on the contrast will then depend on the color of the object compared to that of the background. An object that is darker than the background will have a lesser contrast and an object that is brighter than the background will have a greater contrast. In the case of a black object observed against a clear sky background, air pollution extending beyond the object will decrease its contrast. However, it is not possible to calculate this change in contrast simply and it is necessary to perform a radiative transfer calculation using a numerical model (see Section 5.3).

The assumption that is generally made in the empirical formulas used to calculate visual range is that the Koschmieder relationship applies to the extinction coefficient (i.e., both scattering and absorption). Since the extinction coefficient is generally dominated by scattering rather than by absorption, this approximation is considered acceptable. However, it is not strictly correct:

$$VR = \frac{-\ln(0.02)}{b_e} = \frac{3.9}{b_e} \quad (5.34)$$

For a pristine atmosphere, scattering follows the Rayleigh regime and there is no absorption. The scattering coefficient can be calculated according to the following relationship:

$$b_{Ray}(\lambda) = b_{Ray,\,ref}(\lambda)\, P\frac{273}{T} \quad (5.35)$$

where P is the atmospheric pressure in atm, T is the temperature in K, and $b_{Ray,ref}(\lambda)$ is the Rayleigh scattering coefficient at 1 atm and 0 °C. Some values for the visible range are provided in Table 5.2.

Table 5.2. Rayleigh scattering coefficient at 1 atm and 0 °C as a function of wavelength in the visible range (1 Mm = 10^6 m).

Wavelength (nm)	$b_{Ray,ref}(\lambda)$ (Mm^{-1})
400	45.40
450	27.89
500	18.10
550	12.26
600	8.604
650	6.217
700	4.605

At 1 atm and 15 °C, for a wavelength in the middle of the visible range, i.e., 550 nm (green light):

$$b_{Ray} = 11.6 \text{ Mm}^{-1}$$

This simple formula does not take into account the fact that the refraction index depends on temperature. Nevertheless, it is a good approximation at ambient temperatures (for example, 1 % error at 15 °C).

The calculation of the extinction coefficient requires knowing the concentrations, chemical composition, and size distribution of atmospheric particles and, to a lesser extent, the concentration of NO_2 (which absorbs light in the visible range). In the absence of detailed information on particle size distribution, empirical formulas based on the chemical composition of particulate matter and atmospheric humidity may be used (humidity affects the size of hygroscopic particles, see Chapter 9). For example, the formula used by the U.S. National Park Service (EPA, 2003) is as follows:

$$
\begin{aligned}
b_e = &\, 3f_{as}(RH) \text{ [ammonium sulfate]} + 3f_{as}(RH) \text{ [ammonium nitrate]} \\
&+ 4 \text{ [organic particulate matter]} + 10 \text{ [black carbon]} \\
&+ 1 \text{ [dust]} + 1.7f_{ss}(RH) \text{ [sea salt]} \\
&+ 0.6 \text{ [coarse particles]} + b_{Ray}
\end{aligned}
\tag{5.36}
$$

This equation corresponds to a wavelength of 550 nm (green light), which is in the middle of the visible range. Units are Mm^{-1} (per megameter) for the extinction and scattering coefficients, $\mu\text{g m}^{-3}$ for concentrations, and $\text{m}^2\text{ g}^{-1}$ for the scattering and absorption efficiencies. The numerical values represent the scattering efficiencies of various components of atmospheric particulate matter and the absorption efficiency of black carbon. The first five terms correspond to fine particles (i.e., those particles with a diameter less than 2.5 μm) and the size distribution of those particles is supposed to be lognormal with a mean diameter of 0.3 μm and a standard deviation of 2 (see Chapter 9 for a description of the lognormal distribution). The function $f_{as}(RH)$ represents the growth of inorganic fine particles via absorption of water as atmospheric humidity increases (i.e., hygroscopicity); a larger particle is more efficient at scattering light. This function was obtained by calculating the size of an ammonium sulfate particle as a function of relative humidity (RH). The particle size is an average value between the deliquescence and efflorescence particle sizes (see Chapter 9). For example, $f_{as}(RH)$ is equal to 1 (dry particles) at very low humidity (<30 %), but is equal to 1.5 at 60 %, 2 at 70 %, 2.5 at 80 %, 3 at 85 %, 4 at 90 %, and 10 at 95 % relative humidity. Organic particulate matter is assumed to be hydrophobic and, therefore, its scattering efficiency does not depend on humidity. Black carbon is also hydrophobic; in any case, its absorption efficiency depends on its concentration and not on particle size. The size distributions of coarse soil dust and sea salt particles are assumed to be lognormal, with a mean diameter of 5 μm and a standard deviation of 2. The function $f_{ss}(RH)$ represents the hygroscopic growth of sea salt particles; it differs from that used for ammonium sulfate particles. This type of formula is empirical (i.e., it involves various assumptions on particle size distributions, hygroscopic growth, etc.) and there are other similar formulas available to estimate the extinction coefficient for polluted atmospheres.

The haze index, DHI, which is expressed in deciview (dv), was introduced to quantify atmospheric visibility (Pitchford and Malm, 1994). It is defined by the following equation, by analogy with the decibel unit for sound:

$$
DHI = 10\ln\left(\frac{b_e}{10}\right) \tag{5.37}
$$

where b_e is expressed in Mm^{-1}. A change of 1 dv corresponds to a change of about 10 % in the extinction coefficient and, therefore, in the visual range (10.52 % more precisely). Given that the scattering coefficient for a Rayleigh atmosphere is about 10 Mm^{-1}, such an atmosphere has a haze index of about 0 dv. A visual range of 1 km corresponds to 60 dv,

10 km to 37 dv, and 100 km to 13.6 dv. The advantage of the haze index is that it changes linearly in terms of perception by the human eye.

Example: Calculation of the visual range for a Rayleigh atmosphere (pristine air)

According to Equation 5.35, the Rayleigh scattering coefficient at 550 nm is 11.6 Mm^{-1} at 1 atm and 15 °C. Therefore, in an atmosphere without any particles, the visual range given by the Koschmieder relationship (Equation 5.26) is:

$$VR = \frac{3.9}{0.0116} = 336 \text{ km}$$

Actually, such a visual range should take into account the curvature of the Earth and should integrate the Rayleigh scattering coefficient along an optical path, because this scattering coefficient decreases with altitude. Indeed, looking toward the horizon over a distance of 336 km leads to an altitude located well above sea level at the distance of the visual range. The radius of the Earth is 6,371 km; the angle θ_o between the direction from the center of the Earth to the observer and that from the center of the Earth to the visual range virtual location may be calculated as follows:

$$\tan(\theta_o) = 336/6{,}371; \text{ i.e., } \theta_o = 3°$$

Therefore, the distance from the center of the Earth to the location corresponding to the visual range is:

$$6{,}371 \text{ km}/\cos(\theta_o) = 6{,}380 \text{ km}$$

Therefore, the corresponding altitude of that location is (6,380 − 6,371) = 9 km. This altitude is greater than the highest Himalayan mountain tops. Since the Rayleigh scattering coefficient decreases with atmospheric pressure, the visual range would actually be greater. In any case, no fixed object is located at such an altitude along the sight path of the observer and the concept of visual range seems then difficult to apply.

Example: Estimation of the visual range during an air pollution episode in Paris, France

During the air pollution episode of March 2014 (see Chapter 9), the particulate matter concentrations were measured by the *Laboratoire des sciences du climat et de l'environnement* (LSCE) and by the *Institut national de l'environnement et des risques industriels* (Ineris) in Saclay, a suburban area located southwest of Paris. On March 14 at 10 a.m., these concentrations were approximately as follows (the background air pollution during this episode was fairly uniform and it seems appropriate to use those measurements to represent air pollution over the Paris region):

- Ammonium sulfate concentration: 9 μg m^{-3}
- Ammonium nitrate concentration: 45 μg m^{-3}

- Organic particulate matter concentration: 18 μg m^{-3}
- Black carbon concentration: 1 μg m^{-3}
- Sea salt concentration: negligible
- Soil dust concentration: no available measurements (zero concentration by default)
- Coarse particulate matter concentration: ~70 μg m^{-3} (estimation based on PM$_{10}$ measurements performed by Airparif; see Chapter 9)

A temperature of 15 °C and a relative humidity of 60 % are assumed; thus $f(RH) = 1.5$. The empirical formula given by Equation 5.36 is used the calculate the extinction coefficient:

$$b_e = (3 \times 1.5 \times 9) + (3 \times 1.5 \times 45) + (4 \times 18) + (10 \times 1) + (1 \times 0) \\ + (1.7 \times 0) + (0.6 \times 70) + 11.6$$

Thus: $b_e = 379$ Mm^{-1}

Absorption contributes only 3 % to b_e, which is dominated by light scattering. Applying the Koschmieder relationship (Equation 5.34) to estimate visual range:

$$VR = \frac{3.9}{0.379} = 10.3 \text{ km}$$

This distance corresponds approximately to the distance between Notre-Dame Cathedral and the *Grande Arche* in the *Défense* business district located west of Paris or the distance between Saint-Germain-en-Laye Castle and the *Défense* business district. This result implies that during such an air pollution episode, it would not be possible to see the *Défense* business district from Notre-Dame in downtown Paris or from Saint-Germain-en-Laye (actually, since the *Grande Arche* in the *Défense* business district is not black, it will be even less visible than estimated by the Koschmieder relationship). The corresponding deciview haze index is:

$$DHI = 10 \ln\left(\frac{b_e}{10}\right) = 36 \text{ dv}$$

5.2.5 Colors of Air Pollution

Plumes emitted from tall stacks or from tail pipes may have different colors depending on the pollutants emitted or, in the case of particles, the relative positions of the observer, the Sun, and the plume.

Some gases absorb radiation and, if this absorption occurs in the visible range, the color of the gas will correspond to the wavelengths that are not being absorbed. The main example in air pollution is nitrogen dioxide (NO$_2$), which absorbs radiation in the blue, violet, and UV (Burkholder et al., 2015). Thus, its color results from wavelengths ranging from the green to the red wavelengths and a NO$_2$ plume will appear orange-brown.

Black-carbon particles absorb strongly and, therefore, they appear black, since almost no light is scattered by those particles (Bond and Bergstrom, 2006). For example, some diesel vehicles without any particle filter may lead to a small black puff of black carbon when starting or accelerating.

Ultrafine particles have sizes that are near those of molecules and they scatter radiation preferentially at small wavelengths, i.e., in the visible range, at blue and violet wavelengths. Therefore, a large concentration of ultrafine particles leads to a bluish color of the air. This phenomenon occurs in areas with pristine air where there is a lot of vegetation. Vegetation leads to emissions of volatile organic compounds, terpenes and terpenoids, which, after oxidation in the air, form semi-volatile organic compounds. In the absence of large concentrations of other particles, these compounds nucleate to form new ultrafine particles (see Chapter 9). This phenomenon has led to several terms on various continents:

- The Blue Ridge Mountains in Virginia, United States
- The Blue Mountains in New South Wales, Australia
- The Blue Mountains in Jamaica, Caribbean
- And perhaps also the blue line of the Vosges Mountains, France

Fine and coarse particles and water droplets scatter all wavelengths of the visible spectrum and they appear gray. However, this appearance may vary from white to black depending on the particle/droplet concentration and the angle between the solar radiation direction, the plume, and the line of sight of the observer.

5.3 Numerical Modeling of Atmospheric Radiation

Numerical modeling of atmospheric radiation is needed in air pollution models to calculate the kinetics of photochemical reactions (see Chapter 7). The assumption of a plane atmosphere is generally made to simplify the geometry of the radiative fluxes. The simplest approach consists in considering only two directions for the radiative transfer, upward and downward. This simplification allows one to obtain a simple numerical solution of the radiative transfer equation. The adding and doubling method (or one of its variants) may be used to obtain the numerical solution. If standard atmospheric conditions are used, the calculation of the atmospheric radiative fluxes may be performed as a preprocessing step to an air pollution simulation. It can then be performed using a large number of wavelengths, and the photolytic kinetic constants can be tabulated for future use during the air pollution simulations. However, this approach only applies to optically thin atmospheric conditions, i.e., in cases where the presence of particles and droplets can be neglected.

Since atmospheric radiation may be significantly affected by the presence of clouds and particles that scatter radiation (and in the case where black carbon absorbs it) and since clouds and particle concentrations vary spatially and temporally during an air pollution simulation, it is preferable to perform the radiative transfer calculation online, i.e., concurrently with the air pollution simulation, in order to calculate the photolytic rates based on the more representative atmospheric conditions. Several algorithms have been developed to perform such calculations. For example, the FAST-J algorithm can perform this type of calculation for atmospheres with cloud layers and particles (Wild et al., 2000). FAST-J uses a two-dimensional geometry with eight directions for radiation scattering and a spectral resolution of up to 18 wavelengths.

Modeling the degradation of atmospheric visibility is generally limited to the calculation of the extinction coefficient to assess regional haze (see Section 5.2.4). Nevertheless, models of atmospheric visibility have been developed to calculate the visual impacts (contrast and color) of industrial stack plumes. This type of modeling is significantly more difficult than modeling regional haze, because the position of the observer with respect to the plume and the Sun must be taken into account, in order to calculate light scattering by the particles present in the plume correctly. Atmospheric visibility models give satisfactory results for plumes that absorb the atmospheric light; for example, a nitrogen dioxide plume emitted from a power plant (White et al., 1985). However, their performance is less satisfactory in the case of light-scattering plumes (i.e., particle-laden plumes) (White et al., 1986). The reason is that the phase functions depend strongly on the particle size distribution of the particles or droplets and the simulation of multiple scattering with complex geometries is challenging. If the particle/droplet size distribution is known, numerical methods such as discrete ordinate methods may be used to discretize the radiative transfer equation in different directions over finite volumes. Monte-Carlo methods may also be used to obtain better accuracy; such methods solve the radiative transfer equation with a large array of calculations representing the effect of multiple scattering on the paths of photons that are sampled randomly (e.g., Thomas and Stamnes, 1999). However, Monte-Carlo methods are computationally demanding and they have not been used operationally in air pollution modeling.

Problems

Problem 5.1 Absorption of atmospheric radiation

The absorption efficiency of visible light by black carbon is assumed to be 10 m^2 per g of black carbon. The black-carbon atmospheric concentration in Paris during an air pollution episode is given to be 4 µg m^{-3}. In Paris, an observer is looking at *Arc de Triomphe* from *Place de la Concorde*; the distance between these two locations along *Avenue des Champs-Élysées* is 2 km. Calculate the fraction of light that is absorbed by black carbon along this sight path using Beer-Lambert's law.

Problem 5.2 Visual range

During a spring air pollution episode in Paris, scattering dominates over absorption to degrade atmospheric visibility (b_s = 400 Mm^{-1}). Instead of applying the Koschmieder relationship to calculate the visual range, as done in the example above, it is assumed here that a monument will be perceptible only when its contrast is 5 % (i.e., instead of 2 %).

a. Calculate the distance at which this monument will no longer be visible.
b. By which amount should the ammonium nitrate particulate concentration be reduced in order to double the visual range? One assumes here that the initial ammonium nitrate concentration is 50 µg m^{-3}.

References

Bond, T. C. and R.W. Bergstrom, 2006. Light absorption by carbonaceous particles: An investigative review, *Aerosol Sci. Technol.*, **40**, 27–67.

Burkholder, J.B., S.P. Sander, J.P.D. Abbatt, J.R. Barker, R.E. Huie, C.E. Kolb, M.J. Kurylo, V.L. Orkin, D.M. Wilmouth, and P.H. Wine, 2015. Chemical Kinetics and Photochemical Data for Use in Atmospheric Studies, Evaluation n° 18, *JPL Publication* **15–10**, Jet Propulsion Laboratory, Pasadena, CA, available at http://jpldataeval.jpl.nasa.gov/pdf/JPL_Publication_15–10.pdf.

Chandrasekhar, S., 1960. *Radiative Transfer*, Dover Publications, Inc., New York.

EPA, 2003. *Guidance for Estimating Natural Visibility Conditions under the Regional Haze Program*, Report EPA-454/B-03–005, U.S. Environmental Protection Agency, Office of Air Quality Planning and Standards, Research Triangle Park, NC.

Goody, R.M. and Y.L. Yung, 1989. *Atmospheric Radiation – Theoretical Basis*, Oxford University Press, Oxford, UK.

Kopp, G. and J. Lean, 2011. A new, lower value of total solar irradiance: Evidence and climate significance, *Geophys. Res. Lett.*, **38**, L01706.

Koschmieder, H., 1924. Theorie der horizontalen Sichtweite, *Beitr. Phys. frei. Atmos.*, **12**, 33–53, 171–181.

NASA, 2017. *Solar Fact Sheet*, NASA Space Science Data Coordinated Archive (NSSDCA), National Aeronautics and Space Administration, United States, available at http://nssdc.gsfc.nasa.gov/planetary/factsheet/sunfact.html.

NREL, 2017. Renewable Resource Data Center (RReDC), National Renewable Energy Laboratory (NREL), Golden, CO, available at http://rredc.nrel.gov/solar/spectra/am1.5/.

Pitchford, M.L. and W.C. Malm, 1994. Development and applications of a standard visual index, *Atmos. Environ.*, **28**, 1049–1054.

Thomas, G.E. and K. Stamnes, 1999. *Radiative Transfer in the Atmosphere and Ocean*, Cambridge University Press, Cambridge, UK.

White, W.H., C. Seigneur, D.W. Heinold, M.W. Eltgroth, L.W. Richards, P.T. Roberts, P.S. Bhardwaja, W.D. Conner, and W.E. Wilson, Jr., 1985. Predicting the visibility of chimney plumes: An intercomparison of four models with observations at a well-controlled power plant, *Atmos. Environ.*, **19**, 515–528.

White, W.H., C. Seigneur, D.W. Heinold, L.W. Richards, W.E. Wilson, Jr., and P.T. Roberts, 1986. Radiative transfer budgets for scattering and absorbing plumes: Measurements and model predictions, *Atmos. Environ.*, **20**, 2243–2257.

Wild, O., X. Zhu, and M.J. Prather, 2000. Fast-J: Accurate simulation of in- and below-cloud photolysis in tropospheric chemical models, *J. Atmos. Chem.*, **37**, 245–282.

6 Atmospheric Dispersion

Atmospheric dispersion is a very important process in air pollution. The use of tall stacks to minimize the impacts of air pollutant emissions reflected the saying that "the solution to pollution is dilution." Although this saying turned out to be wrong for several reasons, including the cumulative effect of a large number of individual sources and the formation of secondary pollutants at large regional scales, atmospheric dispersion is nevertheless one of the key processes that govern air pollution levels. This chapter presents the fundamental processes of atmospheric dispersion, their theoretical basis, as well as the advantages and shortcomings of different types of atmospheric dispersion models.

6.1 General Considerations on Atmospheric Dispersion

Pollutants are dispersed, and therefore diluted, in the atmosphere as a result of turbulence. This turbulence covers a wide range of spatial scales, including large-scale eddies, smaller eddies, and so on until the dispersion of kinetic energy at the molecular scale. Lewis Fry Richardson (1881–1953), an English physicist, summarized this cascade of processes in a more poetic fashion:

> "Big whirls have little whirls that feed on their velocity,
> and little whirls have lesser whirls and so on to viscosity."

Since it is not possible to represent turbulence in a mathematical deterministic manner, it is necessary to introduce approximate representations of the main processes involved. Because of the presence of the Earth's surface, which creates an anisotropy in the lower layers of the atmosphere, it is appropriate to differentiate between turbulence in the vertical and horizontal directions.

Vertical turbulence results from processes mentioned in Chapter 4, which include mechanical turbulence and thermal turbulence. The latter depends strongly on the thermal structure of the planetary boundary layer (PBL); it is very large when the atmosphere is unstable and weak when the atmosphere is stable. Convection phenomena, which occur in the PBL as well as in the free troposphere, mix also air parcels (and, therefore, air pollutants) in the vertical direction, but they typically correspond to coherent motions and, therefore, are not treated in detail here.

Horizontal turbulence may be created by mechanical processes in the PBL, as well as by thermal processes. It is more challenging to parameterize the relevant physical processes, compared to those governing vertical motions. Therefore, its mathematical modeling can be difficult, particularly at very large spatial scales.

Turbulent processes can be addressed either through a lagrangian or an eulerian approach. In the lagrangian approach, one uses a reference system that moves with the mean wind and turbulent motions are studied with respect to that moving reference system. Therefore, one refers to relative turbulent diffusion, because it is relative to this reference system. Thus, the meandering of a plume, which may result from large-scale motions affecting the reference system, is not included. Relative diffusion corresponds to the instantaneous plume size, because averaging the plume size over time will correspond to including some of the plume meandering in the plume size. In the eulerian approach, the ensemble of turbulent motions is considered with respect to a fixed reference system. Therefore, one refers to absolute turbulent diffusion. Both approaches become equivalent when the lagrangian approach covers the full range of the turbulent eddies.

6.2 Lagrangian Representation of Turbulence

In the lagrangian representation of turbulence, one follows the statistical evolution of the distance between two particles (or molecules) emitted from the same source (Csanady, 1973). If a particle is present at the location x' at time $t = 0$, its random motion may be represented with the probability density $p(x - x', t)$ such that $p(x - x', t)\,dx$ is the probability that the motion of the particle during time t leads to its arrival in a space dx centered at the location x. Then, the concentration of particles present at time t at the location x (i.e., the mass of particles within a volume dx) is as follows:

$$C(x, t|x') = S_0\, p(x - x', t) \qquad (6.1)$$

where S_0 is the mass of particles emitted at time 0 at the location x'. The assumption of a random turbulent diffusion process leads to a gaussian distribution of the particle concentrations with a standard deviation σ and its maximum at the source location x'. In three dimensions, the standard deviation of the distribution of particles undergoing turbulent diffusion may be defined in one direction (here x) for the entire particle population as follows (assuming that the source is at $x' = y' = z' = 0$):

$$\sigma_x^2 = \frac{1}{S_0}\int\int\int x^2\, C(x,y,z,t)\,dx\,dy\,dz = \int\int\int x^2 p(x,y,z,t)\,dx\,dy\,dz = \overline{x^2} \qquad (6.2)$$

Therefore, the standard deviation of the particle distribution in one direction is equivalent to the root mean square of the motions of the particles in that direction. If turbulence is uniform and isotropic, then: $\sigma_x = \sigma_y = \sigma_z = \sigma$. The distance traveled by a particle at the speed u during time t is:

$$x(t) = \int_0^t u(t')\,dt' \qquad (6.3)$$

6.2 Lagrangian Representation of Turbulence

Therefore, the change in the square of the distance traveled as a function of time is as follows:

$$\frac{dx^2(t)}{dt} = 2x\frac{dx}{dt} = 2\int_0^t u(t)u(t')dt' \qquad (6.4)$$

Taking the ensemble average for both sides of the equation:

$$\frac{\overline{dx^2(t)}}{dt} = 2\int_0^t \overline{u(t)u(t')}dt' \qquad (6.5)$$

The autocorrelation between the velocities of two particles at times t and $t' = t + \tau_L$ is as follows:

$$R(\tau_L) = \frac{\overline{u(t)u(t+\tau_L)}}{\overline{u^2}} \qquad (6.6)$$

Therefore, the change of the standard deviation as a function of time is written as follows (dt' becomes $d\tau_L$):

$$\frac{d\sigma^2(t)}{dt} = 2\overline{u^2}\int_0^t R(\tau_L)d\tau_L \qquad (6.7)$$

This expression is known as Taylor's theorem (1922) (Geoffrey Ingram Taylor was a British physicist and mathematician). After integration:

$$\sigma^2(t) = 2\overline{u^2}\int_0^t\int_0^{t'} R(\tau_L)d\tau_L dt' \qquad (6.8)$$

Integrating by parts provides the solution for the square of the standard deviation of the distribution of particles for a steady-state dispersion process:

$$\sigma^2(t) = 2\overline{u^2}\left[t\int_0^t R(\tau_L)d\tau_L - \int_0^t \tau_L R(\tau_L)d\tau_L\right] \qquad (6.9)$$

Thus:

$$\sigma^2(t) = 2\overline{u^2}\int_0^t (t-\tau_L)R(\tau_L)d\tau_L \qquad (6.10)$$

where σ is the standard deviation of the particle distribution in one direction (σ_x in the mean wind direction, σ_y in the horizontal crosswind direction, σ_z in the vertical direction), t is the time since the emission, u is the wind velocity, and $R(\tau_L)$ is the lagrangian correlation function for a uniform and stationary turbulence. For such an idealized situation (constant in time and space), one may use a Markov process to represent the successive values of the velocity of the dispersed particles. (In a Markov process, the prediction of the future is performed using only information on the present state with no information on earlier states; although it is not entirely correct for a turbulent flow, it is a good approximation except for very short times.) Then, the correlation function may be represented by an exponential function:

$$R(\tau_L) = \exp(-\frac{\tau_L}{t_L}) \qquad (6.11)$$

where t_L is the lagrangian time scale. It is defined as the integral of the correlation function:

$$t_L = \int_0^\infty R(\tau_L) \, d\tau_L \qquad (6.12)$$

The lagrangian time scale is on the order of 60 to 100 s in the planetary boundary layer. Near the source, i.e., as $t \longrightarrow 0$, $R(\tau_L) \longrightarrow 1$ and the solution for the standard deviation is:

$$\sigma^2(t) = 2\overline{u^2} \int_0^t (t - \tau_L) d\tau_L = 2\overline{u^2} \left(t^2 - \frac{t^2}{2} \right) \qquad (6.13)$$

Therefore:

$$\sigma^2(t) = \overline{u^2} \, t^2 \qquad (6.14)$$

The standard deviation σ is proportional to the standard deviation of the instantaneous wind velocities, here σ_u, and to the time elapsed since the emission (or the distance traveled by the particle from the source if the mean wind velocity is constant, $x = u \, t$).

Far from the source, i.e., as $t \longrightarrow \infty$, $R(\tau_L) \longrightarrow 0$ and the solution for the standard deviation is:

$$\sigma^2(t) = 2\overline{u^2} \int_0^\infty (t - \tau_L) R(\tau_L) \, d\tau_L = 2\overline{u^2} \, t \int_0^\infty R(\tau_L) d\tau_L - 2\overline{u^2} \int_0^\infty \tau_L R(\tau_L) d\tau_L \qquad (6.15)$$

Let us define:

$$t_0 = \int_0^\infty R(\tau_L) d\tau_L$$

$$t_1 = \frac{1}{t_0} \int_0^\infty \tau_L R(\tau_L) \, d\tau_L \qquad (6.16)$$

Then, $\sigma^2(t)$ may be written as a function of t, t_0, and t_1:

$$\sigma^2(t) = 2\overline{u^2} \, t_0 (t - t_1) \qquad (6.17)$$

For an exponential correlation function, one obtains:

$$t_0 = t_L \qquad (6.18)$$

The value of t_1 is obtained by partial integration:

$$t_1 = \frac{t_L^2}{t_0} = t_L \tag{6.19}$$

Therefore:

$$\sigma^2(t) = 2\overline{u^2} \, (t_L t - t_L^2) \tag{6.20}$$

And since $t \gg t_L$:

$$\sigma^2(t) = 2\overline{u^2} \, t_L t \tag{6.21}$$

The standard deviation σ is proportional to the square root of the time elapsed since the emission or to the square root of the distance x traveled by the particle from the source if the mean wind velocity is constant ($x = u\,t$). Therefore, the lagrangian representation leads to a dispersion process that increases faster near the source (proportional to t or x) than farther downstream (proportional to $t^{1/2}$ or $x^{1/2}$). In his 1922 article, Taylor mentioned that this result was consistent with observations obtained in the laboratory and field observations of stack plumes. Indeed, the plume near the source displayed a conic shape (i.e., σ proportional to x), whereas farther downstream the shape of the plume resembled that of an elliptic paraboloid (i.e., σ proportional to $x^{1/2}$). The change in the formulation of the standard deviation as a function of distance from the source is discussed further in Section 6.3.2, when the lagrangian and eulerian representations of turbulence are compared.

This lagrangian representation used the implicit assumption that the distribution of the instantaneous velocities about the mean velocity is gaussian. This assumption is generally appropriate for the horizontal directions; however, it is rarely verified in the vertical direction because the Earth's surface leads to an important dissymmetry between upward and downward motions. Upward motions generally occur along narrow columns, whereas downward motions tend to occur with lower velocities and over wider columns of air.

6.3 Eulerian Representation of Turbulence

6.3.1 The Atmospheric Diffusion Equation

The continuity equation for an air pollutant (particle or molecule) is written similarly to that written for temperature in Chapter 4 (here, one does not account for emission, deposition, and transformation):

$$\frac{\partial C}{\partial t} + \frac{\partial (uC)}{\partial x} + \frac{\partial (vC)}{\partial y} + \frac{\partial (wC)}{\partial z} = 0 \tag{6.22}$$

This equation is sometimes referred to as the atmospheric diffusion equation. Molecular diffusion is not taken into account here, because it is negligible in the atmosphere compared to turbulent diffusion, except near surfaces (see Chapter 11). Accounting for the turbulent nature of the flow, which affects not only momentum and temperature (see Chapter 4), but also the pollutant concentration:

$$C = \overline{C} + C'; \quad u = \overline{u} + u'; \quad v = \overline{v} + v'; \quad w = \overline{w} + w' \tag{6.23}$$

Therefore, since $\overline{\overline{u}C'} = 0$; $\overline{u'\overline{C}} = 0$; etc.:

$$\frac{\partial \overline{C}}{\partial t} + \frac{\partial (\overline{u}\,\overline{C} + \overline{u'C'})}{\partial x} + \frac{\partial (\overline{v}\,\overline{C} + \overline{v'C'})}{\partial y} + \frac{\partial (\overline{w}\,\overline{C} + \overline{w'C'})}{\partial z} = 0 \tag{6.24}$$

Using the continuity equation for momentum (see Equation 4.19) leads to the simplified equation:

$$\frac{\partial \overline{C}}{\partial t} + \overline{u}\frac{\partial \overline{C}}{\partial x} + \frac{\partial (\overline{u'C'})}{\partial x} + \overline{v}\frac{\partial \overline{C}}{\partial y} + \frac{\partial (\overline{v'C'})}{\partial y} + \overline{w}\frac{\partial \overline{C}}{\partial z} + \frac{\partial (\overline{w'C'})}{\partial z} = 0 \tag{6.25}$$

Hereafter, as in Chapter 4, the following notations are used: $\overline{C} = C$; $\overline{u} = u$; etc. A first-order closure is performed by analogy with Fick's law for molecular diffusion (Bird et al., 2006). For example, for the term $\overline{u'C'}$:

$$\overline{u'C'} = -K_{xx}\frac{\partial C}{\partial x} - K_{xy}\frac{\partial C}{\partial y} - K_{xz}\frac{\partial C}{\partial z} \tag{6.26}$$

where K_{xx}, K_{xy}, and K_{xz} are the turbulent diffusion coefficients. It is generally assumed that the non-diagonal terms of the matrix of the turbulent diffusion coefficients are zero ($K_{xy} = K_{xz} = 0$). Then:

$$\overline{u'C'} = -K_{xx}\frac{\partial C}{\partial x} \tag{6.27}$$

Similar terms are obtained for $\overline{v'C'}$ and $\overline{w'C'}$. Note that the analogy with molecular diffusion does not have a true physical sense. Molecular diffusion is governed by the concentration gradient (the values of the molecular diffusion coefficient vary little among chemical species). In the case of turbulent diffusion, the term is also proportional to the concentration gradient, but it is the turbulent nature of the flow, which governs the importance of this term in the atmospheric diffusion equation. The term "dispersion" is typically used as synonymous of turbulent diffusion. The atmospheric diffusion equation is then written as follows:

$$\frac{\partial C}{\partial t} + u\frac{\partial C}{\partial x} - \frac{\partial}{\partial x}K_{xx}\frac{\partial C}{\partial x} + v\frac{\partial C}{\partial y} - \frac{\partial}{\partial y}K_{yy}\frac{\partial C}{\partial y} + w\frac{\partial C}{\partial z} - \frac{\partial}{\partial z}K_{zz}\frac{\partial C}{\partial z} = 0 \tag{6.28}$$

Rewriting by grouping the advection and turbulent diffusion terms:

$$\frac{\partial C}{\partial t} + (u\frac{\partial C}{\partial x} + v\frac{\partial C}{\partial y} + w\frac{\partial C}{\partial z}) - (\frac{\partial}{\partial x}K_{xx}\frac{\partial C}{\partial x} + \frac{\partial}{\partial y}K_{yy}\frac{\partial C}{\partial y} + \frac{\partial}{\partial z}K_{zz}\frac{\partial C}{\partial z}) = 0 \tag{6.29}$$

For the sake of simplicity, one assumes hereafter that the wind velocity and direction are constant along the x direction (i.e., $v = w = 0$), thus:

$$\frac{\partial C}{\partial t} + u\frac{\partial C}{\partial x} - \left(\frac{\partial}{\partial x}K_{xx}\frac{\partial C}{\partial x} + \frac{\partial}{\partial y}K_{yy}\frac{\partial C}{\partial y} + \frac{\partial}{\partial z}K_{zz}\frac{\partial C}{\partial z}\right) = 0 \quad (6.30)$$

6.3.2 Instantaneous Point Source

The case study of an instantaneous point source, which was used for the lagrangian representation, is also used here first for the eulerian representation. Therefore, the initial and boundary conditions are as follows:

$$\begin{aligned} C(x,y,z,0) &= S_0\, \delta(x-x_s)\, \delta(y-y_s)\, \delta(z-z_s) \\ C(x,y,z,t) &= 0 \quad \text{for } x,\ y,\ \text{and } z \to \infty \end{aligned} \quad (6.31)$$

where S_0 is the mass (g) emitted from the source, which is located at (x_s, y_s, z_s), and δ is the Dirac function. For example, for $\delta(x-x_s)$:

$$\begin{aligned} \delta(x-x_s) &= 0 \quad \text{for } x \neq x_s \\ \delta(x-x_s) &= +\infty \quad \text{for } x = x_s \\ \int_{-\infty}^{+\infty} \delta(x-x_s)\,\mathrm{d}x &= 1 \end{aligned} \quad (6.32)$$

The solution of the continuity equation is the concentration in a puff emitted by the source:

$$C(x,y,z,t) = \frac{S_0}{8\,\pi^{\frac{3}{2}} t^{\frac{3}{2}} \sqrt{K_{xx}K_{yy}K_{zz}}} \exp\left(-\frac{(x-x_s)^2}{4K_{xx}t} - \frac{(y-y_s)^2}{4K_{yy}t} - \frac{(z-z_s)^2}{4K_{zz}t}\right) \quad (6.33)$$

This equation shows a gaussian distribution of the concentration. Introducing the standard deviations of gaussian distributions, σ_x, σ_y, and σ_z:

$$C(x,y,z,t) = \frac{S_0}{(2\pi)^{\frac{3}{2}} \sigma_x \sigma_y \sigma_z} \exp\left(-\frac{(x-x_s)^2}{2\sigma_x^2} - \frac{(y-y_s)^2}{2\sigma_y^2} - \frac{(z-z_s)^2}{2\sigma_z^2}\right) \quad (6.34)$$

A simple comparison between Equations 6.33 and 6.34 shows that the eulerian turbulent diffusion coefficients are related to the standard deviations of the concentration distributions by the following relationships:

$$\begin{aligned} \sigma_x &= \sqrt{2K_{xx}t} \\ \sigma_y &= \sqrt{2K_{yy}t} \\ \sigma_z &= \sqrt{2K_{zz}t} \end{aligned} \quad (6.35)$$

In this eulerian formulation, the standard deviations are proportional to the square root of the time elapsed since the emission (or the square root of the distance traveled from the source if the mean wind velocity is constant). Therefore, this relationship is the same as that obtained with the lagrangian representation for the case when one is far downwind from the source (Equation 6.21). On the other hand, near the source, the lagrangian representation led to standard deviations that are proportional to the time elapsed since the emission or the distance traveled from the source (Equation 6.14).

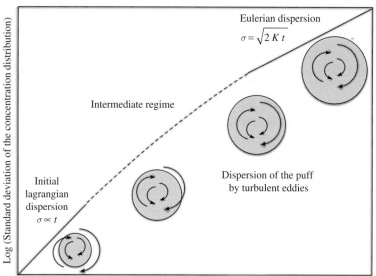

Figure 6.1. Schematic representation of the evolution of the standard deviation of the concentration in a puff as a function of the distance from the source.

The change of the dispersion rate with distance from the source may be explained conceptually as follows.

Near the source, the pollutant puff is very small. The relative dispersion of the puff results only from turbulent eddies that are smaller than the puff. As the puff grows in size due to dispersion, there are more and more eddies that are smaller than the puff. Farther downwind, when the puff size is large enough to cover the entire spectrum of eddy sizes, puff dispersion stabilizes and the eulerian formula applies. There is a transient regime between the near-source dispersion, represented by the near-source lagrangian dispersion, and the far-downwind dispersion, represented by the eulerian dispersion (as well as the lagrangian formula for $t \gg t_L$), where the standard deviations are proportional to t^α where $0.5 < \alpha < 1$. This change of the function governing the standard deviation of the puff concentration distribution is shown schematically in Figure 6.1.

6.3.3 Continuous Point Source

For a continuous point source under stationary and uniform meteorological conditions (i.e., with constant wind speed and direction), the steady-state concentration of a pollutant in the plume is calculated with the following mass conservation equation:

$$u\frac{\partial C}{\partial x} - \frac{\partial}{\partial x}K_{xx}\frac{\partial C}{\partial x} - \frac{\partial}{\partial y}K_{yy}\frac{\partial C}{\partial y} - \frac{\partial}{\partial z}K_{zz}\frac{\partial C}{\partial z} = S\,\delta(x - x_s)\,\delta(y - y_s)\,\delta(z - z_s) \quad (6.36)$$

where S is the source emission rate (g s^{-1}). Turbulence in the x direction may be neglected by assuming that atmospheric transport in that direction is dominated by the mean wind; this is called the slender plume approximation ($K_{xx} = 0$):

$$u\frac{\partial C}{\partial x} - \frac{\partial}{\partial y} K_{yy} \frac{\partial C}{\partial y} - \frac{\partial}{\partial z} K_{zz} \frac{\partial C}{\partial z} = S\,\delta(x-x_s)\,\delta(y-y_s)\,\delta(z-z_s) \qquad (6.37)$$

The solution is as follows:

$$C(x,y,z) = \frac{S}{4\pi\,u\,t\,\sqrt{K_{yy}K_{zz}}} \exp\left(-\frac{(y-y_s)^2}{4K_{yy}t} - \frac{(z-z_s)^2}{4K_{zz}t}\right) \qquad (6.38)$$

where $t = x/u$. This equation may be written according to its gaussian form:

$$C(x,y,z) = \frac{S}{2\pi\,u\,\sigma_y\,\sigma_z} \exp\left(-\frac{(y-y_s)^2}{2\sigma_y^2} - \frac{(z-z_s)^2}{2\sigma_z^2}\right) \qquad (6.39)$$

Thus, the concentration of a gaussian plume emitted from a continuous source is inversely proportional to the wind speed.

6.4 Lagrangian Models of Atmospheric Dispersion

6.4.1 General Considerations

Atmospheric dispersion may be modeled using a lagrangian approach, i.e., with respect to the source, or using an eulerian approach, i.e., without any particular reference to the source locations. In the former case, the model must take into account the distance traveled by the pollutant from the source (or the time elapsed since the emission) and the meteorological conditions and physiographic configuration, which affect turbulence. In the latter case, only the meteorological conditions and physiographic configuration are taken into account.

The lagrangian dispersion models are often referred to as gaussian models, because a gaussian distribution of the concentration around the puff or plume center is assumed. If the meteorological conditions are assumed to be stationary and uniform (i.e., constant in time and space), an analytical solution may be obtained, which is identical to that obtained for the eulerian equation:

$$C(x,y,z) = \frac{S}{2\pi u \sigma_y \sigma_z} \exp\left(-\frac{y^2}{2\sigma_y^2} - \frac{(z-z_s)^2}{2\sigma_z^2}\right) \qquad (6.40)$$

where z_s is the source height, the source is located at $(0, 0, z_s)$, and dispersion along the wind direction is assumed to be negligible (slender plume approximation). Then, the concentration at the plume centerline is obtained for $y = 0$ and $z = z_s$:

$$C(x,y,z) = \frac{S}{2\pi u \sigma_y \sigma_z} \tag{6.41}$$

Near the source, the lagrangian approach leads to standard deviations that are proportional to the distance from the source (Equation 6.14):

$$\sigma_y(x) \propto t = \frac{x}{u} \tag{6.42}$$

and similarly for σ_z. Therefore, the concentration at the plume centerline is related to the distance from the source as follows:

$$C(x,y,z) \propto \frac{S}{x^2} \tag{6.43}$$

The maximum concentration in the plume decreases initially as the inverse of the square of the distance from the source.

6.4.2 Plume Rise

The value of the initial plume height depends not only on the source height, but also on the plume rise with respect to that source. If the effluent at the source has some initial velocity, this momentum leads the plume to rise above the source. Also, if the plume temperature is greater than the ambient temperature, the plume rises due to its buoyancy, up to some height above the source where it reaches equilibrium with the ambient environment. The calculation of the plume rise, Δz_s, may be performed using semi-empirical formulas. The CONCAWE formula is simple:

$$\Delta z_s = 2.71 \frac{Q_h^{0.5}}{u^{0.75}} \tag{6.44}$$

where Q_h is the heat emission rate (J s^{-1}). This heat emission rate may be calculated based on the amount of gases emitted and their temperature:

$$Q_h = n_s\, C_P\, \Delta T \tag{6.45}$$

where n_s is the emission rate in moles emitted per unit time (mol s^{-1}), C_P is the average molar heat capacity of the emitted gases (J mol^{-1} K^{-1}) and $\Delta T = T_s - T$ (K) is the difference between the effluent temperature and the ambient temperature. This CONCAWE formula takes into account the plume rise due to plume buoyancy (i.e., vertical motion of the plume due to the temperature difference between the plume and its environment), but it does not take into account the initial plume momentum.

The Briggs formula (1984) is more complete because it takes into account both effects: buoyancy and initial kinetic energy of the effluent:

$$\Delta z_s = \left(\frac{3 F_a\, x}{0.36\, u^2} + 4.17 \frac{F_b\, x^2}{u^3}\right)^{1/3}$$

$$F_a = \frac{T}{T_s} v_i^2 \frac{d_s^2}{4} \tag{6.46}$$

$$F_b = g\, v_i \frac{d_s^2}{4} \left(\frac{T_s - T}{T_s}\right)$$

where d_s is the diameter of the stack (m), T_s and T are the temperature of the effluent at the source and the ambient temperature (K), respectively, and v_i is the initial vertical velocity of the effluent (i.e., plume or puff) at the source (m s^{-1}), typically at the stack exit. This formula applies up to a distance x_t, which is defined as follows:

$$x_t = 49\, F_b^{5/8} \text{ for } F_b \leq 55 \text{m}^4\text{s}^{-3}$$
$$x_t = 119\, F_b^{2/5} \text{ for } F_b \geq 55 \text{m}^4\text{s}^{-3} \quad (6.47)$$

The final height of the plume above ground level is called the effective source height and it corresponds to the final value of the plume height, $z_{s,f} = z_s + \Delta z_s$.

More detailed formulas that account for atmospheric stability have been developed. They are not given here, but are available in the literature. For example, Carson and Moses (1969) have compared several plume rise formulas with experimental data available at industrial sites. These plume rise formulas were developed initially for gaussian dispersion models. They are now used also in other models, such as eulerian air quality models.

6.4.3 Operational Gaussian Models for Point Sources

A plume emitted into the atmosphere gets into contact with the ground at some distance downwind of the source (immediately if the source is at ground level, $z_s = 0$). Then, the solution for the plume concentration must take into account the fact that concentrations cannot be calculated for $z < 0$. It is generally assumed that the plume is entirely reflected at the ground. Then, the solution is obtained by using a virtual source that is symmetrical of the actual source with respect to the ground to represent this reflection of the plume at the ground. The solution for the plume concentrations is as follows:

$$C(x,y,z,t) = \frac{S}{2\pi u\, \sigma_y \sigma_z} \exp\left(-\frac{y^2}{2\sigma_y^2}\right)\left(\exp\left(-\frac{(z-z_{s,f})^2}{2\sigma_z^2}\right) + \exp\left(-\frac{(z+z_{s,f})^2}{2\sigma_z^2}\right)\right) \quad (6.48)$$

Similarly, if the emission height is below a temperature inversion, there may be reflection of the plume against the base of this inversion layer. The concept of a virtual source may also be used to account for this plume reflection aloft. A virtual source that is symmetrical of the actual source with respect to the inversion base is used to represent the plume reflection against the base of the inversion layer of height z_i above ground level:

$$C(x,y,z,t) = \frac{S}{2\pi u\, \sigma_y \sigma_z} \exp\left(-\frac{y^2}{2\sigma_y^2}\right)\left(\exp\left(-\frac{(z-z_{s,f})^2}{2\sigma_z^2}\right) + \exp\left(-\frac{(z+z_{s,f}-2z_i)^2}{2\sigma_z^2}\right)\right) \quad (6.49)$$

The two equations above may be combined to account for both the reflection of the plume at the ground and against the base of an inversion layer aloft. Then, one must in addition

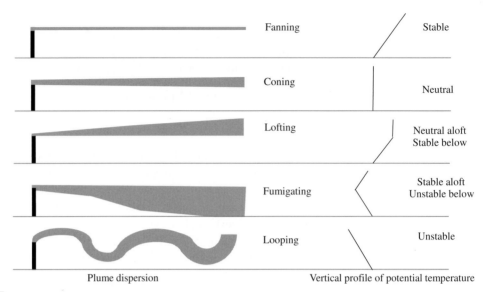

Figure 6.2. Typical profiles of vertical dispersion of plumes under various atmospheric conditions.

account for the reflections of the plumes corresponding to the virtual sources (for example, reflection at ground level of the plume of the virtual source used to represent the reflection against the base of the inversion layer). Therefore, the solution includes a series of terms representing those various reflections:

$$C(x,y,z,t) = \frac{S}{2\pi u \sigma_y \sigma_z} \exp\left(-\frac{y^2}{2\sigma_y^2}\right) \sum_{-N_r}^{+N_r} \left(\exp\left(-\frac{(z - z_{s,f} + 2N_k z_i)^2}{2\sigma_z^2}\right) \right.$$
$$\left. + \exp\left(-\frac{(z + z_{s,f} - 2N_k z_i)^2}{2\sigma_z^2}\right) \right) \tag{6.50}$$

where N_r is the number of reflections ($N_k = 1, \ldots, N_r$). A satisfactory solution is obtained with a single reflection ($N_r = 1$) and five terms ($N_r = 5$) are sufficient to obtain an accurate solution. These formulations are useful to represent the different types of atmospheric dispersion of plumes in the PBL depending on various stability conditions and the presence of inversion layers. Figure 6.2 depicts the major types of vertical dispersion of plumes emitted from a tall stack.

When using the lagrangian gaussian formulation, the standard deviations σ_y and σ_z must be defined. They are usually defined empirically using data from field experiments. Those data correspond typically to averaging times of 5 to 15 min; therefore, they include some plume meandering in addition to relative plume diffusion. These standard deviations are generally called dispersion coefficients. Among the Gaussian dispersion coefficients that are the most widely used, one may cite those of Pasquill–Gifford and those of McElroy–Pooler. The Pasquill-Gifford coefficients are presented in Figure 6.3. They may also be reported as analytical mathematical formulas for direct use in numerical modeling. Table 6.1 lists the analytical formulas for the Pasquill-Gifford and

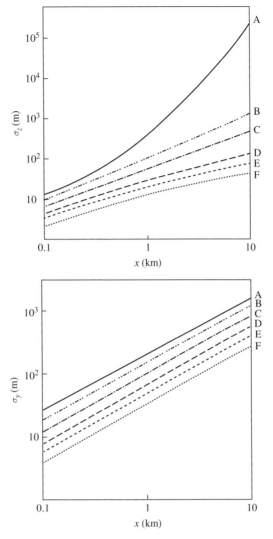

Figure 6.3. Empirical dispersion coefficients of Pasquill-Gifford as a function of distance from the source for different atmospheric stability classes (ranging from A, very unstable, to F, very stable). These coefficients were developed by Pasquill using the data reported by Gifford (1961). Vertical dispersion coefficients, σ_z (top figure), and horizontal dispersion coefficients, σ_y (bottom figure).

McElroy-Pooler coefficients. Other empirical formulas have been developed for dispersion coefficients, such as those of Briggs and those of Doury (the latter apply to longer distances than those applicable for the other formulas), as well as those based on similarity theory (Irwin and Venkatram) and those developed for tall stacks (Hanna and Paine, Gillani and Godowitch).

The assumption of stationary and uniform atmospheric conditions, which is necessary when using the gaussian plume formulas, may be removed by using an ensemble of puffs instead of a single plume. The continuous emission rate is then discretized

Table 6.1. Analytical formulas for the dispersion coefficients of Pasquill-Gifford and Briggs-McElroy-Pooler, σ_y and σ_z (m), as a function of distance from the source, x (m), and atmospheric stability (A to F).

Pasquill-Gifford

Stability	$\sigma_y = \exp(a_y + b_y \ln(x) + c_y (\ln(x))^2)$			$\sigma_z = \exp(a_z + b_z \ln(x) + c_z (\ln(x))^2)$		
	a_y	b_y	c_y	a_z	b_z	c_z
A	−1.104	0.9878	−0.0076	4.679	−1.172	0.2770
B	−1.634	1.0350	−0.0096	−1.999	0.8752	0.0136
C	−2.054	1.0231	−0.0076	−2.341	0.9477	−0.0020
D	−2.555	1.0423	−0.0087	−3.186	1.1737	−0.0316
E	−2.754	1.0106	−0.0064	−3.783	1.3010	−0.0450
F	−3.143	1.0148	−0.0070	−4.490	1.4024	−0.0540

Briggs-McElroy-Pooler

Stability	σ_y	σ_z
Rural conditions		
A	$0.22\,x\,(1 + 0.0001\,x)^{-1/2}$	$0.20\,x$
B	$0.16\,x\,(1 + 0.0001\,x)^{-1/2}$	$0.12\,x$
C	$0.11\,x\,(1 + 0.0001\,x)^{-1/2}$	$0.08\,x\,(1 + 0.0002\,x)^{-1/2}$
D	$0.08\,x\,(1 + 0.0001\,x)^{-1/2}$	$0.06\,x\,(1 + 0.0015\,x)^{-1/2}$
E	$0.06\,x\,(1 + 0.0001\,x)^{-1/2}$	$0.03\,x\,(1 + 0.0003\,x)^{-1}$
F	$0.04\,x\,(1 + 0.0001\,x)^{-1/2}$	$0.016\,x\,(1 + 0.0003\,x)^{-1}$
Urban conditions		
A & B	$0.32\,x\,(1 + 0.0004\,x)^{-1/2}$	$0.24\,x\,(1 + 0.001\,x)^{1/2}$
C	$0.22\,x\,(1 + 0.0004\,x)^{-1/2}$	$0.20\,x$
D	$0.16\,x\,(1 + 0.0004\,x)^{-1/2}$	$0.14\,x\,(1 + 0.0003\,x)^{-1/2}$
E & F	$0.11\,x\,(1 + 0.0004\,x)^{-1/2}$	$0.08\,x\,(1 + 0.0015\,x)^{-1/2}$

with respect to time and each discretized emission rate is assigned to a puff as an instantaneous emission rate. Puffs are emitted at successive (generally, but not necessarily, constant) time steps. Thus, if the wind direction varies in time and/or space, the plume is represented by an ensemble of puffs and will not follow a straight trajectory. Therefore, a puff model provides a more realistic representation of the transport and dispersion or air pollutants emitted from a source than a standard plume model, when the meteorological conditions are not stationary and uniform. The gaussian formulation may be used in puff models. However, since each puff is affected by its own atmospheric conditions (wind speed and direction, atmospheric stability), the plume, which is represented by an ensemble of puffs, is not gaussian. This more realistic representation of plume dispersion involves a larger computational burden, as it is necessary to calculate the transport and dispersion of an ensemble of puffs. There exist several operational lagrangian puff models. Some of the most commonly used include SCIPUFF and its version with atmospheric chemistry SCICHEM, HYSPLIT, and FLEXPART.

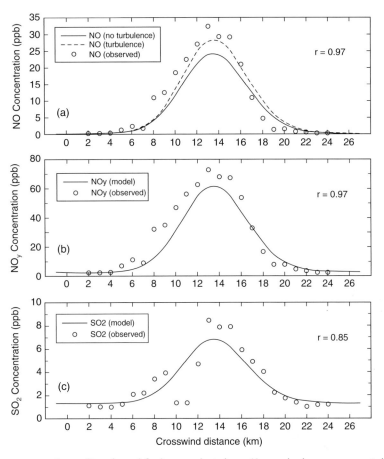

Figure 6.4. Air pollutant concentration profiles of a coal-fired power plant plume. Measured values are represented by circles and simulated values are represented by solid lines. The simulation represented by a dotted line corresponds to the case where the influence of atmospheric turbulence on the chemical kinetics of pollutant reactions was taken into account. Source: Reproduced with permission from Karamchandani, P. et al. (2000).

The gaussian formulation for plume dispersion is theoretical. Nevertheless, it is a good representation of the atmosphere when conditions are relatively stationary. Figure 6.4 shows a comparison of measured and simulated concentrations of nitrogen oxides (nitric oxide, NO, and the ensemble of nitrogen oxides, NO_y, see Chapter 8 for the definition of NO_y) and sulfur dioxide (SO_2) in the plume of a coal-fired power plant. The measured concentrations were obtained with an instrumented aircraft flying across the plume. The simulated concentrations were obtained with the SCICHEM puff model. Although the measured concentration profiles are not exactly gaussian, the model simulation reproduces satisfactorily the observed concentrations.

Figure 6.5 shows the results of ground-level concentrations calculated with a gaussian plume model for a point source (stack). In these simulations, the SO_2 emission rate is 5 g s^{-1}. In the top figure, the stack height (z_s) is 5 m; it is 30 m in the bottom figure. In each

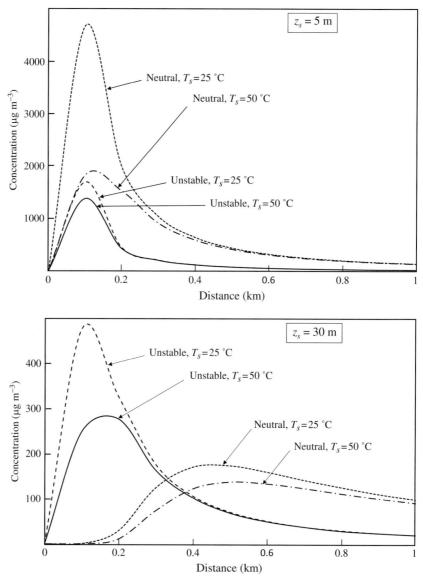

Figure 6.5. Simulation of SO_2 ground-level concentrations (µg m^{-3}) with a gaussian plume model. Top figure: stack height = 5 m; bottom figure: stack height = 30 m.

case, four scenarios were simulated: two separate atmospheric conditions (unstable and neutral) and two initial plume temperatures (25 and 50 °C, with an ambient temperature of 20 °C).

When the stack is only 5 m high, the largest concentrations are obtained under neutral conditions. Although the wind speed is typically greater under neutral and stable conditions compared to unstable conditions (here 2 m s^{-1} for unstable conditions and 4 m s^{-1} for neutral conditions), the dispersion coefficients are much lower for neutral conditions than for unstable conditions, thereby leading to a more concentrated plume. Therefore, the ground-level plume concentrations are greater. If the effluent temperature increases, the plume rise (4 m for neutral conditions and 7 m for unstable conditions) leads to lower ground-level concentrations near the source. The SO_2 hourly concentration standard is about 200 µg m^{-3} (75 ppb) in the U.S. and 350 µg m^{-3} in France (see Chapter 15). Thus, these air quality standards are exceeded for all conditions.

When the stack is 30 m high, the largest concentrations are obtained under unstable conditions. Under neutral conditions, the plume remains aloft and shows ground-level impacts only beyond 100 m with a maximum concentration between 400 and 500 m downwind. On the other hand, unstable conditions are conducive to vertical dispersion, i.e., they dilute the plume concentrations, but mix the plume to the ground at a short distance from the source. An increase in the effluent temperature leads to the same plume rise as above and minimizes the ground-level impacts. Then, the U.S. air quality standard is exceeded only under unstable conditions and the French air quality standard is not exceeded if a high effluent temperature is used.

In summary, stable and neutral conditions lead to the largest ground-level impacts for short stacks, whereas unstable conditions lead to the largest ground-level impacts for tall stacks (see Figure 6.2). Furthermore, plume rise may reduce the ground-level impacts significantly. Plume rise may be obtained either by increasing the effluent initial velocity (i.e., increased momentum) or by increasing the effluent temperature (i.e., increased buoyancy). This type of analysis led to an air quality management approach based on the reasoning that "the solution to pollution is dilution." Thus, many large industrial sources were given tall stacks. Some of those stacks can be as high as 300 m above ground level. Also, under some configurations (stacks located near a building for example), minimum stack height requirements are used to minimize ground-level impacts (for example, "good engineering practice" in the U.S.). However, this approach pertains only to primary pollutants, i.e., those pollutants emitted from the stack. It is not relevant to secondary pollutants, such as acid rain, ozone, and fine particles, because these secondary pollutants are associated with a regional pollution, which does not depend strongly on stack height. Therefore, in the 1990s, emission controls started to be required for some industrial sources that had tall stacks, but contributed to regional pollution. These emission controls of tall stacks were driven first by acid rain regulations (see Chapters 10 and 13) and later by ozone regulations (see Chapter 8).

6.4.4 Operational Gaussian Models for Line Sources

The gaussian equation presented in Section 6.4.3 may be applied to non-point sources, i.e., line, area, and volume sources. However, the integration over a line, area or volume rarely leads to an analytical solution. An exception is the integration over a line source when the wind is perpendicular to the line source:

$$C(x,y,z) = \frac{S_l}{2\sqrt{2\pi}u\sigma_z}\exp\left(-\frac{z^2}{2\sigma_z^2}\right)\left(\text{erf}\left(-\frac{(y-y_1)}{\sqrt{2}\sigma_y}\right) - \text{erf}\left(-\frac{(y-y_2)}{\sqrt{2}\sigma_y}\right)\right) \quad (6.51)$$

where S_l is the line source emission rate (g km^{-1} s^{-1}), y_1 and y_2 are the coordinates of the extremities of the line source and erf is the error function. For other wind directions, it is necessary to either use an approximate analytical solution or to obtain a numerical solution. A numerical solution is usually computationally expensive and, therefore, analytical solutions, even if they are approximate, are preferred. The Horst-Venkatram (HV) model is a good example of an approximate solution of the gaussian equation for a line source (Venkatram and Horst, 2006). It was later improved by Briant et al. (2011) (BKS), who minimized the errors due to the use of analytical functions and combined it with a numerical solution for cases where the wind is almost parallel to the line source (the error of the analytical solution becomes too large for such wind/line source configurations, see the discussion of Equation 6.52).

The analytical form of the HV-BKS model is as follows:

$$C(x,y,z) = \frac{S_l}{2\sqrt{2\pi}\, u \cos(\alpha)\, \sigma_z(d_{\mathit{eff}})} \exp\left(-\frac{z^2}{2\sigma_z^2(d_{\mathit{eff}})}\right)$$
$$\left(\mathrm{erf}\left(-\frac{(y-y_1)\cos(\alpha) - x\sin(\alpha)}{\sqrt{2}\sigma_y(d_1)}\right) - \mathrm{erf}\left(-\frac{(y-y_2)\cos(\alpha) - x\sin(\alpha)}{\sqrt{2}\sigma_y(d_2)}\right)\right)$$
$$\left(\frac{1}{L_{BKS}(x_{wind}) + 1}\right) + E_{BKS}(x_{wind}, y_{wind}, z) \tag{6.52}$$

where α is the angle between the wind direction and the direction perpendicular to the line source, x_{wind} and y_{wind} are the receptor coordinates in the wind direction coordinate system, $d_{\mathit{eff}} = x/\cos(\alpha)$, and $d_i = (x - x_i)\cos(\alpha) + (y - y_i)\sin(\alpha)$, $i = 1$ or 2. The first part of the formula corresponds to the HV model and L_{BKS} and E_{BKS} are the analytical functions added in the BKS version of the model to further minimize the error. When the wind direction becomes parallel to that of the line source, α tends toward 90° and $\cos(\alpha)$ tends toward 0, so that C tends to infinity. Then, a numerical solution is used instead. It consists of discretizing the line source with an ensemble of point sources when the wind direction is within 10° of the direction of the line source. This type of formula may be used, for example, to assess the air quality impacts of on-road traffic emissions.

The application of such models to on-road traffic showed that air pollution due to on-road traffic decreases rapidly with the distance from the roadway. At about 200 to 300 m from the roadway, the air pollution due to the local traffic has decreased enough that the air pollution is commensurate with the background air pollution. Results of simulations conducted with such models are in reasonable agreement with measurements of air pollution conducted in the vicinity of roadways. The modeling of the local impact of on-road traffic on air pollution is discussed in greater detail by Venkatram and Schulte (2018).

6.4.5 Operational Gaussian Models for Area Sources

Two main approaches are used to simulate atmospheric dispersion from area sources. The lagrangian atmospheric dispersion equation may be integrated numerically (for example, using Romberg's method). This approach has been used to take into account the width of a roadway, for example. Then, the formula for a line source presented in Section 6.4.4 is integrated along the width of the roadway. On the other hand, one may use a virtual point source located at a distance upwind of the area source such that the plume has the width of the area source over the area source. This approach is computationally cheaper than the former one because no numerical integration is needed. It presents, however, two shortcomings: (1) the location of the virtual source and the width of the plume over the area source depend on wind speed and atmospheric stability (initial source dimensions may be used instead) and (2) the concentration profile over the source area is gaussian. Nevertheless, this latter approach is widely used, for example for industrial sources, because of its low computational cost (EPA, 1995).

6.4.6 Operational Gaussian Models for Volume Sources

The most commonly used approach for volume sources is similar to the latter one mentioned for area sources. Thus, a virtual source is located at a distance upwind of the volume source such that the plume has the dimensions of the volume source when it reaches that source. This approach has been used, for example, to model fugitive emissions of volatile organic compounds from refineries (Kim et al., 2014).

6.4.7 Gaussian Models with Removal of Air Pollutants in the Plume

The concentrations of an air pollutant may decrease in a plume (or a puff), because of chemical reactions, scavenging by rain or dry deposition on surfaces. In the case where simple parameterizations are used to model such processes, it may be possible to take them into account explicitly in the operational formulas already presented. The addition of these removal terms is presented here in the case of a point source, but the formulas for other source types can be modified similarly.

Chemical kinetics is treated in the following chapters. Here, one considers the case where the air pollutant reacts with a chemical species of constant concentration (for example, an oxidant). Then, its concentration decreases with time according to an exponential function (see Chapters 7 and 8):

$$C(t) = C_0 \exp(-k\,[X]\,t) = C_0 \exp(-k'\,t) \tag{6.53}$$

where C_0 is the initial concentration, k is the rate constant of the chemical reaction, $[X]$ is the oxidant concentration, and k' is the pseudo-first-order rate constant expressed in s^{-1} ($k' = k\,[X]$). The solution of the gaussian equation taking into account this removal by chemical reaction is as follows:

$$C(x,y,z) = \frac{S}{2\pi u \sigma_y \sigma_z} \exp\left(-\frac{y^2}{2\sigma_y^2}\right) \left(\exp\left(-\frac{(z-z_{s,f})^2}{2\sigma_z^2}\right) + \exp\left(-\frac{(z+z_{s,f})^2}{2\sigma_z^2}\right)\right) \exp(-k'\,t) \tag{6.54}$$

where $t = x/u$.

Scavenging by rain (wet deposition) may be parameterized empirically using an exponential function (see Chapter 11):

$$C(t) = C_0 \exp(-\Lambda t) \qquad (6.55)$$

where Λ is the scavenging coefficient expressed in s^{-1}. The solution is similar to that obtained for the removal by chemical reaction:

$$C(x,y,z) = \frac{S}{2\pi u \sigma_y \sigma_z} \exp\left(-\frac{y^2}{2\sigma_y^2}\right) \left(\exp\left(-\frac{(z-z_{s,f})^2}{2\sigma_z^2}\right) + \exp\left(-\frac{(z+z_{s,f})^2}{2\sigma_z^2}\right)\right) \exp(-\Lambda t) \qquad (6.56)$$

where $t = x/u$.

Removal by dry deposition is usually parameterized with a deposition velocity, v_d, expressed in m s^{-1} (see Chapter 11). The solution is more complicated because dry deposition only affects the air pollutants that are in contact with the surface (chemical reaction and rain scavenging affect air pollutants in the entire plume). In other words, a solution that removes the air pollutants uniformly throughout the plume is inappropriate here. A suitable solution is (Overcamp, 1976):

$$C(x,y,z) = \frac{S}{2\pi u \sigma_y \sigma_z} \exp\left(-\frac{y^2}{2\sigma_y^2}\right) \left(\exp\left(-\frac{(z-z_{s,f})^2}{2\sigma_z^2}\right) + \alpha_d \exp\left(-\frac{(z+z_{s,f})^2}{2\sigma_z^2}\right)\right) \qquad (6.57)$$

where α_d is a function of the pollutant dry deposition velocity:

$$\alpha_d = 1 - \frac{2 v_d}{v_d + (u\, z_{s,f}\, \sigma_z^{-1})\left(\frac{d\sigma_z}{dx}\right)} \qquad (6.58)$$

If the dry deposition velocity is zero, then, $\alpha_d = 1$, and the solution becomes identical to the standard gaussian formula. On the other hand, if the dry deposition velocity tends toward infinity, then, α_d tends toward -1 and the surface acts as an irreversible sink for the pollutant in the plume.

6.4.8 Atmospheric Stability Categories

The parameterizations of the dispersion coefficients σ_y and σ_z that are used in operational gaussian plume and puff models depend on atmospheric stability. Therefore, it is useful to be able to determine the atmospheric stability category from simple meteorological data. Table 6.2 provides information to determine the atmospheric stability category given data on wind speed, solar radiation (during daytime) and cloudiness (at night). Table 6.3 provides information to determine the atmospheric stability categories given the vertical temperature gradient, the bulk Richardson number or the Monin-Obukhov length over open terrain.

Table 6.2. Atmospheric stability categories[a] as a function of simple meteorological data, after Turner (1970).

Wind speed (m s^{-1}) at 10 m	Day			Night	
	Solar radiation[b]			Cloudiness[c]	
	Strong	Moderate	Weak	$\geq 4/8$	$\leq 3/8$
<2	A	A or B	B	F	F
2 to 3	A or B	B	C	E	F
3 to 5	B	B or C	C	D	E
5 to 6	C	C or D	D	D	D
>6	C	D	D	D	D

(a) Stability categories: A (very unstable), B (unstable), C (moderately unstable), D (neutral), E (stable), F (very stable).
(b) Solar radiation: strong (>700 W m^{-2}), moderate (between 350 and 700 W m^{-2}), weak (<350 W m^{-2})
(c) Cloudiness: fraction of the sky covered by clouds.

Table 6.3. Atmospheric stability categories as a function of the bulk Richardson number, the vertical temperature gradient, and the Monin-Obukhov length.

Atmospheric stability category	Bulk Richardson number[a]	Vertical temperature gradient[b] (°C m^{-1})	Monin-Obukhov length[c] (m)
A	$Ri_b < -0.86$	$\Delta T/\Delta z < -1.9$	$-10 < L_{MO} < 0$
B	$-0.86 < Ri_b < -0.37$	$-1.9 < \Delta T/\Delta z < -1.7$	$-15 < L_{MO} < -10$
C	$-0.37 < Ri_b < -0.10$	$-1.7 < \Delta T/\Delta z < -1.5$	$-100 < L_{MO} < -15$
D	$-0.10 < Ri_b < 0.053$	$-1.5 < \Delta T/\Delta z < -0.5$	$L_{MO} < -100; 100 < L_{MO}$
E	$0.053 < Ri_b < 0.134$	$-0.5 < \Delta T/\Delta z < 1.5$	$30 < L_{MO} < 100$
F	$0.134 < Ri_b$	$1.5 < \Delta T/\Delta z$	$0 < L_{MO} < 30$

(a) The observed temperature must be corrected using a vertical gradient of -1 °C / 100 m to obtain the potential temperature. See Equation 4.62 for the definition of Ri_b.
(b) The temperature gradient must be measured over a difference in altitude that is significant (at least 100 m)
(c) For a roughness length of 0.1 m (open terrain). For a greater value of the roughness length, the absolute values of the ranges increase, since mechanical turbulence becomes more important. See Equation 4.59 for the definition of L_{MO}.

6.5 Eulerian Models of Atmospheric Dispersion

6.5.1 General Considerations

Lagrangian dispersion models are specific to a single source and, therefore, cannot be applied to a large number of sources, as required when simulating air pollution in an urban area or at a regional, continental or global scale. In such cases, an eulerian chemical-transport model is generally used. Air pollutant concentrations are calculated in a three-dimensional (3D) grid and they are not defined with respect to a source in particular. Therefore, atmospheric dispersion is calculated based on the atmospheric conditions and is independent of the source-based dispersion process. In other words, within a model grid cell, one assumes that the dispersion process includes the full spectrum of the turbulent eddies. The turbulent diffusion coefficients in the horizontal direction, K_h, and vertical direction, K_z, are defined solely as functions of meteorological conditions.

6.5.2 Vertical Dispersion

The vertical dispersion coefficient is important, because it governs the mixing of air pollutants within the PBL. It is a function of meteorological conditions, and several formulations are available. The two main categories of formulations include the local and non-local parameterizations. Local parameterizations consist of a standard first-order closure of turbulent diffusion (so-called K-theory formulations). Thus, vertical dispersion is represented via mass transfer between adjacent model layers. Non-local parameterizations apply to strongly convective conditions (i.e., with important vertical mass transfer), when the atmosphere becomes mixed significantly over large depths. Then, the transfer of air pollutants may occur among non-adjacent layers of the model. Figure 6.6 depicts schematically these different formulations.

The convective dispersion model presented here is the asymmetric convective mixing model (ACM). Mass transfer from the bottom layer occurs upward into all layers above. On the other

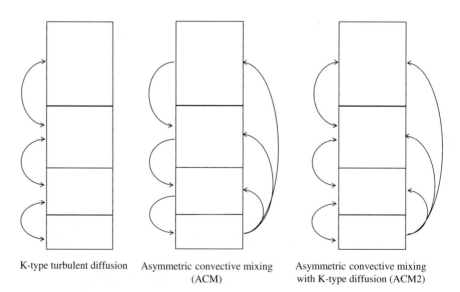

K-type turbulent diffusion   Asymmetric convective mixing (ACM)   Asymmetric convective mixing with K-type diffusion (ACM2)

Figure 6.6. Schematic representations of vertical dispersion in eulerian models. Left figure: standard K-theory turbulent diffusion model; middle figure: convective dispersion model; right figure: model combining convective dispersion and K-theory turbulent diffusion.

hand, downward mass transfer occurs between adjacent layers (some earlier models used formulations allowing downward mass transfer among non-adjacent layers). The assumption used in ACM is that the upward mass transfer is due to strong convection, whereas downward transfer results from a slow subsidence and, therefore, is more gradual. Both approaches (K-theory and ACM) have been combined to provide a more complete representation of vertical mixing processes; the corresponding model is called ACM2 (Pleim, 2007).

Several parameterizations have been developed to represent the vertical turbulent diffusion coefficient, K_z, for models using K-theory for turbulence closure. For example, similarity theory may be used for the PBL to develop parameterizations of K_z (see Chapter 4). By analogy with the parameterization of the turbulent kinematic viscosity (i.e., the coefficient of turbulent transfer of momentum) presented in Equation 4.31, the following equation may be used for K_z:

$$K_z(z) = \kappa \frac{u_*}{\Phi\left(\frac{z'}{L_{MO}}\right)} z \left(1 - \frac{z}{z_i}\right)^2 \qquad (6.59)$$

where κ is the von Kármán constant (0.4), u_* is the friction velocity (see Chapter 4.3), L_{MO} is the Monin-Obukhov length, which characterizes atmospheric stability (see Chapter 4.4), z is altitude, z_i is the PBL height, and z' is equal to z, except for unstable conditions when z' is the surface layer depth ($0.1\ z_i$) for $z > z'$. The PBL height, z_i, is usually obtained from a meteorological numerical simulation; however, parameterizations are also available to estimate it (see for example, Troen and Mahrt, 1986). The function Φ varies depending on the atmospheric stability. Those presented in Equations 4.68 and 4.69 for heat transfer may be used for mass transfer. Figure 6.7 shows the vertical profile of this vertical dispersion coefficient for various atmospheric stability conditions.

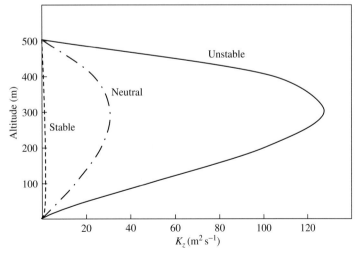

Figure 6.7. Vertical profiles of the eulerian vertical dispersion coefficient (similarity theory formulation). Meteorological conditions: stable (dashed line), neutral (dash-dotted line), and unstable (solid line). The PBL height is 500 m. The friction velocity (u_*) is 0.3, 0.4, and 0.6 m s^{-1} for stable, neutral, and unstable conditions, respectively.

Other parameterizations of vertical dispersion have been developed for the PBL and the free troposphere. For example, the parameterization of Louis (1979) is used in many models for the PBL. Some other parameterizations use the convection velocity, rather than the friction velocity (e.g., Troen and Mart, 1986).

6.5.3 Horizontal Dispersion

Various formulations are available to parameterize the horizontal dispersion coefficient in eulerian chemical-transport models. These formulations depend usually on the horizontal grid size of the eulerian model. However, the dispersion coefficients depend on model grid size in ways that may lead to confusion: one formulation uses a proportional relationship between the dispersion coefficient and the grid surface area, whereas another formulation uses an inversely proportional relationship. The most widely used formulation is that of Smagorinsky (1963):

$$K_{h,Smagorinsky} = 0.16 \, (S_\Gamma^2 + S_\Lambda^2)^{\frac{1}{2}} (\Delta x)^2$$
$$S_\Gamma = \left(\frac{\partial u}{\partial x} - \frac{\partial v}{\partial y}\right) \quad (6.60)$$
$$S_\Lambda = \left(\frac{\partial v}{\partial x} + \frac{\partial u}{\partial y}\right)$$

where S_Γ represents the stretching deformation and S_Λ the shearing deformation of the wind field; Δx is the horizontal grid size (m). This formulation leads to a horizontal dispersion coefficient, which is proportional to the horizontal grid cell surface area. Therefore, horizontal dispersion will become very large for a model simulation where a large horizontal grid size is used (say >10 km). Since numerical diffusion increases with grid size, it is likely that the horizontal dispersion term will become superfluous when the horizontal model resolution is coarse. Thus, it appears that this formulation should be limited to cases where the model horizontal grid size, Δx, is on the order of a few km.

Another parameterization, called Unif, has been proposed to counter numerical diffusion in cases where a coarse horizontal resolution is used (Byun and Schere, 2006):

$$K_{h,Unif} = 2000 \left(\frac{4000}{\Delta x}\right)^2 \quad (6.61)$$

This formulation leads to a horizontal dispersion coefficient, which is inversely proportional to the horizontal grid cell surface area, $(\Delta x)^{-2}$. Therefore, its value decreases when Δx increases, i.e., when numerical diffusion increases. However, this formulation is not applicable to simulations where a very small grid size is used, because K_h tends toward infinity when the grid size decreases (for example, it is 32,000 m^2 s^{-1} for $\Delta x = 1$ km).

An evaluation of these two formulations was performed using a field experiment conducted in 1999 in Texas. Tracer gases were released from different locations and measured downwind with a monitoring network. The model simulation used two imbedded domains with horizontal grid sizes of 12 and 4 km, respectively. The results suggested that, for this

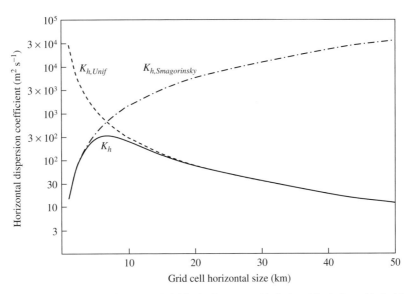

Figure 6.8. Eulerian horizontal dispersion coefficients: K_h (solid line), $K_{h,Smagorinsky}$ (dash-dotted line), $K_{h,Unif}$ (dashed line). A value of 10^{-4} s^{-1} was used for the wind field deformation.

model configuration, the Smagorinsky formulation led to better model performance than the Unif formulation (Pun et al., 2006). Byun and Schere (2006) proposed the following formulation, which combines the two formulations previously described:

$$K_h = \frac{1}{(K_{h,Smagorinsky}^{-1} + K_{h,Unif}^{-1})} \tag{6.62}$$

When Δx tends toward 0, $(K_{h,Unif})^{-1}$ tends toward 0 and K_h tends toward $K_{h,Smagorinsky}$; when Δx tends toward infinity, $(K_{h,Smagorinsky})^{-1}$ tends toward 0 and K_h tends toward $K_{h,Unif}$. Figure 6.8 illustrates how these eulerian horizontal dispersion coefficients depend on the horizontal grid size.

6.6 Street-canyon Models

The application of gaussian models to the urban canopy is strongly limited by the presence of buildings, which affect the air flow and, therefore, the transport and dispersion of air pollutants. In addition, the air pollution in a street canyon results mostly from local sources and it cannot be modeled by eulerian models, which have a grid resolution of at least 1 km, i.e., much greater than the dimension of a street canyon. Computational fluid dynamics (CFD) models can solve the equations that govern the air flow and the dispersion of the pollutants while taking into account the effect of buildings; however, the associated computational costs are too large to make CFD models applicable to large domains for long periods. Therefore, other approaches had to be developed to simulate the transport and dispersion of pollutants in an urban canopy with parameterizations that take into account the presence of buildings.

The most commonly used formulations are currently the Operational Street Pollution Model (OSPM; Berkowicz, 2000) and SIRANE (Soulhac et al., 2011). The latter is the most recent and it is summarized here. This formulation uses the assumption that the air pollutant concentration is uniform within a street canyon (or a segment of a street canyon). The turbulent transfer of air pollutants between the street canyon and the atmosphere above the urban canopy is modeled using a mass transfer coefficient at roof level. The horizontal transport of air pollutants within the street canyon is modeled using a wind that is parallel to the street axis, with a wind speed calculated using an exponential vertical profile. Then, the concentration of an air pollutant within the street canyon, C (g m^{-3}), is calculated via a steady-state mass balance, which takes into account the pollutant emissions in the street as well as deposition processes:

$$u_s h_b W_s C_u + S_l L_s = u_s h_b W_s C + F_d L_s (W_s + 2h_b) + \frac{\sigma_w W_s L_s}{\sqrt{2\pi}} (C - C_b) \qquad (6.63)$$

where u_s is the mean wind speed within the street canyon (m s^{-1}), σ_w is the standard deviation of the vertical wind speed at roof level (m s^{-1}), h_b, W_s, and L_s are the average height of the buildings, the average street width, and the length of the street-canyon segment, respectively (m), S_l is the emission rate of the pollutant in the street-canyon segment (g m^{-1} s^{-1}), C_u is the concentration of the pollutant transported form the upwind street-canyon intersection (g m^{-3}), C_b is the background concentration of the pollutant above the urban canopy (g m^{-3}, concentration obtained either from air monitoring stations or from a model simulation), and F_d (g m^{-2} s^{-1}) is the average dry deposition flux on urban surfaces (street, buildings ...). The average wind speed within the street is calculated using the assumption of an exponential vertical wind profile (see Equation 4.43) and accounting for the angle between the wind direction above the urban canopy and the direction of the street-canyon axis. The first term of the equation represents the pollutant inflow rate from the upwind intersection and the second term represents the emission rate within the street. The third term represents the outflow rate toward the downwind intersection. The fourth term represents dry deposition on surfaces. Scavenging by rain is not included here, but may easily be added if needed. The fifth term represents pollutant transfer between the street canyon and the background atmosphere above the urban canopy. It depends only on atmospheric turbulence (represented by σ_w) and does not account, for example, for traffic-generated turbulence or the effect of vegetation on atmospheric turbulence. Deposition fluxes may be calculated with parameterizations described in Chapter 11; they may be a function of surface type (street, sidewalk, building walls, windows, etc.).

SIRANE simulates the horizontal transport of air pollutants at street intersections by taking into account the variability of the wind direction and by conducting a mass balance of the flows through the intersection. Thus, some turbulent mass transfer at roof level is used to compensate for the horizontal inflows from and outflows to the street canyons connected with the intersection. The set of street-canyon equations is solved numerically, and the pollutant concentrations are calculated for each street-canyon segment.

If the wind is perpendicular to the street-canyon axis, an air recirculation zone may develop on the upwind side of the street while a ventilation zone may appear on the downwind side. Therefore, if an air pollutant is emitted within the recirculation zone, its concentration will be greater than that of an air pollutant emitted within the ventilation zone. The OSPM formulation takes into account the presence of recirculation and ventilation zones within a street-canyon configuration. However, the horizontal configuration of the street-canyon network may also generate recirculation zones and an academic representation of recirculation and ventilation zones may not always apply. Then, a CFD model is needed to simulate the air flow and calculate air pollutant concentrations in cases of complex configurations.

6.7 Numerical Modeling of Pollutant Transport in the Atmosphere

Gaussian plume models and street-canyon models have analytical solutions, which do not present any particular numerical difficulties. However, the calculation must be conducted for each source and receptor (i.e., the location where the air pollutant concentration is calculated) or for each street-canyon segment. Therefore, a large number of sources, receptors, and street-canyon segments may lead to significant computational costs.

Puff models require more computations, because puff trajectories must be tracked according to a wind field obtained from a numerical meteorological model. However, the computation of the pollutant concentrations within the puffs does not lead to any more numerical difficulties than those of gaussian plumes (puffs that treat chemical reactions may lead to numerical difficulties due to fast kinetics, see Chapter 7).

Eulerian chemical-transport models require a numerical solution of the set of the mass conservation equations representing the processes governing the concentrations of the air pollutants. Two distinct terms appear in the partial differential equation (excluding source, sink, and transformation terms): a term representing transport of the air pollutants by the mean wind and a turbulent dispersion term.

The first term (transport by the mean wind) is hyperbolic and presents a numerical difficulty: it leads to numerical diffusion, which artificially spreads concentrations throughout the model gridded system. Numerical diffusion may be reduced by using a finer spatial resolution, but, then, the computational cost increases considerably. Therefore, numerical algorithms have been developed to minimize numerical diffusion while keeping the computation costs manageable. However, some algorithms may have negative secondary effects such as the propagation of concentrations upwind (e.g., Bott's algorithm) or non-conservation of mass (e.g., some semi-lagrangian algorithms based on spatial gradients). Decisions on the choice of a numerical algorithm must be made based on the application at hand. The Smolarkiewicz algorithm is widely used; it presents few shortcomings, however, it is more diffusive than some other algorithms. The piecewise parabolic method is used in several models, because it presents the advantage of being positive definite and monotonous (Colella and Woodward, 1984). The algorithm of Walcek and Aleksic (1998) is also monotonous and mass conserving; in addition, it may be used via operator splitting for both transport terms (mean wind transport and dispersion). Comparisons of numerical algorithms have been conducted, which provide valuable information on the pros and cons of various available algorithms (e.g., Dabdub and Seinfeld, 1994).

The second term (dispersion) is parabolic and, therefore, does not present any particular numerical difficulty. Standard numerical algorithms, such as semi-implicit methods, may be used (e.g., von Rosenberg, 1969).

Problems

Problem 6.1 Dispersion of a gaussian plume

A plume is emitted from an incinerator stack. Atmospheric conditions are assumed to be stationary, and a gaussian plume model is used to simulate the atmospheric dispersion of this plume. The dioxin emission rate is 6×10^{-4} g s^{-1}, the stack height is 20 m, atmospheric stability is neutral, the wind speed is 5 m s^{-1}, and plume rise is assumed to be negligible. Calculate the dioxin concentration, C, at a location 1 km downwind, underneath the plume at a height of 1.5 m corresponding to a person's exposure.

Problem 6.2 Atmospheric dispersion of air pollutants

Air pollutants emitted from a tall stack lead to exceedance of an ambient air quality standard at a nearby monitoring station. Reducing the emissions with some control equipment would be too costly, and increasing the stack height is not allowed by the local regulatory agency. Suggest another approach to increase the dilution of the air pollutants in order to attain the ambient air quality standard.

Problem 6.3 Plume rise

Calculate the plume rise for a power plant stack with the following characteristics and atmospheric conditions: the stack diameter is 7.3 m, the plume initial vertical velocity at the stack exit is 18 m s^{-1}, the plume initial temperature is 77 °C, the ambient temperature is 20 °C, and the horizontal wind speed is 5 m s^{-1}.

Problem 6.4 Lagrangian and eulerian representations of atmospheric dispersion

The gaussian dispersion coefficients of McElroy-Pooler under rural conditions and neutral atmospheric stability are used here (Table 6.1) with a wind speed of 10 m s^{-1}.

a. Calculate the corresponding eulerian coefficients.
b. At which downwind distance from the source can these eulerian dispersion coefficients be used?
c. Check whether the values of these eulerian dispersion coefficients are consistent with the values given in this chapter.

References

Berkowicz, R., 2000. OSPM – A parameterized street pollution model, *Environ. Monitoring Assessment*, **65**, 323–332.

Bird, R.B., W.E. Stewart, and E.N. Lightfoot, 2006. *Transport Phenomena*, John Wiley & Sons, Inc., USA.

Briant, R., I. Korsakissok, and C. Seigneur, 2011. An improved line source model for air pollutant dispersion from roadway traffic, *Atmos. Environ.*, **45**, 4099–4107.

Briggs, G.A., 1984. Plume rise and buoyancy effects, *Atmospheric Science and Power Production*, U.S. Department of Energy, USA.

Byun, D. and K.L. Schere, 2006. Review of the governing equations, computational algorithms, and other components of the Models-3 Community Multiscale Air Quality (CMAQ) Modeling System, *Appl. Mechanics Rev.*, **59**, 51–77.

Carson, J.E. and H. Moses, 1969. The validity of several plume rise formulas, *J. Air Pollut. Control Assoc.*, **19**, 862–866.

Colella, P. and P.R. Woodward, 1984. The piecewise parabolic method (PPM) for gas-dynamical simulations, *J. Comp. Phys.*, **54**, 174–201.

Csanady, G.T., 1973. *Turbulent Diffusion in the Environment*, D. Reidel Publishing Company, Dordrecht, The Netherlands.

Dabdub, D. and J.H. Seinfeld, 1994. Numerical advective schemes used in air quality models – Sequential and parallel implementation, *Atmos. Environ.*, **28**, 3369–3385.

EPA, 1995. User's guide for the industrial source complex (ISC3) dispersion models, U.S. Environmental Protection Agency, Research Triangle Park, NC, USA; available at: www.epa.gov/ttn/scram/userg/regmod/isc3v2.pdf.

Gifford, F.A., 1961. Use of routine meteorological observations for estimating atmospheric dispersion, *Nucl. Safety*, **2**, 47–51.

Karamchandani, P., L. Santos, I. Sykes, Y. Zhang, C. Tonne, and C. Seigneur, 2000. Development and evaluation of a state-of-the-science reactive plume model, *Environ. Sci. Technol.*, **34**, 870–880.

Kim, Y., C. Seigneur, and O. Duclaux, 2014. Development of a plume-in-grid model for industrial point and volume sources: Application to power plant and refinery sources in the Paris region, *Geosci. Model Dev.*, **7**, 569–585.

Louis, J.-F., 1979. A parametric model of vertical eddy fluxes in the atmosphere, *Bound.-Lay. Meteor.*, **17**, 187–202.

Overcamp, T.J., 1976. A general Gaussian diffusion-deposition model for elevated point sources, *J. Appl. Meteor.*, **15**, 1167–1171.

Pleim, J.E., 2007. A combined local and non-local closure model for the atmospheric boundary layer. Part 1: Model description and testing, *J. Appl. Meteor. Climatol.*, **46**, 1383–1395.

Pun, B., C. Seigneur, K. Vijayaraghavan, S.-Y. Wu, S.-Y. Chen, E. Knipping, and N. Kumar, 2006. Modeling regional haze in the BRAVO study using CMAQ-MADRID. 1. Model evaluation, *J. Geophys. Res.*, **111**, D06302, doi:10.1029/2004JD005608.

von Rosenberg D., 1969. *Methods for the Numerical Solution of Partial Differential Equations*, Elsevier, New York.

Smagorinsky J., 1963. General circulation experiments with the primitive equations: 1. The basic experiment, *Mon. Wea. Rev.*, **91**, 99–164.

Soulhac, L., P. Salizzoni, F.-X. Cierco, and R.J. Perkins, 2011. The model SIRANE for atmospheric urban air pollution dispersion: Part I, presentation of the model, *Atmos. Environ.*, **45**, 7379–7395.

Taylor, G.I., 1922. Diffusion by continuous movements, *Proc. London Math. Soc.*, **A20**, 196–212.

Troen, I.B. and L. Mahrt, 1986. A simple model of the atmospheric boundary layer: Sensitivity to surface evaporation, *Bond.-Lay. Meteor.*, **37**, 129–148.

Turner, J.S., 1970. *Workbook of Atmospheric Dispersion Estimates*, U.S. Environmental Protection Agency, Research Triangle Park, NC, USA.

Venkatram, A. and T. Horst, 2006. Approximating dispersion from a finite line source. *Atmos. Environ.*, **40**, 2401–2408.

Venkatram, A. and N. Schulte, 2018. *Urban Transportation and Air Pollution*, Elsevier, New York.

Walcek, C.J. and N.M. Aleksic, 1998. A simple but accurate mass conservative, peak-preserving, mixing ratio bounded advection algorithm with Fortran code, *Atmos. Environ.*, **32**, 3863–3880.

7 The Stratospheric Ozone Layer

The stratospheric ozone layer results from the photolysis of molecular oxygen by ultraviolet (UV) solar radiation in the high atmosphere. This large atmospheric layer is stable and, therefore, affects the general atmospheric circulation by decreasing significantly the vertical motions of air parcels. In addition, ozone protects the Earth from harmful UV radiation. Therefore, its destruction by anthropogenic activities may lead to public health impacts. This chapter presents first some fundamentals of atmospheric chemical kinetics (i.e., the speed at which chemical reactions occur in the atmosphere), which are needed to understand the processes leading to the presence of the ozone layer. These notions are also needed to understand the formation of gaseous and particulate pollutants, which is presented in the following chapters. Next, the processes that govern the ozone layer are described in terms of its natural formation and its destruction by man-made substances. Finally, the public policies introduced to address the protection of the stratospheric ozone layer are summarized.

7.1 Fundamentals of Chemical Kinetics

7.1.1 Atmospheric Concentration Units

The ideal gas law was shown in Chapter 4 to apply to the atmosphere. Therefore, at an atmospheric pressure $P = 1$ atm, i.e., at sea level, and a temperature $T = 298$ K (i.e., 25 °C), the air contains 40.9 moles m^{-3}. Atmospheric concentrations may be expressed in various units, and the following ones are generally used in atmospheric chemistry:

– Moles per unit volume of air
– Molecules per unit volume of air
– Molar fraction (also called mixing ratio)
– Mass per unit volume of air

The conversion of moles per unit volume of air into molecules per unit volume of air is done using the Avogadro number, N, which is the number of molecules per mole ($N = 6.02 \times 10^{23}$ molecules mole^{-1}).

There are 40.9 moles of air per m^3 at 1 atm and 25 °C, therefore:

$$n/V = 40.9 \text{ moles m}^{-3} \times 6.02 \times 10^{23} \text{molecules mole}^{-1}$$
$$= 2.46 \times 10^{25} \text{molecules m}^{-3} \quad (7.1)$$
$$= 2.46 \times 10^{19} \text{molecules cm}^{-3}$$

Molecular concentrations are usually given in molecules per cm^3 (i.e., molec cm^{-3}).

The molar fraction (also referred to as mixing ratio) is the ratio of the number of moles (or molecules) of a chemical species and of the number of moles (or molecules) of air (i.e., nitrogen + oxygen + argon + other minor constituents). Therefore, the molar fraction of pure air is 1. For a gaseous pollutant, it is of course much less than 1. Accordingly, the molar fraction is usually expressed as ppm (parts per million), ppb (parts per billion), or ppt (parts per trillion). If there is a molecule of a chemical species per million molecules of air, its molar fraction is 1 ppm. If there is a molecule of a chemical species per billion molecules of air, its molar fraction is 1 ppb. The molar fraction of pure air is by definition 10^6 ppm, i.e., 10^9 ppb, or 10^{12} ppt.

The conversion of the molar (or molecular) concentration of a species into a molar fraction (or vice versa) is done using the ideal gas law. Therefore, this conversion depends on pressure and temperature. For example, let C_{molec} be the molecular concentration expressed in molec cm^{-3}. The molar fraction expressed in ppm, C_{ppm}, is calculated as follows (V is expressed here in m^3):

$$C_{ppm} = \frac{10^{12} \, C_{molec}}{N \frac{n}{V}} = \frac{10^{12} \, C_{molec} \, R \, T}{N \, P} \quad (7.2)$$

where $R = 8.206 \times 10^{-5}$ atm m^3 mole^{-1} K^{-1}. If a chemical species is uniformly mixed within the atmosphere, its molar fraction is constant. On the other hand, its molar or molecular concentration decreases with altitude, since pressure decreases with altitude (temperature also decreases with altitude in the troposphere, but the absolute temperature gradient is less than the pressure gradient).

The mass concentration is derived from the molar concentration (or from the molecular concentration) using the molar mass of the chemical species, MM. Let C_{mass} be the mass concentration of a chemical species (expressed here in µg m^{-3}). It is calculated from the molecular concentration, C_{molec} (expressed here in molec cm^{-3}), as follows:

$$C_{mass} = \frac{10^{12} \, C_{molec} \, MM}{N} \quad (7.3)$$

where the factor 10^{12} corresponds to conversions from g to µg and from cm^3 to m^3.

The conversion of the mass concentration into molar fraction (or vice versa) is done using the ideal gas law and, therefore, depends on pressure and temperature. For example, to obtain the molar fraction in ppb from a mass concentration in µg m^{-3}:

$$C_{ppb} = \frac{10^3 \, C_{mass}}{MM \, \frac{n}{V}} = \frac{10^3 \, C_{mass} \, R \, T}{MM \, P} \quad (7.4)$$

where the factor 10^3 corresponds to conversions of molar fraction to ppb and of mass from µg to g.

Example: Conversion in ppm of 4 µg m^{-3} of sulfuric acid, H_2SO_4

Conditions are 1 atm and 25 °C. The molar masses of H, S, and O are 1, 32, and 16 g mole^{-1}, respectively.

The molar mass of sulfuric acid is:

$$MM_{H2SO4} = (2 \times 1) + 32 + (4 \times 16) = 98 \text{ g mole}^{-1}$$

Thus, $\quad C_{ppm} = 4 \times 8.2 \times 10^{-5} \times 298 \,/\, 98$

$\quad\quad\quad C_{ppm} = 10^{-3} \text{ ppm}$

7.1.2 Chemical Species and Chemical Reactions

The following types of chemical species are present in the atmosphere:

- Molecules such as molecular nitrogen (N_2) and molecular oxygen (O_2). Molecules do not have any free electrons and are chemically stable. Their chemical reactions with other species involve breaking a bond between two atoms, which requires energy.
- Radicals such as the hydroxyl radical (OH), the hydroperoxyl radical (HO_2, also called perhydroxyl radical; the prefix "per" indicates saturation, here in terms of oxygen), and the nitrate radical (NO_3). Radicals have one or more free electrons. Therefore, they are very chemically reactive, because they aim to stabilize their electron cloud by adding one or more electrons. The presence of a free electron may be represented by a dot next to the chemical formula, for example $HO_2^{\bullet}$; this notation is not used here for the sake of simplicity, except in some figures of chemical mechanisms presented in Chapter 8.
- Atoms, which may be chemically stable or to the contrary very reactive. Some atoms such as elementary mercury (Hg°) do not have any free electron and they are, therefore, stable. Others such as the oxygen atom (O) have a free electron and are very reactive.
- Excited chemical species such as the excited oxygen atom $O(^1D)$, which has more energy than the standard oxygen atom $O(^3P)$. The excited state results from the absorption of energy (from solar radiation via a photolytic reaction, for example); the excited species then seeks to lose that extra energy through a collision with other species (typically molecules present in high concentrations such as nitrogen, oxygen or water).
- In the aqueous phase, ions such as H^+, OH^-, and NO_3^-. In an aqueous solution, the positive and negative electrical charges must balance each other to give a neutral solution: this is called electroneutrality (see Chapter 10).

Reactions may be categorized as photochemical and chemical reactions. Photochemical reactions (also called photolytic reactions) correspond to the absorption by a molecule of energy from solar radiation. Chemical reactions concern one (rarely, this is called thermal decomposition), two (in most cases), or three chemical species.

7.1.3 Photochemical Reactions

Solar radiation energy is carried by photons (see Chapter 5). A photon carries an energy quantity, E_p, equal to $h\nu$, where h is the Planck constant (6.626×10^{-34} m^2 kg s^{-1}) and ν is the frequency of the radiation (s^{-1}). For a given wavelength, the frequency is equal to the speed of light divided by the wavelength:

$$\nu = \frac{c}{\lambda} \tag{7.5}$$

Therefore, a photon of wavelength $\lambda = 400$ nm (violet light), i.e., 4×10^{-7} m, has a frequency equal to 7.5×10^{14} s^{-1} and the energy per photon at that wavelength is about 5×10^{-19} m^2 kg s^{-2}, i.e., 5×10^{-19} J.

The dissociation energy of a bond of the oxygen molecule (O-O), $E_{b,O2}$, is 500 kJ mole^{-1}, i.e., $5 \times 10^5 / N$ J molec^{-1}, where N is Avogadro's number.

$$E_{b,O2} = 500 \text{ kJ mole}^{-1} = (5 \times 10^5 / 6.02 \times 10^{23}) \text{ J molec}^{-1}$$
$$= 8.3 \times 10^{-19} \text{ J molec}^{-1}$$

Therefore, a photon corresponding to the wavelength of violet light does not have enough energy to break the bond of molecular oxygen. Visible light ranges from 400 nm (violet) to 700 nm (red). Since the energy of photons decreases when the wavelength increases, red light has less energy than violet light. Therefore, oxygen is not photolyzed by visible light.

Example: Below which wavelength can a molecule of oxygen be photolyzed?

$$E_{b,O_2} = 8.3 \times 10^{-19} \text{ J molec}^{-1} < \frac{h\,c}{\lambda}$$

$$8.3 \times 10^{-19} < \frac{6.6 \times 10^{-34} \times 3 \times 10^8}{\lambda}$$

Thus: $\quad \lambda < 2.4 \times 10^{-7}$ m $= 240$ nm

However, a photon with enough energy may not necessarily break a chemical bond. The rate constant of a photolytic reaction depends on several characteristics of the molecule that absorbs the radiation at a given wavelength, λ. Thus, a photochemical rate constant (also called photolysis rate constant or photolysis rate coefficient), J, is equal to the product of three terms:

$$J = \sigma_J(\lambda)\, I_J(\lambda)\, \phi_J(\lambda) \tag{7.6}$$

where $\sigma_J(\lambda)$ is the absorption cross-section of the molecule, which represents its capacity to absorb the radiation (in cm^2 per molecule), $I_J(\lambda)$ is the actinic flux, which represents the amount of radiation received from all directions (in photons per cm^2 per s), and $\phi_J(\lambda)$ is the quantum yield, which represents the probability that the molecule will be photolyzed when a photon is absorbed by the molecule (molecule per photon).

The actinic flux is maximum when the Sun is at its zenith and it is zero at night. Therefore, there are no photochemical reactions at night. This results from the fact that infrared radiation emitted by the Earth does not have enough energy, since it corresponds to large wavelengths, >700 nm, i.e., to low energy.

The zenith angle of the Sun, θ_z, depends on latitude, date, and hour. It is calculated with the following formula (Jacobson, 2005):

$$\cos(\theta_z) = \sin(\phi)\sin(\delta_s) + \cos(\phi)\cos(\delta_s)\cos(\theta_h) \tag{7.7}$$

where ϕ is the latitude, δ_s is the Sun's declination angle (which depends on the date), and θ_h is the angle of the local hour. The declination angle is given by the following formula:

$$\delta_s = \arcsin(\sin(\varepsilon_{ob})\sin(\lambda_{ec})) \tag{7.8}$$

where ε_{ob} is the inclination (or obliquity) of the ecliptic and λ_{ec} is the ecliptic longitude of the Sun. The ecliptic is the mean plane of the orbit of the Earth around the Sun; it passes through both tropics and the equator. The inclination of the ecliptic is the angle between this plane and the equatorial plane; it is about 23.44 °, however, it decreases slightly every year by about 0.468," i.e., 1.3×10^{-4} °. The ecliptic longitude of the Sun may be calculated as follows:

$$\begin{aligned}\lambda_{ec} &= L_{ec} + 1.915° \sin(g_{ec}) + 0.020° \sin(2g_{ec}) \\ L_{ec} &= 280.460° + 0.9856474° N_J \\ g_{ec} &= 357.528° + 0.9856003° N_J \\ N_J &= D_J - 2451545\end{aligned} \tag{7.9}$$

where D_J is the Julian day, in a chronology starting on January 1 of year −4,713 BC, at noon on the Greenwich meridian. $D_J = 2{,}457{,}388.5$ for January 1, 2017.

The angle of the local hour is given by the following formula:

$$\theta_h = \frac{2\pi t_s}{86400} \tag{7.10}$$

where t_s is time in seconds starting at noon. Therefore, $\theta_h = 0$ at noon when the zenith angle is minimum during the day. For example, at noon on June 21, 2016 in Paris (48 ° 51 ' 12 " N), the correction on solar radiation is $\cos(\theta_z) = 0.9$.

It is straightforward to calculate that at the tropics (latitude = 23.44 °) on June 21 (solstice) at noon: $\theta_z = 0$ (the Sun is at the zenith). At the equator (latitude = 0 °) on March 20 (equinox) at noon: $\theta_z = 0$. (Actually, $\theta_z \approx 0$, because the solstice and the equinox do not occur exactly at noon.)

The quantum yield is zero if the energy of the photon is less than the dissociation energy of the bond (it is an approximation because the formation of an excited molecule may occur after absorption of the photon, with subsequent reaction of the excited species with another species resulting in bond dissociation; the energy of the excited molecule must then be taken into account; this is the case, for example, for the photolysis of nitrogen dioxide, NO_2). For some molecules, the quantum yield may be equal or close to 1 at some wavelengths.

Since chemical species with free electrons are more reactive than molecular species and stable atoms, photolysis (which generates radicals and atoms with free electrons, as well as excited species) leads to an atmosphere that is more chemically reactive. Therefore, the formation of secondary pollutants, which are produced via chemical reactions in the atmosphere, is more important when photolysis occurs, i.e., during the day. At night, the atmosphere is not very chemically reactive, because photochemical reactions do not take place.

Data (absorption cross-sections and quantum yields) for the main atmospheric photochemical reactions are available in the evaluations of the Jet Propulsion Laboratory (Burkholder et al., 2015).

7.1.4 Chemical Reactions

General Considerations

Atmospheric gas-phase chemical reactions may be grouped in three main categories.

– Unimolecular reactions lead to the dissociation of a molecule via absorption of thermal energy. In the atmosphere, the amount of thermal energy that may be absorbed by a molecule is much smaller than the energy available from solar radiation. Nevertheless, some molecules may undergo thermal dissociation. An important thermal dissociation reaction in air pollution is the dissociation of peroxyacetyl nitrate (PAN), which may dissociate into nitrogen dioxide (NO_2) and a peroxyacetyl radical (CH_3COO_2). PAN is formed from those two species and is, therefore, considered to be a reservoir species, since PAN can form NO_2 back by dissociation when the ambient temperature increases.
– Bimolecular reactions are the most common. Two chemical species undergo a collision. Note that such a bimolecular reaction does not necessarily involve two molecules, but may also correspond to a reaction between a molecule and a radical, a molecule and an atom, two radicals or a radical and an atom. With some probability, this collision may lead to the dissociation of some chemical bonds and the formation of new chemical species. The species that collide and react are called the reactants. The species that are produced by a chemical reaction are called the products. A reaction may lead to one or more products.
– Trimolecular (or termolecular) reactions, require the presence of a third molecule for the reaction to take place. In the atmosphere, this third molecule is N_2 or O_2; it is represented by M. As for bimolecular reactions, a trimolecular reaction may involve a radical or an atom as one of the three chemical species involved.

For a reaction to occur, a chemical bond must be broken and, therefore, a quantity of energy greater than the bond dissociation energy must be added. One defines for each reaction an activation energy, E_a, which is the energy required for the reaction to occur. Once the reaction occurs, the products have different chemical bonds than the reactants. The difference between the energies of the products and reactants may be calculated from the energies of their chemical bonds (for a molecule with only two atoms, the bond energy is equal to the bond dissociation energy; for molecules with more than two atoms, the energy of a chemical bond is an average value, which differs from the dissociation energy of that bond). If the total energy of the products is greater than that of the reactants, the reaction is endothermic (the net budget requires adding thermal energy); if the total energy of the products is less than that of the reactants, the reaction is exothermic (the net budget leads to a release of thermal energy). For example, combustion is exothermic.

The mean thermal energy of a gas, E_t (J mole^{-1}), is usually too low compared to the activation energy of a chemical reaction. It is related to the kinetic energy of its molecules and is given by the following equation:

$$E_t = \frac{3}{2} N k_B T \qquad (7.11)$$

where k_B is the Boltzmann constant (1.38 × 10^{-23} J K^{-1}). Thus, E_t = 3.7 kJ mole^{-1} at T = 25 °C. However, the kinetic theory of gases implies that the velocities of the atoms and molecules (and, therefore, their kinetic energy) are not uniform, but follow a distribution, which is called the Maxwell–Boltzmann distribution. Thus, some molecules or atoms may have velocities sufficiently high to exceed the activation energy of the chemical reaction so that the reaction may occur.

The kinetics of most reactions increases with temperature, because (1) the random motion of molecules, atoms, and radicals in the gas increases with temperature (and, therefore, the number of collisions increases) and (2) the velocity distribution of molecules, atoms, and radicals changes in such a way that the probability of high velocities increases.

The kinetics of a chemical reaction is limited by the diffusion of molecules, atoms, and radicals in the gas, since two chemical species must collide for the reaction to occur. The kinetic theory of gases implies that there is a maximum value of the kinetics of bimolecular reactions that corresponds to the case where each collision results in a reaction between the two chemical species; this maximum value of the rate constant is about 4.3 × 10^{-10} molec^{-1} cm^3 s^{-1}.

Chemical Kinetics of Unimolecular Thermal Dissociation Reactions

There are few unimolecular thermal dissociation reactions. A unimolecular reaction is usually a reaction consisting of two elementary reactions. The first reaction is bimolecular and leads to an excited state of the molecule of interest. The second reaction corresponds to the decomposition of the excited molecule:

$$A + M \underset{k_2}{\overset{k_1}{\longleftrightarrow}} A^* + M \qquad (R7.1)$$

$$A^* \overset{k_3}{\to} B + C \qquad (R7.2)$$

where M represents the air molecules, either molecular oxygen (O_2) or molecular nitrogen (N_2). The excited molecule may either dissociate into two distinct chemical species due to its extra energy or return to its initial state. Therefore, there are actually three elementary reactions, which correspond to what is perceived as a unimolecular dissociation:

$$A \rightarrow B + C \tag{R7.3}$$

The kinetics of this type of reaction has been studied by Lindemann, Hinshelwood, and Troe. Two extreme regimes may be considered: one at low pressure (low concentration of M) and the other at high pressure (high concentration of M).

Assuming that the excited molecule A* is at steady-state, due to its unstable excited energy state (brackets, [ ], indicate concentrations):

$$[A^*] = \frac{k_1 [A][M]}{k_2 [M] + k_3} \tag{7.12}$$

where k_1, k_2, and k_3 are the reaction rate constants. Therefore, the kinetics of the chemical reaction (also called the reaction rate), r_r (molec cm^{-3} s^{-1}) leading to the formation of B and C is as follows:

$$r_r = k_3 [A^*] = k_3 \frac{k_1 [A][M]}{k_2 [M] + k_3} \tag{7.13}$$

Thus, the rate constant expressed under its unimolecular form is as follows:

$$A \xrightarrow{(+M)} B + C \tag{R7.4}$$

$$k = k_3 \frac{k_1 [M]}{k_2 [M] + k_3}; \quad r_r = k[A] \tag{7.14}$$

The rate constant may be expressed in terms of two constants representing the extreme cases at low pressure (low concentration of M) and high pressure (high concentration of M):

$$k = \frac{1}{\frac{k_2}{k_1 k_3} + \frac{1}{k_1 [M]}} = \frac{1}{\frac{1}{k_\infty} + \frac{1}{k_0 [M]}}; \quad k_0 = k_1; \quad k_\infty = \frac{k_1}{k_2} k_3 \tag{7.15}$$

In the original formulation, $k_0 = k_1 [M]$ (Troe, 1983); however, the formulation used in Equation 7.15 is now used, so that k_0 depends only on temperature and does not depend on pressure (Finlayson-Pitts and Pitts, 2000; Jacobson, 2005). At low pressure, the concentration of air molecules is low and the probability of dissociation is greater than that of deexcitation by collision with M. Therefore, $k_2 [M] \ll k_3$, and $k \approx k_1 [M]$. Then, the rate constant k is proportional to pressure, i.e., [M]. At high pressure, there is saturation of air molecules; the limiting step of the kinetics is the dissociation of the excited molecule A* and the rate constant is simply proportional to k_3 and the equilibrium constant k_1/k_2. The rate constant may be written in terms of these two rate constants, k_0 and k_∞, which may be estimated experimentally at low and high pressures. This is called the Lindemann-Hinshelwood expression (after the names of the English scientists who introduced and developed the concept of these elementary steps involving an excited molecule). A correction was proposed by Troe (1983) for the intermediate pressure regime:

$$k = \frac{k_\infty k_0 [M]}{k_\infty + k_0 [M]} F_c^{\left[1 + \left(\log \frac{k_0 [M]}{k_\infty}\right)^2\right]^{-1}} \tag{7.16}$$

where k_0 and k_∞ are the rate constants for the low and high pressure cases, respectively, and the Troe correction uses the factor F_c (sometimes called the falloff factor), which is estimated theoretically and tends toward 1 when the pressure tends toward 0 or infinity. The rate constants k_0 and k_∞ are expressed in cm^3 molec^{-1} s^{-1} and s^{-1}, respectively, and [M] is expressed in molec cm^{-3}. The constants k_0 and k_∞ depend on temperature, as described in the section on bimolecular reactions. When [M] tends toward 0, k tends toward k_0 [M] and when [M] tends toward infinity, k tends toward k_∞ and does not depend on [M]. For atmospheric reactions, the Troe correction factor, F_c, ranges between 0.3 and 0.9 depending on the reaction.

Chemical Kinetics of Bimolecular Reactions

Chemical kinetics consists of quantifying the rate at which a chemical reaction occurs. It is governed by the law of mass action. Let us consider the following generic bimolecular reaction:

$$\alpha A + \beta B \to \gamma C + \delta D \quad \text{(R7.5)}$$

The coefficients α, β, γ, and δ are the stoichiometric coefficients. They are typically equal to 1, although in some cases some may be equal to 2.

The reaction rate is as follows:

$$r_r = k\,[A]^\alpha\,[B]^\beta \quad (7.17)$$

where k is the rate constant and the brackets indicate concentrations.

The changes in chemical concentrations with respect to time are defined as follows:

$$r_r = -\frac{1}{\alpha}\frac{d[A]}{dt} = -\frac{1}{\beta}\frac{d[B]}{dt} = +\frac{1}{\gamma}\frac{d[C]}{dt} = +\frac{1}{\delta}\frac{d[D]}{dt} \quad (7.18)$$

The negative signs correspond to the consumption of a reactant, and the positive signs correspond to the formation of a product. This equation conserves the total mass of reactants and products.

The rate constant depends on temperature and is defined in its most general form as follows:

$$k = A_T\,T^{B_T}\exp\left(-\frac{E_a}{RT}\right) \quad (7.19)$$

where A_T is a pre-exponential factor, B_T is the exponent of the temperature term, E_a is the activation energy of the reaction (J mole^{-1}), R is the ideal gas law constant (8.314 J mole^{-1} K^{-1}), and T is temperature (K).

The most commonly used expression is the Arrhenius expression, which is a simplified version where the pre-exponential term does not depend on temperature:

$$k = A_T\exp\left(-\frac{E_a}{RT}\right) \quad (7.20)$$

However, when the activation energy is very small, e.g., on the order of 1 kJ mole^{-1}, the exponential term tends toward 1 and the temperature dependence is then governed by the pre-exponential term T^{B_T}.

Chemical Kinetics of Trimolecular Reactions

In the case of a trimolecular reaction in the atmosphere, the third molecule is a molecule of air, i.e., an oxygen molecule (O_2) or a nitrogen molecule (N_2). Therefore, this type of reaction depends on the atmospheric pressure. The trimolecular reaction process is not an elementary reaction, because the probability that three molecules collide simultaneously is very low. It represents a series of reactions with the formation of an intermediate excited species (i.e., a species with excess energy), which returns to a stable energy level either by dissociation into the original reactants, or by colliding with a third molecule (O_2 or N_2). Thus:

$$A + B \underset{k_2}{\overset{k_1}{\longleftrightarrow}} AB^* \overset{k_3\ (+M)}{\longrightarrow} C \tag{R7.6}$$

An approach similar to that used for unimolecular reactions is used to express the rate constant as a function of its extreme values under low- and high-pressure regimes. Assuming steady state for the excited molecule AB^*, due to its highly unstable nature:

$$[AB^*] = \frac{k_1[A][B]}{k_2 + k_3[M]} \tag{7.21}$$

The kinetics of the formation of C is given by the following expression:

$$r_r = k_3[AB^*][M] = k_3 \frac{k_1[A][B]}{k_2 + k_3[M]}[M] \tag{7.22}$$

Therefore, a rate constant corresponding to the formulation of the trimolecular reaction expressed as a bimolecular reaction may be written as follows:

$$A + B \overset{(+M)}{\longrightarrow} C \tag{R7.7}$$

$$k = k_3 \frac{k_1[M]}{k_2 + k_3[M]}; \quad r_r = k[A][B] \tag{7.23}$$

The rate constant may be written as a function of two constants representing the extreme cases, i.e., at low and high pressure:

$$k = \frac{1}{\frac{1}{\frac{k_1}{k_2}k_3[M]} + \frac{1}{k_1}} = \frac{1}{\frac{1}{k_0[M]} + \frac{1}{k_\infty}}; \quad k_0 = \frac{k_1}{k_2}k_3; \quad k_\infty = k_1 \tag{7.24}$$

At low pressure, the concentration of the air molecules is limiting the formation of C and, therefore, the rate constant is proportional to pressure, i.e., $[M]$, as well as to the equilibrium constant between the two reactants and the excited molecule, k_1/k_2. Thus, the reaction appears as being a trimolecular reaction. At high pressure, there is saturation of air molecules. The limiting step is the formation of the excited molecule AB^* and the rate constant is simply k_1 (i.e., the kinetics of the dissociation of the excited molecule into the original reactants becomes negligible) and the reaction appears to be bimolecular. The rate constant can be written as a function of these two rate constants, k_0 and k_∞, which depend only on temperature. These two constants may be estimated experimentally under low- and high-pressure conditions, respectively. The expression of the rate constants, taking into account the Troe correction is as follows:

$$k = \frac{k_\infty k_0 [M]}{k_\infty + k_0 [M]} F_c^{\left[1 + \left(\log \frac{k_0 [M]}{k_\infty}\right)^2\right]^{-1}}$$ (7.25)

where k_0 and k_∞ are the rate constants for the cases at low and high pressures, respectively, and [M] is the total concentration of N_2 and O_2 (i.e., corresponding to the atmospheric pressure). Therefore, k_0 is expressed in cm^6 $molec^{-2}$ s^{-1}, k_∞ is expressed in cm^3 $molec^{-1}$ s^{-1}, and [M] is expressed in molec cm^{-3}. When [M] tends toward 0, k tends toward k_0 [M] and when [M] tends toward infinity, k tends toward k_∞ and no longer depends on [M]. For atmospheric reactions, the Troe factor, F_c, ranges between 0.3 and 0.9 depending on the reaction.

The constants k_0 and k_∞ depend on temperature, as described previously in the section on bimolecular reactions. However, the temperature dependence of k_1 and k_3 is low and the temperature dependence of k_2 dominates (at high temperature, the vibrational energy of AB* is greater, which favors the backward reaction to form A and B). Therefore, the overall rate constant, k, tends to decrease as temperature increases.

Kinetic data for inorganic atmospheric reactions are available in the evaluations of the Jet Propulsion Laboratory (Burkholder et al., 2015). For organic reactions, kinetic data are available in the references provided in Chapter 8.

7.1.5 Principle of Microreversibility and Chemical Equilibrium

All elementary reactions are reversible. This is called the principle of microreversibility. An elementary reaction is a single-step reaction. However, it is possible to write "reactions" that lump several steps, i.e., several elementary reactions occurring sequentially, in order to simplify the representation of chemical reactions. The microreversibility principle is based on the fact that the reaction process can be reversed in time so that the elementary reaction may occur in the reverse (backward) direction (the products become the reactants and the reactants become the products). In most cases, the kinetics of the reverse reaction is much slower than that of the forward reaction; therefore, only the forward reaction is of interest. In some case, both kinetics are commensurate. Then, there is chemical equilibrium:

$$A + B \xrightarrow{k_f} C + D$$ (R7.8)

$$A + B \xleftarrow{k_r} C + D$$ (R7.9)

At equilibrium, the rates of the forward and reverse reactions are equal:

$$k_f [A][B] = k_r [C][D]$$ (7.26)

The equilibrium relationship is defined as follows:

$$K_{eq} = \frac{[C][D]}{[A][B]} = \frac{k_f}{k_r} \tag{7.27}$$

where K_{eq} is the equilibrium constant. It depends on temperature, since the rate constants k_f and k_r depend on temperature. If the forward reaction dominates, the equilibrium is displaced toward the chemical species C and D, i.e., the products of that reaction. If the reverse reaction dominates, the equilibrium is displaced toward the chemical species A and B, i.e., the reactants of the forward reaction.

7.2 Chemistry of the Stratospheric Ozone Layer

The stratosphere is the atmospheric layer where the temperature increases with altitude (see Chapter 3). Therefore, it is a stable atmospheric layer, i.e., with limited vertical air motions. The increase in temperature with altitude is due to the partial absorption of solar radiation by oxygen (O_2) and ozone (O_3) molecules. The actinic flux is defined as the solar radiation flux, including both direct and scattered radiation, integrated over all directions. It is pertinent to the photolysis of molecules since they may absorb radiation from any direction. The actinic flux at different altitudes ranging from 50 km to sea level is illustrated in Figure 7.1. It appears that oxygen and ozone absorb solar radiation very efficiently in the ultraviolet and, as a result, the corresponding actinic flux is almost zero at sea level below 290 nm.

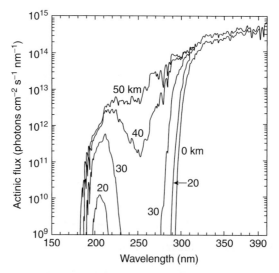

Figure 7.1. Actinic flux in the ultraviolet range at different altitudes. Conditions are a zenith angle of 30 ° and an albedo of 0.3. Source: DeMore et al. (1997).

7.2.1 Chapman Cycle

As mentioned in Section 7.1.3, solar radiation at wavelengths less than 240 nm has enough energy to dissociate oxygen molecules and to lead to the formation of oxygen atoms:

$$O_2 + h\nu \rightarrow 2\,O \quad \lambda < 242\,\text{nm} \tag{R7.10}$$

These oxygen atoms may react with oxygen molecules to form ozone. This reaction requires an air molecule:

$$O_2 + O + M \rightarrow O_3 + M \tag{R7.11}$$

Ozone may be photolyzed:

$$O_3 + h\nu \rightarrow O(^1D) + O_2 \quad 240 < \lambda < 320\,\text{nm} \tag{R7.12a}$$

$O(^1D)$ is an excited oxygen atom, which may lose its excess energy by reaction with air molecules (N_2 or O_2), which are represented by M (the oxygen atom at its lowest energy level is $O(^3P)$, which is represented here by O for the sake of simplicity):

$$O(^1D) + M \rightarrow O(^3P) + M \tag{R7.12b}$$

In addition, ozone may react with oxygen atoms to form molecular oxygen:

$$O_3 + O \rightarrow 2\,O_2 \tag{R7.13}$$

This set of reactions is called the Chapman cycle, after the British mathematician and physicist Sydney Chapman, who first proposed it in 1930 to explain the high ozone concentrations observed in the stratosphere.

In this reaction cycle, reactions R7.12a and R7.12b may be combined into a single reaction by assuming that $O(^1D)$ reacts only with air molecules:

$$O_3 + h\nu \rightarrow O + O_2 \quad 240 < \lambda < 320\,\text{nm} \tag{R7.12}$$

Then, the Chapman cycle includes four reactions. Here, the rate constants k_1, k_2, k_3, and k_4 are used for reactions R7.10, R7.11, R7.12, and R7.13, respectively. Reactions R7.10 and R7.13 are slow, whereas reactions R7.11 and R7.12 are fast. Therefore, the ozone concentration in the stratosphere may be calculated by considering that the Chapman cycle consists of two sets of two reactions each. First, two reactions with slow kinetics (with a characteristic time on the order of one day to several years depending on altitude and latitude):

$$O_2 + h\nu \rightarrow 2\,O \quad \lambda < 242\,\text{nm} \tag{R7.10}$$

$$O_3 + O \rightarrow 2\,O_2 \tag{R7.13}$$

Second, two reactions with fast kinetics (with a characteristic time on the order of a minute):

$$O_3 + h\nu \rightarrow O + O_2 \quad 240 < \lambda < 320\,\text{nm} \tag{R7.12}$$

$$O_2 + O + M \rightarrow O_3 + M \tag{R7.11}$$

Since these two reactions are fast, steady-state may be assumed and at equilibrium their rates are equal:

$$k_3 [O_3] = k_2 [O_2] [O] [M] \tag{7.28}$$

Then, the concentration of oxygen atoms may be calculated from the ozone concentration:

$$[O] = [O_3] \frac{k_3}{k_2 [O_2] [M]} \tag{7.29}$$

Equilibrium between the two reactions with slow kinetics may also be assumed for large time scales:

$$k_1 [O_2] = 2 k_4 [O_3] [O] \tag{7.30}$$

Substituting [O] by its expression as a function of [O_3]:

$$k_1 [O_2] = 2 k_4 [O_3]^2 \frac{k_3}{k_2 [O_2] [M]} \tag{7.31}$$

Thus, [O_3] may be expressed as a function of [O_2] and [M]:

$$[O_3] = [O_2] \left(\frac{k_1 k_2 [M]}{2 k_4 k_3} \right)^{1/2} \tag{7.32}$$

Example: What is the ozone concentration at an altitude of 25 km?

At an altitude of 25 km, atmospheric conditions are assumed to be: $P = 0.025$ atm and $T = 221$ K, i.e., -52 °C.

The rate constants are as follows:

$$k_1 = 1.2 \times 10^{-11} \text{ s}^{-1} \text{ at 25 km altitude}$$
$$k_2 = 6 \times 10^{-34} \times (300/T)^{2.3} \text{ molec}^{-2} \text{cm}^6 \text{ s}^{-1}$$
$$k_3 = 6 \times 10^{-4} \text{ s}^{-1} \text{ at 25 km altitude}$$
$$k_4 = 8 \times 10^{-12} \exp(-2060/T) \text{ molec}^{-1} \text{ cm}^3 \text{ s}^{-1}$$

At $T = 221$ K, $k_2 = 1.21 \times 10^{-33}$ molec^{-2} cm^6 s^{-1} and $k_4 = 7.16 \times 10^{-16}$ molec^{-1} cm^3 s^{-1}. Also, [M] = 8.3×10^{17} molec cm^{-3}. Then, the ozone concentration is calculated as follows:

$$[O_3] = 1.18 \times 10^{-4} [O_2]$$
$$[O_2] = 0.21 [M] = 1.74 \times 10^{17} \text{ molec cm}^{-3}$$

Thus: $\quad [O_3] = 2.05 \times 10^{13}$ molec cm^{-3}

Actually, the ozone concentrations calculated with the Chapman cycle overestimate the measured concentrations. The observed concentrations may be simulated better by taking into account the reaction of the excited oxygen atom with water vapor:

$$O(^1D) + H_2O \rightarrow 2 \text{ OH} \tag{R7.14}$$

where OH is the hydroxyl radical. This radical is very reactive and leads to other reactions:

$$O_3 + OH \rightarrow O_2 + HO_2 \qquad (R7.15)$$

$$O_3 + HO_2 \rightarrow 2\ O_2 + OH \qquad (R7.16)$$

$$OH + HO_2 \rightarrow H_2O + O_2 \qquad (R7.17)$$

where HO_2 is the hydroperoxyl radical. As long as the OH and HO_2 radicals have not reacted with each other, they convert ozone into molecular oxygen, which explains the overestimation of ozone when these reactions are not taken into account.

7.2.2 Ultraviolet Radiation

The presence of oxygen and ozone in the stratosphere leads to an absorption of ultraviolet (UV) radiation. Therefore, in the troposphere, this radiative flux is significantly attenuated. UV radiation may be classified according to wavelengths as follows, starting from the visible light: UV A (400 to 315 nm), B (315 to 280 nm), and C (280 to 153 nm). This classification corresponds to the health hazards of UV radiation interactions with human skin. UV A radiation penetrates deeply and leads to suntan. However, it also leads to skin aging and may be harmful because it may trigger skin diseases such as cancers (carcinoma and melanoma). UV B radiation does not penetrate as deeply. It leads to sunburns and is harmful to the eyes, because it is not stopped by the eye lens. It leads to skin aging and skin cancer. UV C radiation has the most energy and may penetrate deeply; however, it is entirely stopped in the stratosphere. The ozone layer protects from UV radiation, in particular, from UV B radiation. Therefore, its destruction may lead to a greater flux of UV radiation reaching the Earth's surface, which may lead to a public health problem due to an increase in skin cancer cases.

7.2.3 Depletion of the Stratospheric Ozone Layer

The first theory concerning the potential depletion of the stratospheric ozone layer was proposed in the early 1970s. Harold Johnston, a chemistry professor at the University of California at Berkeley, suggested that nitrogen oxide emissions (NO_x) from supersonic aircraft (such as the Franco-British Concorde airplane) could react in the stratosphere and lead to a cycle of reactions that would deplete ozone concentrations (Johnston, 1971). Nitrogen oxide emissions consist mostly of nitric oxide (NO). The following reactions occur:

$$NO + O_3 \rightarrow NO_2 + O_2 \qquad (R7.18)$$

$$NO_2 + O \rightarrow NO + O_2 \qquad (R7.19)$$

Thus, NO is converted into NO_2, which is then converted back to NO. This reaction cycle may continue for a while, leading to the following net reaction budget:

$$O_3 + O \rightarrow 2\ O_2 \qquad (R7.20)$$

This cycle ends when, for example, NO_2 reacts with OH to form nitric acid (HNO_3). Overall, there is a reduction of O_3 concentrations, which is similar to that obtained in Section 7.2.1 via the cycle involving the OH and HO_2 radicals of reactions R7.15 and

R7.16. (NO_2 also gets photolyzed, but the corresponding cycle leads to no net change in the O_3 and O_2 budget; see Chapter 8.) Reactions with NO and NO_2 could have been an issue if NO emissions in the stratosphere had become important. However, stratospheric commercial aircraft flights have been limited and their impact on the stratospheric ozone has, therefore, been negligible.

Then, in the 1970s, another category of chemical species, the chlorofluorocarbons (CFC, also called "freons," which was their commercial name given by the American company Dupont de Nemours), was identified as potentially leading to a depletion of the stratospheric ozone layer. CFC were used, for example, as refrigerants and in some aerosol spray products. The most common CFC were $CFCl_3$ and CF_2Cl_2, also called CFC 11 and CFC 12 according to the nomenclature of Dupont de Nemours (this CFC nomenclature uses the abc numbering system, where a is the number of carbon atoms − 1 (omitted if zero), b is the number of hydrogen atoms + 1, and c is the number of fluorine atoms). These substances are chemically very stable and, therefore, they have a long tropospheric lifetime. Chemistry professor Sherwood Rowland and his postdoctoral student Mario Molina of the University of California at Irvine suggested that, because of their chemical stability, these chemical species would be transported up to the stratosphere, where they would be photolyzed at wavelengths that are not available in the troposphere, because of the UV radiation filtering effect of O_2 and O_3 (Molina and Rowland, 1974). The photolysis of chlorofluorocarbons leads to the production of chlorine atoms, because the carbon-chlorine bonds are much weaker than the carbon-fluorine bonds. They proposed the following chemical mechanism for the stratosphere:

Initiation

$$CFC + h\nu \rightarrow Cl + \text{radical } (CFCl_2, CF_2Cl) \qquad (R7.21)$$

Propagation

$$Cl + O_3 \rightarrow ClO + O_2 \qquad (R7.22)$$

$$ClO + O \rightarrow Cl + O_2 \qquad (R7.23)$$

The net budget is a conversion of ozone and oxygen atoms (which are equivalent to ozone since there is a quasi-steady state, see reactions R7.11 and R7.12 in Section 7.2.1) into molecular oxygen. Therefore, there is partial depletion of the stratospheric ozone layer. This cycle ends when the chlorine radicals are stabilized:

Termination

$$Cl + CH_4 \rightarrow HCl + CH_3 \qquad (R7.24)$$

$$ClO + NO_2 + M \rightarrow ClNO_3 + M \qquad (R7.25)$$

Paul Crutzen, a Dutch meteorologist, then at the National Center for Atmospheric Research (NCAR) in Boulder, Colorado, who had been investigating stratospheric chemistry, confirmed the impact of the CFC industry on the stratospheric ozone layer (Crutzen et al., 1978). In the 1980s, measurements confirmed that the CFC were depleting the ozone layer and the chemistry Nobel Prize was awarded in 1995 to Paul Crutzen, Sherwood

Rowland, and Mario Molina (only three people may share a Nobel Prize; therefore, Harold Johnston was not officially recognized for his theory concerning the potential impact of NO_x from supersonic commercial flights on the ozone layer, which inspired the theory of the impact of CFC).

The depletion of the ozone layer is greater at the South Pole, because of processes involving heterogeneous reactions on ice crystals in polar stratospheric clouds (PSC) (e.g., Solomon, 1988). The South Pole is characterized by an important vortex, which isolates this cold region during winter (the isolation is less important at the North Pole, because the presence of a greater land mass in the northern hemisphere leads to more dynamic perturbations of the atmosphere). The following heterogeneous reactions occur during the polar winter night (the "psc" subscript indicates molecules present in ice crystals of polar stratospheric clouds):

$$ClNO_3 + H_2O_{psc} \rightarrow HOCl + HNO_3 \qquad (R7.26)$$

$$ClNO_3 + HCl_{psc} \rightarrow Cl_2 + HNO_3 \qquad (R7.27)$$

$$HOCl + HCl_{psc} \rightarrow Cl_2 + H_2O \qquad (R7.28)$$

These heterogeneous reactions are faster when the temperature is very low (Burkholder et al., 2015), i.e., when the polar air masses are isolated from other (warmer) air masses. When spring arrives, solar radiation leads to the photolysis of the molecular chlorine molecules that have accumulated during the cold and dark winter:

$$Cl_2 + h\nu \rightarrow 2\ Cl \qquad (R7.29)$$

Then, the destruction cycle of the ozone layer is triggered with the cycle presented previously in this section. It leads to the so-called ozone "hole" at the South Pole.

Chlorofluorocarbons that contain a hydrogen atom (HCFC) are chemically reactive in the troposphere, because hydroxyl radicals, OH, react with the hydrogen atom to form water vapor, thereby leading to the formation of a carbonaceous radical (see Chapter 8 for a description of the chemistry of organic compounds with OH). Chemical-transport simulations showed that these HCFC did not reach the stratosphere in significant amounts and, therefore, had little impact on the stratospheric ozone layer (e.g., Seigneur et al., 1977). Therefore, such commercial products were used in substitution of CFC 11 and 12. However, these compounds are greenhouse gases (see Chapter 14) and their commercial use is now being reduced (see Figure 7.2).

Bromine-containing compounds (for example, halons) also play a role in the depletion of the stratospheric ozone layer, because bromine plays a role equivalent to that of chlorine. Compounds containing carbon and one or more halogen atoms (i.e., chlorine, bromine, and/or fluorine) are called halocarbons (including, therefore, freons and halons).

The potential of a halocarbon to deplete the stratospheric ozone layer may be characterized with an indicator called the ozone depletion potential (ODP). It depends in great part on the atmospheric lifetime of the halocarbon. ODP are calculated with respect to $CFCl_3$ (which has by definition an ODP equal to 1). ODP may be determined with a numerical model or via a combination of observations and modeling. The second

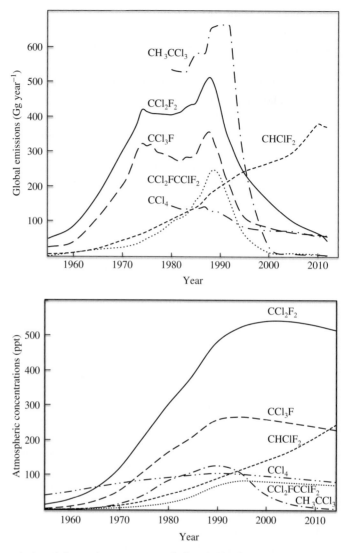

Figure 7.2. Timeline of atmospheric emissions and concentrations of selected chlorofluorocarbons (CFC), hydrochlorofluorocarbons (HCFC), and chlorocarbons. Top figure: global annual emissions (Gg y^{-1}); bottom figure: atmospheric concentrations (ppt). Source of the data: WMO (2014); AFEAS (1998) (for emissions prior to 1980). Emissions after 1980 were obtained by inverse modeling from ambient concentration measurements.

approach is called a semi-empirical method; it is currently the most commonly used method. Both approaches lead to comparable results. Table 7.1 summarizes the ODP of the halocarbons evaluated in a recent report by the World Meteorological Organization (WMO, 2014). Halons (chemical species containing bromine) have ODP, which are greater than those of CFCl$_3$, because of the large chemical reactivity of bromine. HCFC have ODP, which are smaller than those of CFC, because of their tropospheric chemical reactivity.

Table 7.1. Ozone depletion potentials (ODP) of selected chemical species.

Chemical species		
Commercial name	Chemical formula	ODP (semi-empirical method)
Chlorofluorocarbons		
CFC-11	CCl_3F	1
CFC-12	CCl_2F_2	0.73
CFC-113	$CClF_2CCl_2F$	0.81
CFC-114	$CClF_2CClF_2$	0.50
CFC-115	CF_3CClF_2	0.26
Hydrochlorofluorocarbons		
HCFC-22	$CHClF_2$	0.034
HCFC-123	$CHCl_2CF_3$	–
HCFC-124	$CHFClCF_3$	–
HCFC-141b	CCl_2FCH_3	0.102
HCFC-142b	$CClF_2CH_3$	0.057
HCFC-225ca	$CF_3CF_2CHCl_2$	–
HCFC-225cb	$CClF_2CF_2CHClF$	–
Halons		
Halon-1202	CBr_2F_2	1.7
Halon-1211	$CBrClF_2$	6.9
Halon-1301	$CBrF_3$	15.2
Halon-2402	$CBrF_2CBrF_2$	15.7
Other halocarbons		
Methyl bromide	CH_3Br	0.57
Methyl chloride	CH_3Cl	0.015
Carbon tetrachloride	CCl_4	0.72
Methyl chloroform	CH_3CCl_3	0.14

Source: WMO (2014). ODP are given with respect to that of CCl_3F, which is 1 by definition.

The reduction of halocarbon emissions has been addressed in several international protocols. The first one was the Montreal protocol in 1987 (originally signed by 24 countries and the European Economic Community). It was later amended, in particular in 1990 (London), 1992 (Copenhagen), 1997 (Montreal), and 1999 (Beijing). As of 2015, all 197 countries of the United Nations had ratified the protocol and its amendments. Measurements that are now performed continuously (for example, via satellite measurements of the atmospheric ozone column) show that the stratospheric ozone "hole," which deepened during the 20th century, now starts to recover. Therefore, the halocarbon emission controls start to show benefits (see Figure 7.2). However, the long lifetimes of CFC and of some other halocarbons imply that the recovery of the stratospheric ozone layer to its original state (i.e., pre-CFC) will happen slowly.

7.3 Numerical Modeling of Atmospheric Chemical Kinetics

A gas-phase chemical kinetic mechanism includes a number of chemical and photochemical reactions among a variety of chemical species (molecules, atoms, and radicals). The simulation of the time-dependent chemical species concentrations is governed by a system of ordinary differential equations (ODE). A major difficulty for the numerical solution of this ODE system is due to the fact that some chemical species (radicals and atoms) have very short chemical lifetimes, whereas other species (molecules) may have long lifetimes. Therefore, the solution of such a system with a standard numerical algorithm for ODE (for example, a Runge–Kutta algorithm) requires very small time steps to correctly address the change in concentrations of species with very short lifetimes. Using large time steps would lead to numerical instabilities in the solution, i.e., the concentrations of some species would oscillate and lead to unrealistic values. Since it is generally not feasible to use extremely small time steps, because of the enormous associated computational cost, other approaches must be considered.

Two general approaches are possible. It is possible to assume pseudo-steady state for some chemical species with very short lifetimes. Then, the derivative of their concentration with respect to time is set to zero and their concentration is calculated by assuming that the rate of formation is equal to their rate of consumption (see, for example, Equation 7.28). This approach allows one to eliminate from the ODE system the chemical species with very short lifetimes. However, although this approach is appropriate for simple chemical kinetic mechanisms, it may not be applicable to large ODE systems involving a large number of chemical species. First, the steady-state approximation may not be verified for all selected species all the time. Second, an analytical solution to the steady-state equations may not always be available, because a large number of species assumed to be at steady state may lead to complex relationships among those species. Then, another approach must be considered.

Numerical algorithms have been developed to solve stiff ODE systems: an ODE system is numerically stiff if the variables require integration steps ranging over several orders of magnitude. Gear (1971) developed an algorithm to solve such stiff ODE systems. Other algorithms have been developed since then. Sandu et al. (1997a, 1997b) have compared several such algorithms. In addition, the SWGEAR algorithm (Jacobson and Turco, 1994) is very efficient for parallelized air pollution models. It is possible to combine some assumptions of steady state for some species with the use of a numerical algorithm for stiff ODE systems to further optimize the computational time.

Problems

Problem 7.1 Unit conversion

A concentration of 10 ppb of nitrogen dioxide (NO_2) is measured. Atmospheric conditions are a pressure of 1 atm and a temperature of 15 °C. The ideal gas law constant is R = 8.2×10^{-5} m^3 atm mole^{-1} K^{-1}. The molar mass of oxygen (O) is 16 g mole^{-1} and that of nitrogen (N) is 14 g mole^{-1}. What is the concentration of NO$_2$ in μg m^{-3}?

Problem 7.2 Photolysis

Formaldehyde (HCHO) is photolyzed by solar radiation. The dissociation energy of the first C-H bond is 369 kJ mole^{-1}. Calculate the maximum wavelength at which there could be direct dissociation of HCHO by photolysis to form H and HCO.

References

AFEAS, 1998. *Production, Sales and Atmospheric Release of Fluorocarbons through 1997. Alternative Fluorocarbons Environmental Acceptability Study*, Washington, DC.

Burkholder, J.B., S.P. Sander, J.P.D. Abbatt, J.R. Barker, R.E. Huie, C.E. Kolb, M.J. Kurylo, V.L. Orkin, D.M. Wilmouth, and P.H. Wine, 2015. *Chemical Kinetics and Photochemical Data for Use in Atmospheric Studies*, Evaluation n° 18, JPL Publication 15–10, Jet Propulsion Laboratory, Pasadena, CA, available at http://jpldataeval.jpl.nasa.gov/pdf/JPL_Publication_15–10.pdf.

Chapman, S., 1930. A theory of upper atmospheric ozone, *Mem. R. Met. Soc.*, **3**, 103–125.

Crutzen, P.J., I.S.A. Isaksen, and J.R. McAfee, 1978. The impact of the chlorocarbon industry on the ozone layer, *J. Geophys. Res.*, **83**, 345–363.

DeMore, W.B. et al., 1997. *Chemical Kinetics and Photochemical Data for Use in Stratospheric Modeling*, JPL Publication 97–4, Jet Propulsion Laboratory, Pasadena, CA.

Finlayson-Pitts, B.J. and J.N. Pitts, Jr., 2000. *Chemistry of the Upper and Lower Atmosphere: Theory, Experiments, and Applications*, Academic Press, New York.

Gear, C.W., 1971. *Numerical Initial Value Problems in Ordinary Differential Equations*, Prentice-Hall, Englewood Cliffs, NJ.

Jacobson, M.Z., 2005. *Fundamentals of Atmospheric Modeling*, Cambridge University Press, Cambridge, UK.

Jacobson, M. Z. and R.P. Turco, 1994. SMVGEAR: A sparse-matrix, vectorized Gear code for atmospheric models. *Atmos. Environ.*, **17**, 273–284.

Johnston, H., 1971. Reduction of stratospheric ozone by nitrogen oxide catalysts from supersonic transport exhaust, *Science*, **173**, 517–522.

Molina, M.J. and F.S. Rowland, 1974. Stratospheric sink for chlorofluoromethanes: Chlorine atom-catalysed destruction of ozone, *Nature*, **249**, 810–812.

Sandu, A., J.G. Verwer, M. van Loon, G.R. Carmichael, F.A. Potrat, D. Dabdub, and J.H. Seinfeld, 1997a. Benchmarking stiff ODE solvers for atmospheric chemistry problems – I. Implicit vs explicit, *Atmos. Environ.*, **31**, 3151–3166.

Sandu, A., J.G. Verwer, J.G. Bloom, E.J. Spee, G.R. Carmichael, and F.A. Potrat, 1997b. Benchmarking stiff ODE solvers for atmospheric chemistry problems – II. Rosenbrock solvers, *Atmos. Environ.*, **31**, 3459–3472.

Seigneur, C., H. Caram, and R.W. Carr, Jr., 1977. Atmospheric diffusion and chemical reaction of the chlorofluoromethanes CHFCl$_2$ and CHF$_2$Cl, *Atmos. Environ.*, **11**, 205–214.

Solomon, S., 1988. The mystery of the Antarctic ozone "hole," *Rev. Geophys.*, **26**, 131–148.

Troe, J., 1983. Theory of thermal unimolecular reactions in the fall-off range. I. Strong collision rate constants, *Ber. Bunsenges. Phys. Chem.*, **87**, 161–169.

WMO, 2014. Scientific Assessment of Ozone Depletion: 2014. Global Ozone Research and Monitoring Project – Report N° 55, World Meteorological Organization, Geneva, Switzerland.

8 Gaseous Pollutants

Several gaseous chemical species may lead to adverse health effects and, therefore, several of those are regulated. Brief descriptions of those chemical species, including their major sources and atmospheric fate, are presented. Next, the focus of this chapter is on urban and regional pollution, since it corresponds to most of the population exposure to ambient air pollution. The gaseous pollutants that are currently the most relevant at the urban/regional scale in terms of adverse health effects are ozone and nitrogen dioxide. These pollutants are major components of photochemical smog, which results from chemical reactions between nitrogen oxides (NO_x) and volatile organic compounds (VOC) in the presence of sunlight. The fact that photochemical smog precursors such as NO_x and some VOC (alkenes) are both producers and destructors of ozone makes the development of efficient strategies to reduce photochemical smog difficult. Therefore, this chapter addresses gaseous air pollutants with a focus on the complex processes leading to the formation of photochemical smog. Following a presentation of the main reactions that govern the formation of ozone, nitrogen dioxide, and other gaseous pollutants, the different chemical regimes are analyzed in order to understand how efficient approaches to reduce the concentrations of the main constituents of photochemical smog can be developed.

8.1 General Considerations on Gaseous Pollutants

Gaseous pollutants include primary pollutants, secondary pollutants, and precursors of secondary pollutants. Some chemical species may belong to more than one category. The main gaseous pollutants that are regulated or are precursors of regulated pollutants are briefly presented in the following sections.

8.1.1 Sulfur Dioxide

Sulfur dioxide (SO_2) is a primary pollutant. SO_2 is the main pollutant of air pollution episodes such as that of 1952 in London. It is regulated because it leads to respiratory problems. As a primary pollutant, its impacts occur near its sources. In North America and Europe, those impacts are now mostly limited to some industrial sites and maritime traffic, because sulfur content in fuels used by road traffic is now regulated.

SO_2 is oxidized to sulfuric acid in the atmosphere (see Chapter 10). Sulfuric acid contributes to fine particle air pollution, because its low volatility implies that it is preferentially present in the particulate phase, typically as ammonium salts. In addition, sulfuric acid is an important component of acid rain. SO_2 is slightly soluble in water, and it may be removed from the atmosphere by dry and wet deposition.

8.1.2 Carbon Monoxide

Carbon monoxide (CO) is an inorganic carbonaceous compound, because it does not contain any hydrogen atoms. CO is regulated because it combines with hemoglobin in the blood stream and leads to anoxia (lack of oxygen), when the blood stream cannot carry enough oxygen. CO is a primary pollutant, which is mostly emitted by combustion processes, including internal combustion engines. Historically, CO concentrations were, therefore, high near roadways. CO was the first pollutant to be regulated for road traffic in North America and Europe, using catalytic converters, which convert it to carbon dioxide, CO_2 (see Chapter 2). Consequently, ambient concentrations of CO are now fairly low in these regions.

CO is oxidized slowly to CO_2 in the atmosphere. Its oxidation, which is described in Section 8.3.3, leads also to the formation of ozone, the main gaseous pollutant of photochemical smog. However, the contribution of CO to photochemical air pollution is less than that of other carbonaceous compounds because of its low chemical reactivity and the fact that its emissions are currently regulated.

8.1.3 Ozone and Gaseous Photochemical Oxidants

Photochemical pollution is generally called "photochemical smog." Smog is the contraction of smoke and fog, because its appearance falls between these two phenomena. Photochemical pollution was initially identified in the Los Angeles Basin in the 1950s. Heavy pollution was present in that region of southern California, because of a large number of anthropogenic pollution sources, such as road traffic, fossil-fuel fired power plants, refineries, and other industrial sources, as well as meteorological conditions that were conducive to air pollution (strong sunlight and low atmospheric dispersion). Arie Haagen-Smit, a biochemistry professor at the California Institute of Technology (Caltech), was the first to identify the processes that lead to the formation of photochemical air pollution, and in particular ozone (O_3), which is its main gaseous pollutant. He showed that photochemical air pollution results from atmospheric chemical reactions occurring among precursor gases that include nitrogen oxides (NO_x) and volatile organic compounds (VOC), in the presence of sunlight. Hence, the adjective "photochemical" was attributed to that form of air pollution, since photolytic reactions induced by sunlight (see Chapter 7) initiate the set of photochemical reactions that lead to ozone formation.

Ozone is not emitted in the atmosphere and is, therefore, a secondary air pollutant. Its precursors are NO_x, VOC, and, to a lesser extent, CO. Although O_3 is not the only photochemical oxidant, it is the main gaseous constituent of photochemical smog in terms of ambient concentrations. Therefore, O_3 is targeted for regulations pertaining to the gaseous fraction of photochemical smog. Since O_3 is produced by photochemical reactions, its concentrations are highest during spring and summer.

8.1.4 Nitrogen Oxides

NO_x include by definition nitric oxide (NO) and nitrogen dioxide (NO_2). Their emissions result mostly from combustion processes. These two compounds represent in terms of ambient concentrations the majority of nitrogen oxides present in the urban atmosphere. However, there are other nitrogen oxides, such as nitrogen protoxide (N_2O), which is a greenhouse gas, nitric acid (HNO_3), nitrogen pentoxide (N_2O_5), the nitrate radical (NO_3), and a large number of organic nitrates. By convention, NO_y represent all nitrogen oxides, with the exception of N_2O, and NO_z represent the difference between NO_y and NO_x. Thus, in summary:

$$NO_x = NO + NO_2$$
$$NO_y = NO + NO_2 + HNO_3 + N_2O_5 + NO_3 + \text{organic nitrates}$$
$$NO_z = HNO_3 + N_2O_5 + NO_3 + \text{organic nitrates}$$

NO_2 is the only nitrogen oxide that is regulated in terms of its ambient concentrations. It leads to respiratory problems. NO_2 is both a primary and a secondary pollutant, because (1) it constitutes a fraction of NO_x emissions and (2) it is a product of the atmospheric oxidation of NO.

NO_x are rapidly oxidized in the atmosphere. They are precursors of a large number of secondary pollutants, such as ozone, nitric acid, and organic nitrates. Nitric acid may contribute significantly to secondary particulate matter formation. It is also a major constituent of acid rain. Some organic nitrates contribute to the secondary fraction of organic particulate matter. In addition, inorganic and organic nitrates play a major role in the eutrophication of ecosystems.

8.1.5 Volatile and Semi-volatile Organic Compounds

Volatile organic compounds (VOC) and semi-volatile organic compounds (SVOC) are mostly emitted from combustion processes and also from the evaporation of liquid fuels and some organic-containing products (e.g., some paints, solvents, and cleaning products). They play a major role in the formation of photochemical smog. VOC are precursors of ozone. SVOC and some VOC are also important precursors of the secondary fraction of fine particulate matter. In addition, some VOC, such as benzene, 1,3-butadiene, and formaldehyde, are carcinogenic. These carcinogenic VOC may be regulated individually as such (as it is the case in Europe for benzene) or may be regulated indirectly via regulatory approaches that target carcinogenic compounds as a whole (as it is the case in the United States).

VOC include mostly alkanes (hydrocarbons with only single bonds, also called paraffins), alkenes (hydrocarbons with one or more double bonds, also called olefins), aromatic compounds (compounds with one or more phenyl rings), and aldehydes (compounds with a carbonyl group, HC=O). Alcohols (compounds with a C-OH group), alkynes (hydrocarbons with a triple bond), ethers (compounds with a C-O-C group), and other organic compounds are typically present in the atmosphere to a lesser extent. However, alcohols are becoming more prominent due to the use of biofuels and their increased use in gasoline.

The term hydrocarbon (HC) refers to organic compounds that contain only carbon and hydrogen atoms (therefore, alkanes, alkenes, alkynes, and some aromatic compounds).

We will use hereafter the term VOC to refer to all organic compounds, thus including aldehydes and alcohols. Among the alkanes, methane has a very low atmospheric reactivity (atmospheric lifetime on the order of 10 years) and, therefore, it is typically not included among the precursors of photochemical air pollution. Thus, when referring to VOC that are precursors of photochemical pollution, one should use the term non-methane VOC. However, for the sake of simplicity, we will use the term VOC to mean non-methane VOC hereafter.

SVOC are organic compounds with a saturation vapor pressure that is such that they can be present in the atmosphere in both the gas phase and the particulate phase. They play a major role in the formation of particulate organic compounds and may also be involved in the formation of gaseous air pollutants.

The ultimate chemical fate of VOC and SVOC is CO_2, via atmospheric oxidation. However, organic compounds may be removed from the atmosphere before being converted to CO or CO_2. Their removal may occur either as gaseous or particulate compounds, via dry and wet deposition.

8.1.6 Ammonia

Ammonia (NH_3) is the reduced form of nitrogen in the atmosphere. It is mostly emitted by agricultural activities. NH_3 does not present any adverse health effects at the ambient concentrations usually observed in the atmosphere. However, it contributes to the formation of sulfate and nitrate ammonium salts, which may constitute a significant fraction of fine particulate matter. In addition, NH_3 contributes to nitrogen deposition and, therefore, may lead to the eutrophication of ecosystems.

8.2 Oxidizing Power of the Atmosphere and Chemical Reactivity

The atmosphere is an oxidizing environment because of the presence of 21 % of oxygen. However, oxygen is not the main oxidizing species in the atmosphere and atmospheric oxidation processes are due mostly to other chemical species that contain oxygen atoms and are formed photochemically in the atmosphere. The main atmospheric oxidants are:

- The hydroxyl radical, OH
- The nitrate radical, NO_3
- Ozone, O_3

Their formation in the atmosphere is described in Section 8.3, along with the chemistry of photochemical air pollution.

The oxidation of chemical species, such as VOC, NO_x, CO, and SO_2 by these oxidants can occur more or less rapidly depending on the oxidant concentrations and the reactivity of the chemical species toward those oxidants. For a given chemical reaction, two terms are generally used to characterize this chemical reactivity:

- The half-life
- The lifetime (also called residence time)

In the case of a reaction with a constant oxidant concentration, the half-life corresponds to the median of the times needed for all the individual molecules of the chemical species initially present to react. The lifetime corresponds to the mean of those reaction times.

The half-life, $t_{1/2}$, is the time needed for half of the molecules initially present to react. Let $[X]_0$ be the initial concentration of chemical species X, which here will be oxidized by OH (with a constant concentration), as an example:

$$X + OH \rightarrow \text{Oxidation products} \tag{R8.1}$$

The change with time of the concentration of X is given by the following equation (see Chapter 7):

$$\frac{d[X]}{dt} = -k[X][OH] \tag{8.1}$$

where k is the rate constant of the chemical reaction. Integrating this equation between times 0 and t leads to the following solution:

$$[X] = [X]_0 \exp(-k[OH]t) \tag{8.2}$$

where $[X]_0$ is the initial concentration, i.e., at $t = 0$. Therefore, when:

$$[X] = \frac{[X]_0}{2}; \quad \exp(-k[OH]t_{1/2}) = \frac{1}{2} \tag{8.3}$$

Thus, the half-life of X is:

$$t_{1/2} = \frac{\ln(2)}{k[OH]} = \frac{0.7}{k[OH]} \tag{8.4}$$

where ln is the natural logarithm. The lifetime, t_l, is the characteristic time of the chemical reaction and is, therefore, defined simply via a dimensional analysis as:

$$t_l = \frac{1}{k[OH]} \tag{8.5}$$

It can be shown that t_l corresponds to the mean of the reaction times of all X molecules initially present. Let $p(t)$ be the normalized distribution of the individual reaction times of all X molecules (some will react right away or almost right away, whereas others will react after a long, or very long, time). Since all molecules have the same probability of reacting, the number of molecules that will react is proportional to the concentration of these molecules and, by definition, this function, $p(t)$, is proportional to the concentration of X molecules:

$$p(t) = A_p[X]_0 \exp(-k[OH]t) \tag{8.6}$$

This function is normalized and its integration over time must be equal to 1. Therefore, the pre-exponential factor is calculated to be equal to (k [OH]):

$$p(t) = k[OH] \exp(-k[OH]t) \tag{8.7}$$

The mean of the reaction times of all X molecules can be obtained by integrating the reaction time weighted by the distribution of those reaction times:

$$t_{mean} = \int_0^\infty t\, p(t)dt = \int_0^\infty k[\text{OH}]\, t\, \exp(-k[\text{OH}]t)dt \tag{8.8}$$

Integrating by parts:

$$t_{mean} = \frac{1}{k[\text{OH}]} = t_l \tag{8.9}$$

The half-life is related to the lifetime (i.e., mean reaction time) as follows:

$$t_{1/2} = 0.7 t_l \tag{8.10}$$

If the chemical species X undergoes several oxidation reactions, its overall half-life and overall lifetime can be calculated by considering all the oxidation reactions. For example, if X is oxidized by OH, NO_3, and O_3:

$$\frac{d[X]}{dt} = -k_{OH}[X][\text{OH}] - k_{NO_3}[X][NO_3] - k_{O_3}[X][O_3] \tag{8.11}$$

where k_{OH}, k_{NO3}, and k_{O3} are the rate constants of the different oxidation reactions. Integrating this equation leads to the following solution:

$$[X] = [X]_0 \exp(-(k_{OH}[\text{OH}] + k_{NO_3}[NO_3] + k_{O_3}[O_3])t) \tag{8.12}$$

The half-life and the lifetime are, respectively, as follows:

$$t_{1/2} = \frac{\ln(2)}{(k_{OH}[\text{OH}] + k_{NO_3}[NO_3] + k_{O_3}[O_3])}$$

$$t_l = \frac{1}{(k_{OH}[\text{OH}] + k_{NO_3}[NO_3] + k_{O_3}[O_3])} \tag{8.13}$$

The overall half-life and lifetime can be expressed in terms of the half-lives and lifetimes of the individual reactions:

$$t_{1/2} = \left(\frac{1}{t_{1/2,OH}} + \frac{1}{t_{1/2,NO_3}} + \frac{1}{t_{1/2,O_3}}\right)^{-1}$$

$$t_l = \left(\frac{1}{t_{l,OH}} + \frac{1}{t_{l,NO_3}} + \frac{1}{t_{l,O_3}}\right)^{-1} \tag{8.14}$$

The lifetimes and half-lives of the main atmospheric pollutants vary greatly, ranging from a few hours for species such as NO_x and propane (C_3H_8) to about 10 years for methane. The chemical reactivity of VOC is related to their ozone formation potential as discussed in Section 8.5.3. Table 8.1 lists typical atmospheric chemical lifetimes of selected chemical species undergoing oxidation by OH, NO_3, and O_3, as well as photolysis.

Table 8.1. Lifetimes of selected chemical species in the atmosphere at 1 atm and 25 °C for various oxidation reactions and photolysis. Sources of the rate constants: Calvert et al., 2000, 2002, 2008, 2011; Finlayson-Pitts and Pitts, 2000; Mollner et al., 2010.

Chemical species	Photolysis[a]	OH[a]	NO$_3$[a]	O$_3$[a]
NO$_2$	(b)	30 h	(b)	(b)
SO$_2$	–	12 d	–	–
CO	–	48 d	–	–
Methane[c]	–	5 a	>300 a	–
Propane	–	11 d	>4 a	–
n-Butane	–	5 d	7 a	–
Hexane	–	2 d	3 a	–
Octane	–	36 h	20 mo	–
Ethylene	–	33 h	19 mo	7 d
Propylene	–	11 h	12 d	28 h
trans-2-Butene	–	4 h	7 h	90 min
1,3-Butadiene	–	4 h	28 h	44 h
1-Hexene	–	8 h	(d)	25 h
trans-3-Hexene	–	(d)	(d)	100 min
trans-4-Octene	–	4 h	(d)	2 h
Benzene	–	8 d	11 a	–
Toluene	–	2 d	5 a	–
o-Xylene	–	20 h	9 mo	–
Formaldehyde	18 h	33 h	7 mo	–
Acetaldehyde	9 d	18 h	45 d	–
Isoprene	–	3 h	4 h	22 h
MBO[e]	–	4 h	10 d	31 h
α-Pinene	–	5 h	27 min	3 h
Δ^3-Carene	–	3 h	18 min	8 h
Humulene	–	1 h	5 min	1 min
Longifolene	–	6 h	4 h	>23 d

(a) Concentrations: [OH] = 2 × 10^6 cm^{-3} over 12 h per day (daytime); [NO$_3$] = 2 × 10^8 cm^{-3} over 12 h per day (nighttime); [O$_3$] = 40 ppb over 24 h per day. Photolysis for the spring equinox (March 20) in Paris calculated over 24 h; cos(θ_s) = 0.39 on average over 12 h during daytime (see Equations 7.7 to 7.10). Lifetimes are calculated with these values (concentrations or sunlight) averaged over 24 h; therefore, for species with lifetimes less than 24 h, the half-lives are shorter during daytime for photolytic reactions and reactions with OH and O$_3$; they are shorter during nighttime for reactions with NO$_3$.

(b) These reactions were not taken into account because they produce NO, which can subsequently be converted back to NO$_2$ (see Section 8.3); the reaction with OH is the only terminal reaction (see R8.44).

(c) The atmospheric lifetime of methane is actually longer because the kinetics depends on temperature (<25 °C on average) and [OH] decreases with altitude.

(d) No data available on the rates of these reactions.

(e) 2-Methyl-3-buten-2-ol.

8.3 Gas-phase Chemistry of Photochemical Air Pollution

8.3.1 Oxidants

The reactions leading to the formation of the three main oxidant species of photochemical air pollution, OH, NO_3, and O_3, are described in this section.

As mentioned in Chapter 7, hydroxyl radicals can be formed via the photolysis of ozone:

$$O_3 + hv \rightarrow O(^1D) + O_2 \tag{R8.2}$$

$$O(^1D) + H_2O \rightarrow 2\ OH \tag{R8.3}$$

A chemical kinetic mechanism will also need to take into account the fact that only a fraction of the $O(^1D)$ excited oxygen atoms reacts with water vapor, because most of them lose their excess energy by collision with air molecules (N_2 or O_2) and produce O_3 back by reacting next with O_2. (At 100 % relative humidity and 25 °C, the reaction with air molecules is about five times faster than that with water vapor.)

In addition to this OH formation pathway, there are two other important photolytic reactions that lead to OH formation in the troposphere: (1) the photolysis of hydrogen peroxide (H_2O_2) and (2) the photolysis of nitrous acid (HNO_2):

$$H_2O_2 + hv \rightarrow 2\ OH \tag{R8.4}$$

$$HNO_2 + hv \rightarrow NO + OH \tag{R8.5}$$

In addition, as discussed in Section 8.3.3, OH is also formed by reaction of hydroperoxyl radicals, HO_2. They originate mostly from the photolysis of aldehydes. Aldehyde photolysis leads to the production of a hydrogen atom, H, which is then oxidized rapidly by O_2 to form HO_2. Since the OH radical is formed by photolytic reactions, it is present mostly during daytime. There are, however, some formation pathways that do not require photolysis, such as the decomposition of peroxyacetylnitrate (PAN) in presence of NO_x (see the chemistry of PAN in Section 8.3.4) and the oxidation of alkenes by O_3 (see Section 8.3.6). However, these reactions are very limited sources of OH and nighttime OH concentrations are negligible.

The nitrate radical (not to be confused with the nitrate ion, NO_3^-, which is present in the aqueous phase, see Chapter 10) is formed via the reaction of nitrogen dioxide with ozone:

$$NO_2 + O_3 \rightarrow NO_3 + O_2 \tag{R8.6}$$

This radical is rapidly photolyzed:

$$NO_3 + hv \rightarrow NO_2 + O \tag{R8.7}$$

$$NO_3 + hv \rightarrow NO + O_2 \tag{R8.8}$$

Its formation does not require any photochemical reaction; therefore, it can be formed either at night or during the day. However, its rapid photolysis diminishes significantly its concentration during daytime. Thus, this oxidant plays a role mostly at night. Since the kinetics of the reaction of O_3 with NO is about 1,000 times faster than that with NO_2, NO concentrations are negligible when NO_3 is present, because NO will have almost entirely been oxidized into NO_2.

Ozone is formed in the stratosphere by photolysis of oxygen molecules (see Chapter 7). Solar radiation that leads to oxygen photolysis is filtered in the stratosphere and is, therefore, unavailable in the troposphere to lead to ozone formation there. However, the photolysis of nitrogen dioxide, which takes place in the visible and the near ultraviolet (UV) range, takes place in the troposphere:

$$NO_2 + h\nu \rightarrow NO + O \qquad (R8.9)$$

$$O + O_2 + M \rightarrow O_3 + M \qquad (R8.10)$$

Therefore, ozone formation takes place in the presence of sunlight, i.e., during daytime. However, the lifetime of ozone ranges from several hours to a few days. Thus, its oxidizing power can also occur at night.

In summary, the atmospheric gaseous oxidants are the following:

– During daytime: OH and O_3
– During nighttime: NO_3 and O_3

8.3.2 The Photostationary State of Leighton

Ozone formation is balanced by its destruction by nitric oxide:

$$NO + O_3 \rightarrow NO_2 + O_2 \qquad (R8.11)$$

This reaction is very fast and can, therefore, be called a titration reaction when considered in isolation, i.e., it stops when one of the two reactants (NO or O_3) has been entirely depleted. In the atmosphere, in the presence of sunlight, O_3 can be continuously regenerated by NO_2 photolysis. Thus, a system of three reactions that are at equilibrium occurs, i.e., the rates of these three reactions are identical. This set of three reactions, which are at steady state, is called the photostationary state of Leighton, named after the Stanford chemistry professor, Philip A. Leighton.

These three reactions are as follows:

$$NO + O_3 \rightarrow NO_2 + O_2 \quad k_1 = 0.027 \text{ ppb}^{-1} \text{ min}^{-1} \qquad (R8.11)$$

$$NO_2 + h\nu \rightarrow NO + O \quad k_2 = 0.3 \text{ min}^{-1} \qquad (R8.9)$$

$$O + O_2 \xrightarrow{(+M)} O_3 \quad k_3 = 0.022 \text{ ppb}^{-1} \text{ min}^{-1} \qquad (R8.10)$$

The rate constants are given here at 1 atm and 25 °C. A typical average daytime photolysis rate is used. At steady state, the rates of the three reactions are equal:

$$k_1 [NO] [O_3] = k_2 [NO_2] = k_3 [O] [O_2] \qquad (8.15)$$

8.3 Gas-phase Chemistry of Photochemical Air Pollution

The first equation leads to:

$$[O_3] = \frac{k_2 \, [NO_2]}{k_1 \, [NO]} \qquad (8.16)$$

In addition, the sum of the concentrations of nitrogen oxides must remain constant:

$$[NO] + [NO_2] = [NO]_0 + [NO_2]_0 \qquad (8.17)$$

where the subscript 0 indicates the initial concentration. Each ozone molecule reacting with NO leads to a molecule of NO_2, and each photolyzed NO_2 molecule leads to a molecule of ozone; therefore, the sum of the concentrations of O_3 and NO_2 remains constant:

$$[O_3] + [NO_2] = [O_3]_0 + [NO_2]_0 \qquad (8.18)$$

Thus, the concentrations of nitrogen oxides can be calculated as a function of the initial concentrations and the ozone concentration:

$$\begin{aligned}[NO_2] &= [O_3]_0 + [NO_2]_0 - [O_3] \\ [NO] &= [NO]_0 + [NO_2]_0 - [NO_2] = [NO]_0 - [O_3]_0 + [O_3]\end{aligned} \qquad (8.19)$$

Replacing $[NO_2]$ and $[NO]$ in Equation 8.16:

$$[O_3] = \frac{k_2 \, ([O_3]_0 + [NO_2]_0 - [O_3])}{k_1 \, ([NO]_0 - [O_3]_0 + [O_3])} \qquad (8.20)$$

The ozone concentration is then the solution of a quadratic equation:

$$k_1 \, [O_3]^2 + (k_1 \, ([NO]_0 - [O_3]_0) + k_2) \, [O_3] - k_2 \, ([O_3]_0 + [NO_2]_0) = 0 \qquad (8.21)$$

$$[O_3] = \frac{-(k_1([NO]_0 - [O_3]_0) + k_2) + \left((k_1([NO]_0 - [O_3]_0) + k_2)^2 + 4 \, k_1 \, k_2 \, ([O_3]_0 + [NO_2]_0)\right)^{1/2}}{2k_1} \qquad (8.22)$$

Example: Calculation of the ozone concentration produced from nitrogen oxides

The initial nitrogen oxide concentrations are as follows: $[NO]_0 = 100$ ppb, $[NO_2]_0 = 5$ ppb. There is no ozone present initially: $[O_3]_0 = 0$ ppb.
 The solution is: $[O_3] = 0{:}5$ ppb
 The ozone concentration produced from only nitrogen oxides is, therefore, very low.

8.3.3 Oxidation of Carbon Monoxide (CO)

The chemistry of CO is simple and is, therefore, convenient to explain ozone formation when volatile carbonaceous species (CO or VOC) are present. CO is oxidized by OH radicals:

$$CO + OH \rightarrow CO_2 + H \qquad (R8.12)$$

Hydrogen atoms are not stable, and they recombine rapidly with molecular oxygen to form hydroperoxyl radicals (HO_2):

$$H + O_2 \rightarrow HO_2 \qquad (R8.13)$$

Then, these radicals react rapidly to oxidize NO into NO_2:

$$NO + HO_2 \rightarrow NO_2 + OH \qquad (R8.14)$$

The OH radical has been regenerated, and the total budget of these three reactions is then as follows:

$$CO + NO + O_2 \rightarrow CO_2 + NO_2 \qquad (R8.15)$$

Therefore, the oxidation of CO into CO_2 leads to the conversion of NO into NO_2. NO_2 can be photolyzed to form NO and O_3. Since NO is converted to NO_2 without consumption of O_3 (unlike what happens in the Leighton photostationary state), there is formation of a molecule of O_3 for each molecule of CO that is oxidized. This yield is theoretical because all HO_2 radicals do not react with NO and some OH radicals may react with NO_2. The actual yield is, therefore, less than 1. The presence of a carbonaceous species (here CO, but VOC play a similar role, see Sections 8.3.4 to 8.3.8) leads to a perturbation of the photostationary equilibrium, thereby allowing the formation of NO_2 without O_3 consumption and leading, therefore, to O_3 formation.

Example: Calculation of the ozone concentration produced from carbon monoxide in presence of nitrogen oxides

The initial concentration of CO is 1 ppm and its oxidation occurs over an 8-hour period. The OH radical concentration is assumed to be 10^6 cm^{-3}.

The rate constant of the oxidation of CO by OH is 0.35 ppb^{-1} min^{-1} at 1 atm and 25 °C. The OH concentration in ppb is: $10^6 / (2.46 \times 10^{10}) = 4 \times 10^{-5}$ ppb. Formation of O_3 over eight hours is theoretically equivalent to the amount of CO that has reacted. Therefore:

$$[O_3] = [CO]_0 \, (1 - \exp(-k \, [OH] \, t))$$
$$[O_3] = 1000 \times (1 - \exp(-0.35 \times 4 \times 10^{-5} \times 8 \times 60))$$
$$[O_3] = 6.7 \text{ ppb}$$

Thus, 1 ppm of CO (which is not very reactive) has formed 6 ppb of O_3 in eight hours. Therefore, it is the presence of volatile carbonaceous compounds (here CO, but also reactive VOC), which leads to ozone formation, as correctly identified originally by Haagen-Smit.

8.3.4 Photolysis and Oxidation of Aldehydes

The simplest aldehyde (i.e., the aldehyde with only one carbon atom) is formaldehyde (HCHO). Its oxidation occurs by photolysis or by reaction with OH. There are two pathways for the photolysis of formaldehyde. On one hand:

$$\text{HCHO} + h\nu \rightarrow \text{H} + \text{HCO} \tag{R8.16}$$

$$\text{H} + \text{O}_2 \rightarrow \text{HO}_2 \tag{R8.13}$$

$$\text{HCO} + \text{O}_2 \rightarrow \text{HO}_2 + \text{CO} \tag{R8.17}$$

Thus, the overall budget is:

$$\text{HCHO} + h\nu \; (+\; 2\; \text{O}_2) \rightarrow 2\; \text{HO}_2 + \text{CO} \tag{R8.18}$$

On the other hand:

$$\text{HCHO} + h\nu \rightarrow \text{H}_2 + \text{CO} \tag{R8.19}$$

The two products of this reaction are stable molecules (CO will of course be oxidized slowly as described in Section 8.3.3). These two photolysis reactions have similar kinetics. Therefore, one may write a simple overall budget as being the average of these two reactions:

$$\text{HCHO} + h\nu \; (+\; \text{O}_2) \rightarrow \text{HO}_2 + \text{CO} + 1/2\; \text{H}_2 \tag{R8.20}$$

Oxidation by OH leads to the following reactions:

$$\text{HCHO} + \text{OH} \rightarrow \text{HCO} + \text{H}_2\text{O} \tag{R8.21}$$

$$\text{HCO} + \text{O}_2 \rightarrow \text{HO}_2 + \text{CO} \tag{R8.17}$$

The OH radical abstracts a hydrogen atom from the formaldehyde molecule to form a stable molecule (water vapor, H_2O) and an unstable organic radical. Therefore, one obtains an overall budget that is similar to that obtained with the photolytic reactions:

$$\text{HCHO} + \text{OH} \; (+\; \text{O}_2) \rightarrow \text{HO}_2 + \text{CO} + \text{H}_2\text{O} \tag{R8.22}$$

Thus, the oxidation of HCHO, whether it occurs by reaction with OH or results from photolytic reactions leads to one molecule of CO and one HO_2 radical. As shown in Section 8.3.3, the HO_2 radical can later oxidize NO into NO_2 and form an OH radical. The photolysis of NO_2 leads to the formation of a molecule of O_3. Since the oxidation of a molecule of CO leads also to the formation of a molecule of O_3, the oxidation of HCHO can theoretically lead to the formation of two molecules of O_3.

The oxidation of higher aldehydes (i.e., aldehydes with more than one carbon atom) follows the same conceptual scheme as that of formaldehyde, but leads to more complex products due to the greater number of carbon atoms. For example, the reactions of acetaldehyde (two carbon atoms, CH_3CHO) are as follows:

$$\text{CH}_3\text{CHO} + h\nu \rightarrow \text{CH}_3 + \text{HCO} \tag{R8.23}$$

$$\text{CH}_3 + \text{O}_2 \rightarrow \text{CH}_3\text{O}_2 \tag{R8.24}$$

$$\text{HCO} + \text{O}_2 \rightarrow \text{HO}_2 + \text{CO} \tag{R8.17}$$

The overall budget is as follows:

$$\text{CH}_3\text{CHO} + h\nu \; (+\; 2\; \text{O}_2) \rightarrow \text{CH}_3\text{O}_2 + \text{HO}_2 + \text{CO} \tag{R8.25}$$

In this first oxidation step, H originating from formaldehyde has been replaced by CH_3 originating from acetaldehyde. The methylperoxyl radical (also called peroxymethyl), CH_3O_2, behaves similarly to HO_2, that is to say that it can oxidize NO into NO_2 and form a methoxy radical, CH_3O:

$$NO + CH_3O_2 \rightarrow NO_2 + CH_3O \tag{R8.26}$$

The methoxy radical reacts next with O_2 to form formaldehyde and the hydroperoxyl radical:

$$CH_3O + O_2 \rightarrow HCHO + HO_2 \tag{R8.27}$$

The HO_2 radical can subsequently oxidize another NO molecule into NO_2, thereby potentially leading to the formation of another O_3 molecule. Therefore, there may be formation of up to six molecules of O_3 from the photolysis of acetaldehyde (one from the oxidation of CO, one due to the formation of NO_2, two from the HO_2 radicals, and two from the oxidation of HCHO).

The other photolytic pathway is also possible:

$$CH_3CHO + h\nu \rightarrow CH_4 + CO \tag{R8.28}$$

As shown in Section 8.3.5, methane can lead to the formation of four molecules of O_3, but its lifetime is very long; therefore, it is considered to be chemically inert at the time scales of photochemical air pollution.

The oxidation of acetaldehyde by OH occurs according to the following pathway:

$$CH_3CHO + OH \rightarrow CH_3CO + H_2O \tag{R8.29}$$

$$CH_3CO + O_2 \xrightarrow{(+M)} CH_3C(O)O_2 \tag{R8.30}$$

$$CH_3C(O)O_2 + NO_2 \rightarrow CH_3C(O)O_2NO_2 \tag{R8.31}$$

$$CH_3C(O)O_2 + NO \rightarrow CH_3C(O)O + NO_2 \tag{R8.32}$$

$$CH_3C(O)O + O_2 \rightarrow CH_3O_2 + CO_2 \tag{R8.33}$$

Peroxyacetyl nitrate, $CH_3C(O)O_2NO_2$, is commonly called PAN. PAN-type compounds may also be formed from higher aldehydes (propanal, butanal, etc.), in which case PAN refers more generally to peroxyacyl nitrates. Thus, the oxidation of VOC with several carbon atoms leads to more complex products. PAN is an important product because it is the organic nitrate species that is present at the highest ambient concentrations in the atmosphere. PAN can undergo a thermal decomposition reaction in the atmosphere:

$$CH_3C(O)O_2NO_2 \rightarrow CH_3C(O)O_2 + NO_2 \tag{R8.34}$$

PAN, when formed at a low temperature, can be transported over long distances and, when it encounters higher temperatures, it can decompose and form NO_2 back, and potentially O_3. PAN is, therefore, a "reservoir" species, because it acts as a reservoir of O_3 precursor.

Aldehydes are the only VOC that have significant photolytic rates. Calvert et al. (2011) provide a detailed state of the science of the atmospheric chemistry of aldehydes.

8.3.5 Oxidation of Alkanes

The simplest alkane (one carbon atom) is methane. Methane is a greenhouse gas, which reacts slowly with OH (lifetime on the order of several years). Therefore, it is not included in photochemical air pollution studies. Nevertheless, its oxidation mechanism being simple, it provides a useful conceptual model of the oxidation of alkanes in general:

$$CH_4 + OH \rightarrow CH_3 + H_2O \qquad (R8.35)$$

$$CH_3 + O_2 \rightarrow CH_3O_2 \qquad (R8.24)$$

$$CH_3O_2 + NO \rightarrow CH_3O + NO_2 \qquad (R8.26)$$

$$CH_3O + O_2 \rightarrow HCHO + HO_2 \qquad (R8.27)$$

As with aldehydes (see Section 8.3.4), the OH radical abstracts an H atom from the alkane molecule to form a stable H_2O molecule, and generates at the same time an organic radical (CH_3), which is rapidly oxidized by reaction with O_2 into a methylperoxyl radical (CH_3O_2). As discussed, this peroxyl radical and HO_2 may react with NO to form NO_2. The oxidation of methane leads theoretically to four molecules of O_3 (one via the formation of a molecule of NO_2 from CH_3O_2, two by the oxidation of HCHO and one from HO_2).

Higher alkanes, i.e., those with more carbon atoms (ethane, propane, butane, pentane, etc.), follow similar oxidation pathways. Let RH be a generic alkane (thus, R = CH_3 for methane, R = C_2H_5 for ethane, etc.):

$$RH + OH \rightarrow R + H_2O \qquad (R8.36)$$

$$R + O_2 \rightarrow RO_2 \qquad (R8.37)$$

$$RO_2 + NO \rightarrow RO + NO_2 \qquad (R8.38)$$

$$RO + O_2 \rightarrow R'CHO + HO_2 \qquad (R8.39)$$

where R' has one less carbon atom than R. The mechanism is identical to that of methane, except that instead of formaldehyde being the product, a higher aldehyde is obtained (for example, acetaldehyde for ethane). In addition, minor reaction pathways lead to the formation of organic nitrates ($RONO_2$) and peroxynitrates (RO_2NO_2):

$$RO_2 + NO \xrightarrow{(+M)} RONO_2 \qquad (R8.40)$$

$$RO_2 + NO_2 \xrightarrow{(+M)} RO_2NO_2 \qquad (R8.41)$$

For RO radicals, other chemical pathways include decomposition into an aldehyde and a peroxyl radical and, when R > C4, isomerization, which leads to a peroxyl radical and an alcohol group.

Figure 8.1 shows a partial reaction mechanism of the oxidation of n-pentane (an alkane with five carbon atoms) by OH radicals. There are five possible sites for the abstraction of a hydrogen atom by OH; however, for symmetry reasons, this corresponds to only three possible products with formation of a peroxyl radical in positions 1, 2, or 3 (from left to right in Figure 8.1). The yields of these three oxidation pathways are about 8, 57, and 35 %, respectively. Therefore, the probability of abstraction of a hydrogen atom is greater in an internal position than in a terminal position. The following steps are shown in Figure 8.1 only for the 2-pentyl-peroxyl radical, because (1) it is the radical that is formed in greater amount and (2) it provides the most complete reaction mechanism. The peroxyl radical can react with other peroxyl radicals, such as HO_2 and CH_3O_2 (i.e., those peroxyl radicals present in greater concentrations in the atmosphere), to form a peroxide when reacting with HO_2 or various oxygenated organic compounds when reacting with CH_3O_2. When NO_x concentrations are high, the other chemical pathways are favored. The reaction of NO_2 to form an organic peroxynitrate is fast, but reversible. The reactions with NO can lead either to the formation of an organic nitrate or to the oxidation of NO into NO_2 with formation of an alkoxy radical. This alkoxy radical can react along three distinct pathways: isomerization, decomposition, and oxidation by O_2. All these pathways lead to the formation of HO_2 and/or NO_2, and, therefore, to the potential formation of O_3.

As shown in Figure 8.2, alkanes with several carbon atoms are more reactive when the number of carbon atoms increases, because in first approximation an OH radical has a greater probability to encounter an hydrogen atom and abstract it from the alkane molecule. Therefore, ethane (two carbon atoms) is more reactive than methane (one carbon atom), butane (four carbon atoms) is more reactive than propane (three carbon atoms), etc.

Alkanes can also be oxidized by NO_3 radicals. The chemical mechanism is similar to that of the oxidation by OH radicals, because the NO_3 radical also abstracts a hydrogen atom from the alkane molecule to form a stable molecule of nitric acid, HNO_3:

$$RH + NO_3 \rightarrow R + HNO_3 \quad (R8.42)$$

However, the subsequent reactions differ from those resulting from the oxidation by OH, because NO concentrations are negligible when those of NO_3 are high. These differences are shown in Section 8.3.6 for alkenes, which have faster NO_3 oxidation kinetics than alkanes.

Alkanes are oxidized by OH during daytime and by NO_3 at night. However, the oxidation kinetics by NO_3 is much slower than that by OH (see Table 8.1). Calvert et al. (2008) provide a detailed state of the science of the atmospheric chemistry of the alkanes.

8.3.6 Oxidation of Alkenes

The oxidation of alkenes differs from that of alkanes because of their double bond, which is a more favorable site for attack by the OH and NO_3 radicals, and also by ozone. Therefore,

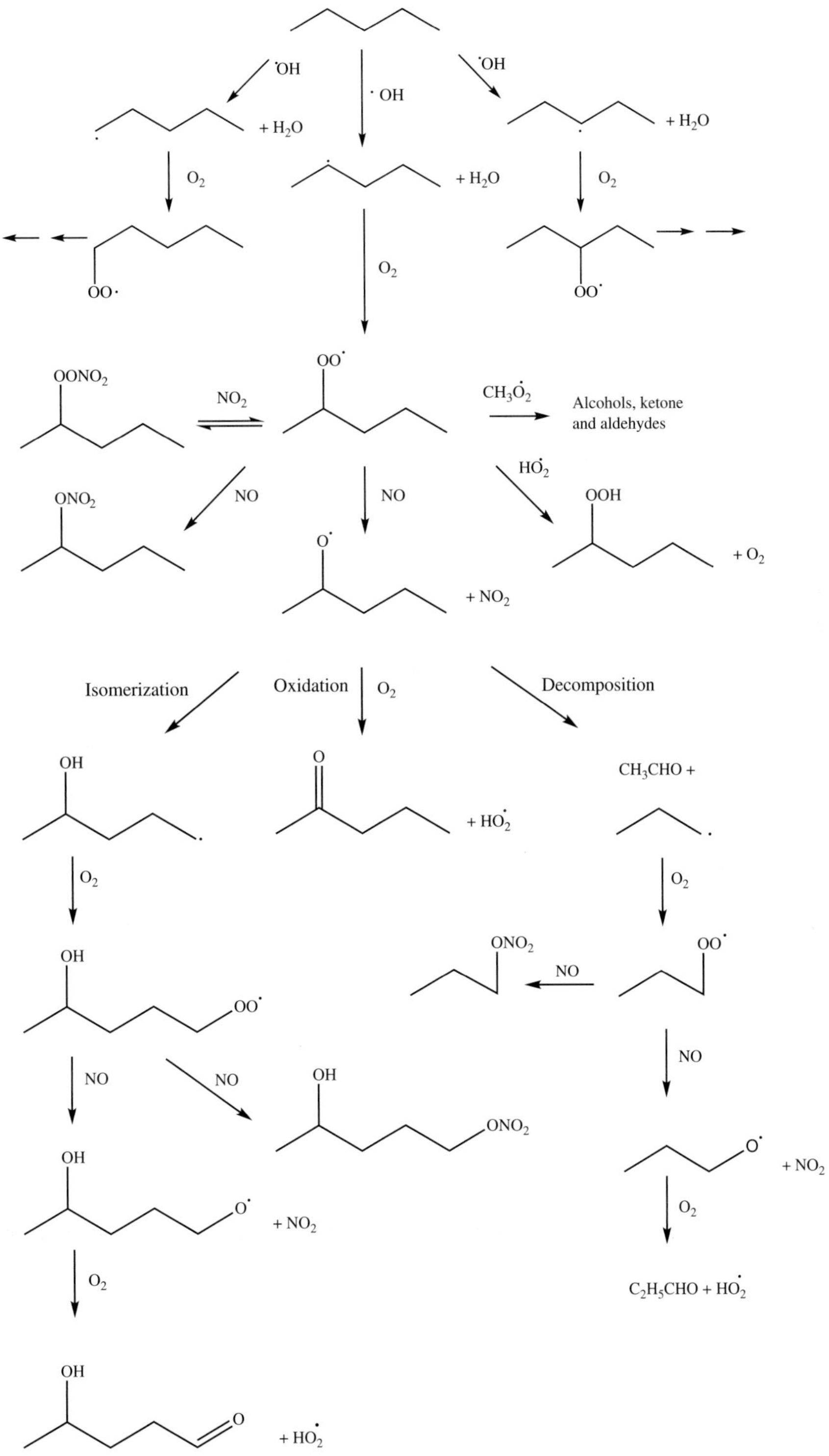

8.3 Gas-phase Chemistry of Photochemical Air Pollution

Figure 8.1. Oxidation of n-pentane: oxidation by OH leading to the formation of peroxyl radicals, and reactions of the 2-pentyl-peroxyl radical.

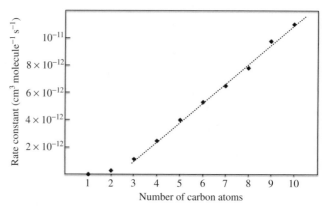

Figure 8.2. Kinetics of oxidation by OH of selected n-alkanes. Source of the data: Calvert et al. (2008).

instead of abstracting a hydrogen atom from the organic molecule, the OH and NO_3 radicals attach to one of the carbon atoms of the double bond. Figure 8.3 shows the chemical mechanism of the oxidation of propene (three carbon atoms, also called propylene) by OH. The OH radical can attach to either one of the carbon atoms of the double bond, but it tends to favor the formation of the radical on the secondary carbon ($CH_3CHC(OH)H_2$), i.e., the carbon atom that is linked to two other carbon atoms (65 % for that pathway, compared to 35 % for the other pathway). Next, these radicals react rapidly with molecular oxygen to form peroxyl radicals. These peroxyl radicals can then react with NO, either to form NO_2 and alkoxy radicals, or to form organic nitrates. However, the formation of nitrates is a relatively minor pathway (<2 %) and the formation of NO_2 prevails. Then, the alkoxy radicals undergo decomposition by reaction with O_2 to form aldehydes. In the case of propene, both chemical reaction pathways lead to the same products, which are formaldehyde and acetaldehyde. Then, the aldehydes lead to ozone formation (for example, two molecules of ozone in the case of formaldehyde, see Section 8.3.4). For alkenes with a greater number of carbon atoms, the mechanism is similar, but it leads to more complex oxo products (aldehydes and ketones).

The NO_3 radicals follow the same general scheme in principle and attach to one of the carbon atoms of the double bond. However, the following steps differ, because the NO concentrations are negligible when NO_3 concentrations are significant. Therefore, the organic peroxyl radical does not react with NO, but instead reacts with other peroxyl radicals, such as HO_2 and CH_3O_2. Figure 8.4 shows the main oxidation pathways for the reaction of propene with NO_3. The oxidation products are mostly organic nitrates with other functional groups, such as peroxide, alcohol or oxo. In the case of propene, products include formaldehyde and acetaldehyde, as well as NO_2 via decomposition of the nitrated alkoxy radical. Unlike alkanes, alkenes can have oxidation kinetics with NO_3 that are faster than those with OH, particularly in the case of biogenic compounds.

Alkenes are the only VOC that react with ozone. Figure 8.5 shows the chemical mechanism of the oxidation of propene by O_3. Ozone adds to the double bond creating a bridge between the two carbon atoms. This unstable species is called an ozonide.

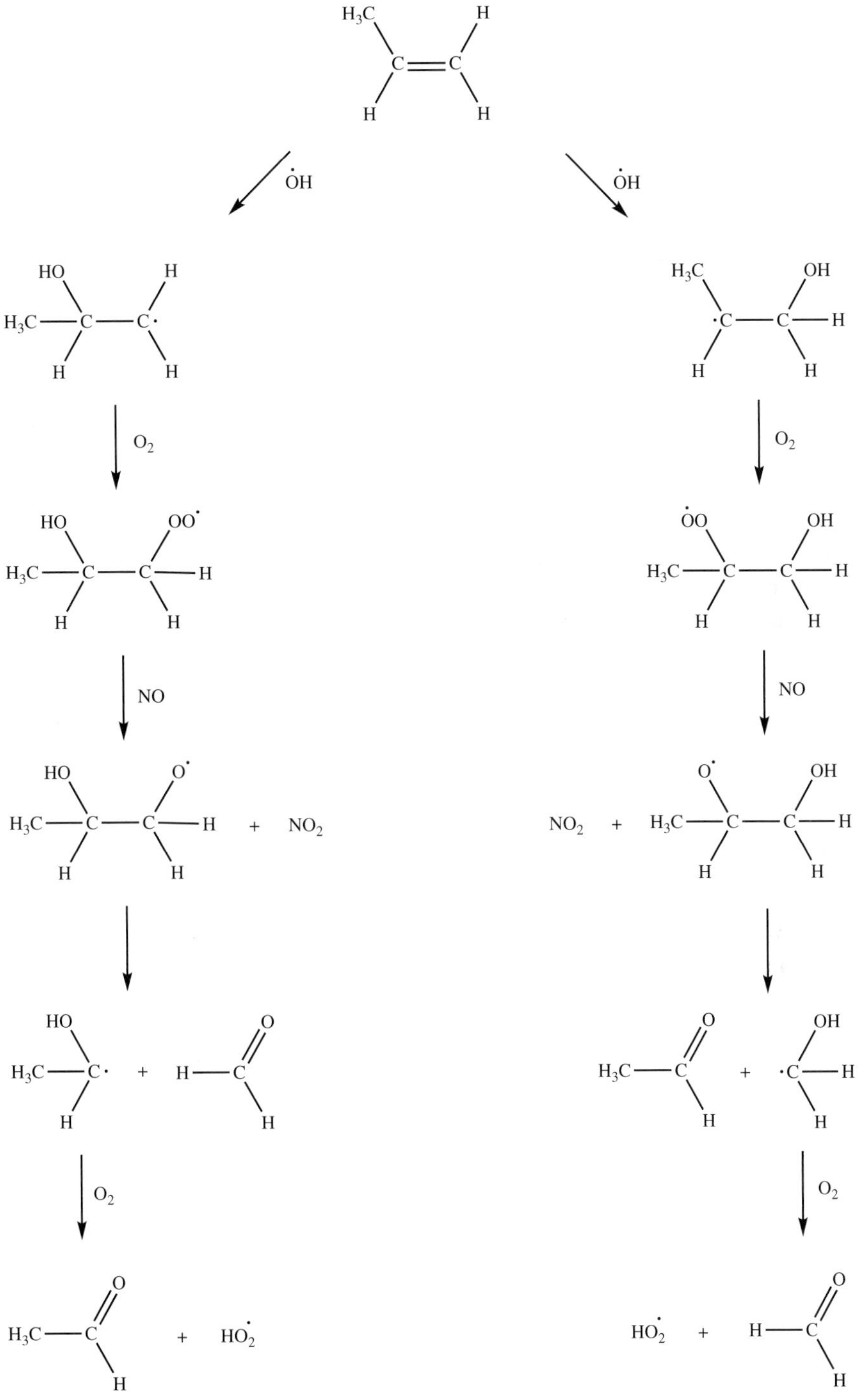

Figure 8.3. Oxidation of propene by OH.

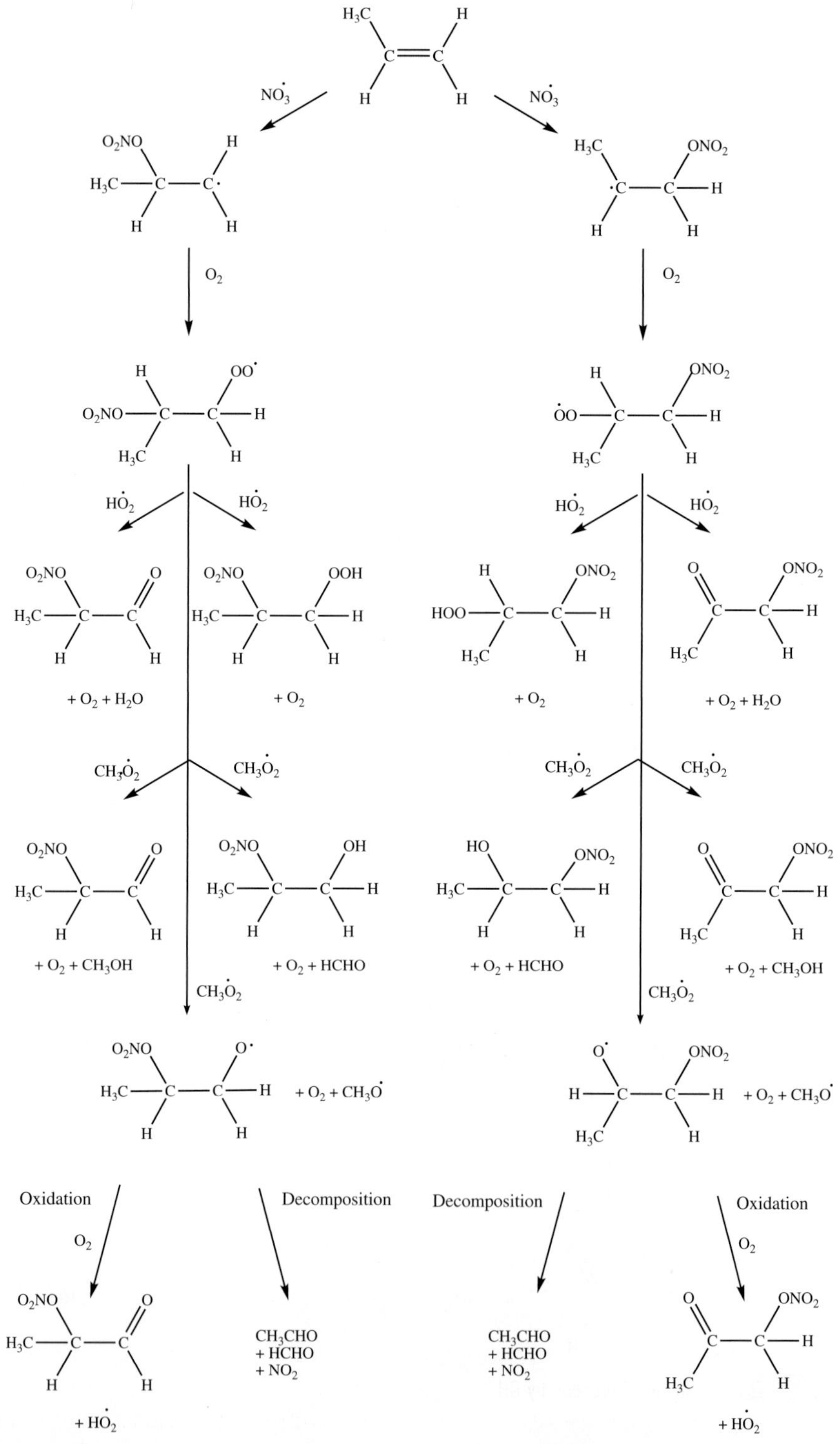

Figure 8.4. Oxidation of propene by NO_3.

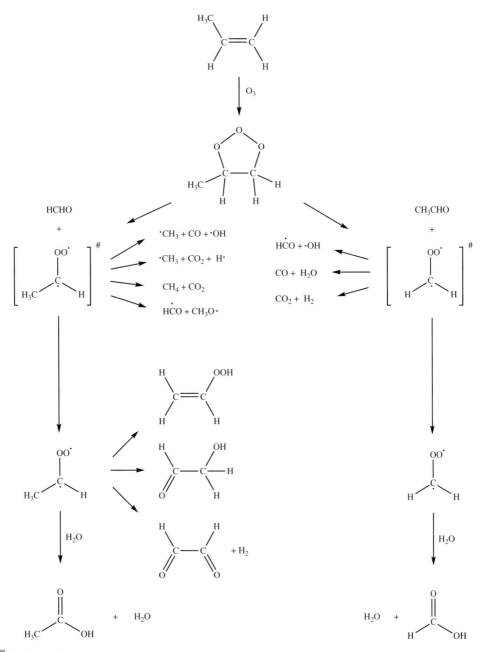

Figure 8.5. Oxidation of propene by ozone.

The products of the ozonide are oxo compounds (aldehydes and ketones) and biradicals (which, therefore, include two reactive sites with free electrons). In the case of propene, formaldehyde and acetaldehyde are formed. The biradicals, which are called Criegee radicals (named after Rudolf Criegee, who first identified them), are extremely reactive and lead to the formation of stable products (in the case of propene, CO, CO_2, CH_4, as well as formic acid and acetic acid) and radicals (OH, H, CH_3, HCO, CH_3O_2) that lead to ozone formation. In particular, one notes that OH radicals are formed. The yield of these OH radicals has been estimated to be 12 % from the H_2CO_2 biradical and 54 % from the $CH_3 C(H)O_2$ biradical. If both reaction pathways of the ozonide are about equivalent, the overall estimated OH yield is 33 %, which is consistent with the overall estimate. The OH yields can be significant; they vary from about 6 % for β-caryophyllene to nearly 120 % for cyclopentene. For some alkenes, the oxidation by O_3 is competitive with that by the OH and NO_3 radicals. The oxidation rate by O_3 is much greater for an internal alkene (internal double bond) than for a terminal alkene (double bond located at the end of the molecule); see Table 8.1.

Calvert et al. (2000) provide a detailed state of the science for the atmospheric chemistry of the alkenes.

8.3.7 Oxidation of Aromatic Compounds

The oxidation of aromatic compounds differs also from that of alkanes. The oxidation by OH can follow the same pathway as that of the alkanes with the abstraction of a hydrogen atom from the molecule to form an organic radical and a water molecule. However, this oxidation pathway is minor (<10 %) and it is the addition of the OH radical to the aromatic ring that prevails, because this addition requires less energy. There are two possibilities following this addition of OH to the ring:

– The aromatic ring is conserved (for example, in the case of toluene, formation of cresol and its derivative products)
– The aromatic ring is broken and species with fewer carbon atoms are formed

Figure 8.6 shows some of the chemical oxidation pathways and products of the oxidation of toluene (seven carbon atoms with a methyl group attached to the aromatic ring) by OH.

The abstraction of a hydrogen atom from the methyl group leads to the formation of benzaldehyde, but with a benzaldehyde yield estimated to be about 6 %. Next, benzaldehyde reacts to form mostly nitrophenol.

The second oxidation pathway accounts for more than 90 % and leads to the formation of a radical that reacts with O_2, which either abstracts a hydrogen atom to form cresols and HO_2 (a potential source of O_3) or adds to the aromatic ring to form a peroxyl radical. The formation of cresols (mostly ortho-cresol, i.e., the CH_3 and OH groups are adjacent on the aromatic ring) accounts for about 18 % of the oxidation of toluene (about 12 % of ortho-cresol, 3 % of meta-cresol, and 3 % of para-cresol). The oxidation of cresols leads to the formation of peroxyl radicals that can lead to O_3 formation via the conversion of NO to NO_2 or can react with NO_x to form nitrocresols. In the latter case of the formation of an aromatic peroxyl radical, which accounts for >70 % of the toluene oxidation, there is mostly an opening of the aromatic ring. The products include aldehydes with fewer than six carbon atoms (glyoxal, methylglyoxal, methylbutenedial, 1,4-butenedial, 2-methyl-2,4-hexadiene-1,6-dial). Among those species, glyoxal and methylglyoxal are the most important ones with yields on the order of 4 to 17 % and 4 to 24 %, respectively.

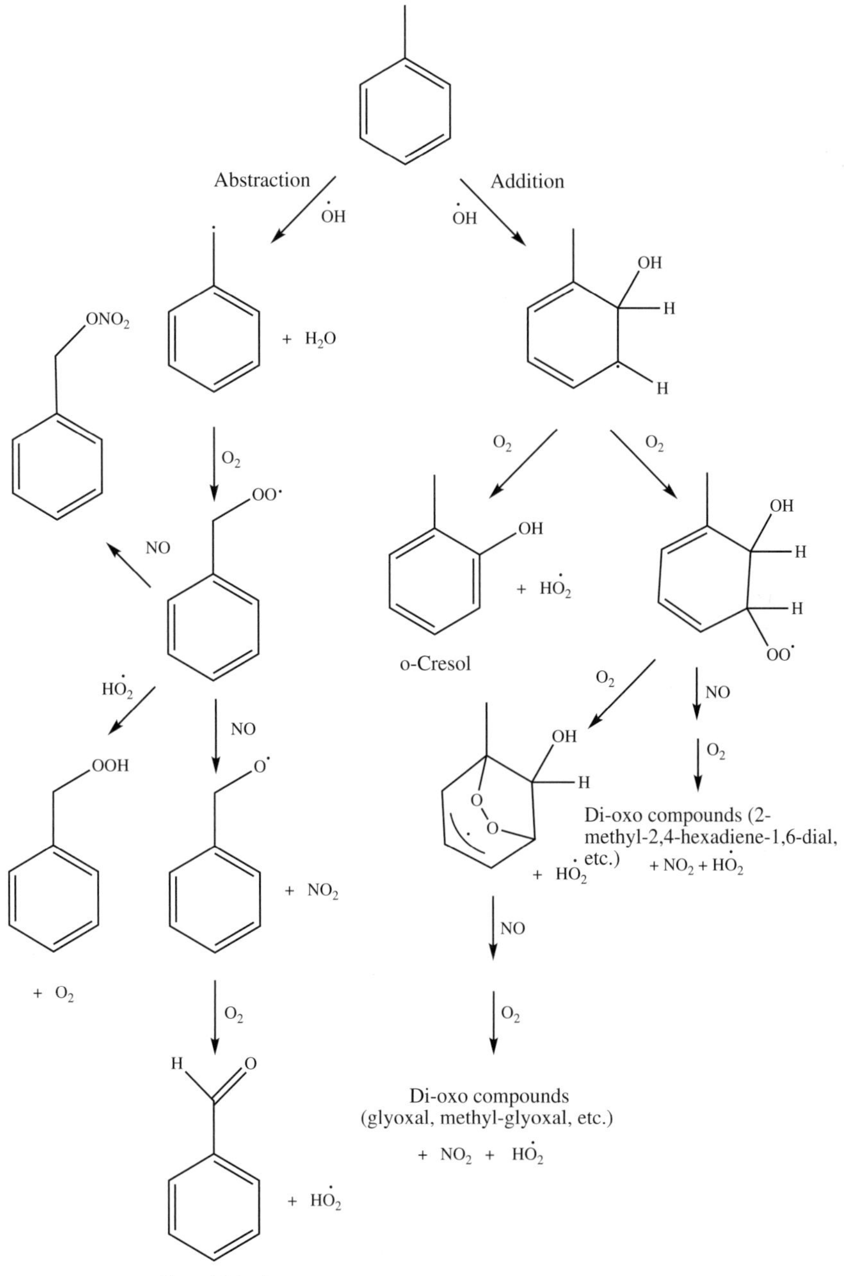

Figure 8.6. Oxidation of toluene by OH: major oxidation pathways and formation of major products.

The oxidation pathways where the aromatic ring is conserved lead to stable species (benzaldehyde, cresol, nitrated derivatives) and, therefore, little ozone formation. However, these species are not very volatile because of their high number of carbon atoms and they lead to particulate matter formation via the formation of particulate organic compounds (see Chapter 9). In addition, the formation of nitrated organic compounds leads to the elimination of nitrogen oxides (NO and NO_2) from the chemical system and, therefore, potentially less ozone. The oxidation pathways where the aromatic ring is opened lead to reactive species (dioxo compounds, i.e., species with two carbonyl groups, and in particular, dialdehydes) leading rapidly to ozone formation. On the other hand, these species have few carbon atoms and are, therefore, unlikely to lead to the formation of particulate organic compounds (however, their aqueous chemistry may lead to species with low volatility). The relative importance of these two main oxidation pathways (ring-retaining versus ring-opening) has long been poorly understood and still remains a source of uncertainty in photochemical air pollution.

Aromatic compounds can be oxidized by NO_3; however, the kinetics is much slower than that with OH radicals (see Table 8.1). Calvert et al. (2002) provide a detailed state of the science of the atmospheric chemistry of aromatic compounds.

8.3.8 Oxidation of Biogenic Compounds

Biogenic compounds include hemiterpenes, monoterpenes, sesquiterpenes, and terpenoids. The most important biogenic compound in terms of atmospheric emissions is isoprene (the hemiterpene), which is a hydrocarbon with five carbon atoms and two double bonds. The term "terpene" refers to a compound that is formed from a combination of several isoprene molecules (two for a monoterpene and three for a sesquiterpene) and, therefore, contains only carbon and hydrogen atoms. The term "terpenoid" is used when methyl groups have been removed or moved or when the compound contains one or more oxygen atoms. For simplicity, we use hereafter the term terpene to refer to all terpenes and terpenoids.

Isoprene is the biogenic compound that has been studied the most in terms of its atmospheric chemistry. The two double bonds are attack sites for OH and NO_3 radicals, as well as for O_3. The oxidation by OH leads to the formation of aldehydes and ketones. The most important ones are methacrolein (MACR) and methyl vinyl ketone (MVK). Isoprene is an important precursor of ozone.

Monoterpenes and other terpenes contribute also to the formation of ozone, but to a lesser extent, because their emissions are less important than those of isoprene (see Chapter 2). Accordingly, their chemistry has been studied more in terms of their contribution to the formation of organic particulate matter. Indeed, their number of carbon atoms (10 for monoterpenes and 15 for sesquiterpenes) favors the formation of compounds with low saturation vapor pressures, which can condense readily on particles. Therefore, the chemistry of biogenic compounds is discussed further in Chapter 9. One notes, however, that these compounds have short lifetimes (see Table 8.1), because they are oxidized rapidly by several oxidant species (OH, NO_3, and O_3).

8.3.9 Termination of the Oxidation Cycles

The formation of ozone occurs when NO is converted to NO_2 by peroxyl radicals (hydroperoxyl, HO_2, or organic peroxyl, RO_2), i.e., without any consumption of O_3. Thus, NO_2

forms O_3 during the day by photolysis. This formation takes place as long as peroxyl radicals are formed by the oxidation of VOC and CO. It can stop (or at least it can be slowed down) when one of the key species of this mechanism is consumed by a reaction that leads to a stable product. There are two main possibilities to stop the formation of ozone.

On one hand, the peroxyl radicals can react without converting NO to NO_2. They can react among themselves to form a peroxide or some other stable products. For example, the hydroperoxyl radicals can form hydrogen peroxide:

$$HO_2 + HO_2 \rightarrow H_2O_2 + O_2 \quad \quad (R8.43)$$

Another possibility for peroxyl radicals is the formation of organic nitrates or peroxynitrates (nitric acid is the inorganic nitrate), when NO and NO_2 react with organic peroxyl radicals. For example, PAN may be formed following the oxidation of acetaldehyde (see Section 8.3.4).

On the other hand, NO_2 can be oxidized into a stable product instead of being photolyzed. Its main oxidation pathway is by reaction with OH radicals to form nitric acid:

$$NO_2 + OH + M \rightarrow HNO_3 + M \quad \quad (R8.44)$$

Therefore, there is competition in the former case for the reaction of peroxyl radicals either with NO to form NO_2 or with other peroxyl radicals or NO_x to form stable products such as peroxides and organic nitrates. In the latter case, there is consumption of NO_2 either by photolysis leading to O_3 formation or by reaction with OH to form HNO_3. We will see in Section 8.4 that the relative importance of these reaction pathways for peroxyl radicals and for NO_2 depends on the chemical regime of the atmosphere.

8.4 Chemical Kinetic Mechanisms of Photochemical Air Pollution

To simulate the formation of gaseous photochemical pollutants such as O_3 and NO_2 requires setting up a chemical kinetic mechanism that takes into account the major reactions involved in the formation of those gaseous air pollutants.

8.4.1 Conceptual Mechanism of Photochemical Air Pollution

Figure 8.7 presents a schematic description of the oxidation of an alkane, RH, by OH. In the first step, the oxidation of RH by OH leads to the formation of an organic peroxyl radical, RO_2, which reacts with NO to form NO_2 and RO. The organic alkoxy radical, RO, undergoes oxidation leading to the formation of an aldehyde (here called R'CHO, where R' contains one less carbon atom than RH) and a hydrogen atom, which is rapidly oxidized into HO_2. The HO_2 radical leads to another oxidation of NO to NO_2 and the regeneration of the OH radical. Therefore, this oxidation of the alkane leads to the formation of two ozone molecules, since each NO_2 molecule is photolyzed during daytime and leads to the formation of NO and O_3. In addition, the aldehyde will be either oxidized by OH or

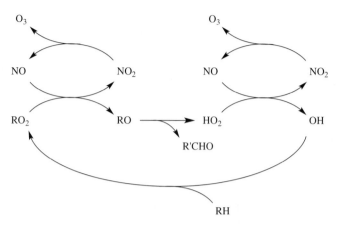

Figure 8.7. Schematic representation of the chemistry of ozone formation from an alkane, RH.

photolyzed, which will lead to the formation of other O_3 molecules. Therefore, the oxidation of an alkane leads to the formation of several O_3 molecules via (1) the formation of peroxyl radicals and (2) the formation of an aldehyde, which via photolysis or oxidation will lead to the formation of other peroxyl radicals. As discussed, the oxidations of alkenes and aromatic compounds follow different reaction schemes, but the general conceptual approach is similar.

8.4.2 Simple Chemical Kinetic Mechanism of a Hydrocarbon

Table 8.2 lists the reactions corresponding to the oxidation mechanism of a generic hydrocarbon and its aldehyde product associated with the first oxidation step, as shown in Figure 8.7. This list of 20 reactions can be reduced during the development of the set of the ordinary differential equations representing the kinetics of the chemical species involved, because the reactions with molecular oxygen, O_2, are very fast. Thus, the reactions of the oxygen atom, O, and of the organic alkoxy radical, RO, with O_2 have kinetics that are not limiting and, accordingly, they can be lumped with the reactions leading to their formation. Then, this mechanism may be reduced to 18 reactions, after the lumping of Reaction 3 with Reactions 2 and 5 and Reaction 11 with Reaction 10. We can write the set of the corresponding differential equations as follows, using the stationary state approximation for the radicals (the numbering of the reactions and their rate constants corresponds to those of Table 8.2).

The radicals and atom of this reacting system are $O(^1D)$, OH, HO_2, RO_2, and $R'O_2$ (the RO radical and the O atom are ignored following the lumping of their formation and destruction reactions). In addition, the RO_2 and $R'O_2$ radicals are lumped here as organic peroxyl radicals and the sum of their concentrations is represented by $[RO_2]$. Thus, there are five species at stationary state and ten species with time-dependent concentrations. The concentrations of the stationary-state species can be obtained with algebraic equations, and the time-dependent concentrations are governed by ordinary differential equations.

The stationary-state solution for $[O(^1D)]$ is as follows:

8.4 Chemical Kinetic Mechanisms of Photochemical Air Pollution

Table 8.2 Simplified chemical kinetic mechanism of the oxidation of a generic hydrocarbon.

Reaction	Rate constant[a]
1. $NO + O_3 \rightarrow NO_2 + O_2$	1.8×10^{-14} cm^3 molecule^{-1} s^{-1}
2. $NO_2 + h\nu \rightarrow NO + O$	7.9×10^{-3} s^{-1}
3. $O + O_2 + M \rightarrow O_3 + M$	6×10^{-34} cm^6 molecule^{-2} s^{-1}
4. $O_3 + h\nu \rightarrow O_2 + O(^1D)$	4.5×10^{-5} s^{-1}
5. $O(^1D) + M \rightarrow O + M$	2.9×10^{-11} cm^3 molecule^{-1} s^{-1}
6. $O(^1D) + H_2O \rightarrow 2\ OH$	2.2×10^{-10} cm^3 molecule^{-1} s^{-1}
7. $NO + HO_2 \rightarrow NO_2 + OH$	8.3×10^{-12} cm^3 molecule^{-1} s^{-1}
8. $O_3 + HO_2 \rightarrow OH + 2\ O_2$	1.9×10^{-15} cm^3 molecule^{-1} s^{-1}
9. $RH + OH\ (+O_2) \rightarrow RO_2 + H_2O$	2.6×10^{-11} cm^3 molecule^{-1} s^{-1}
10. $NO + RO_2 \rightarrow NO_2 + RO$	7.7×10^{-12} cm^3 molecule^{-1} s^{-1}
11. $RO + O_2 \rightarrow R'CHO + HO_2$	9×10^{-15} cm^3 molecule^{-1} s^{-1}
12. $R'CHO + OH\ (+O_2) \rightarrow R'C(O)O_2 + H_2O$	1.6×10^{-11} cm^3 molecule^{-1} s^{-1}
13. $NO + R'C(O)O_2\ (+O_2) \rightarrow NO_2 + R'O_2 + CO_2$	2×10^{-11} cm^3 molecule^{-1} s^{-1}
14. $NO_2 + R'C(O)O_2 \rightarrow PAN$	1×10^{-11} cm^3 molecule^{-1} s^{-1}
15. $PAN \rightarrow NO_2 + R'C(O)O_2$	3.2×10^{-4} s^{-1}
16. $R'CHO + h\nu \rightarrow R'H + CO$	5.2×10^{-6} s^{-1}
17. $R'CHO + h\nu\ (+2\ O_2) \rightarrow R'O_2 + HO_2 + CO$	5.7×10^{-6} s^{-1}
18. $NO_2 + OH\ (+M) \rightarrow HNO_3\ (+M)$	9.2×10^{-12} cm^3 molecule^{-1} s^{-1}
19. $HO_2 + HO_2 \rightarrow H_2O_2 + O_2$	1.5×10^{-12} cm^3 molecule^{-1} s^{-1}
20. $RO_2 + HO_2 \rightarrow$ Products	5.6×10^{-12} cm^3 molecule^{-1} s^{-1}

(a) Values given at 1 atm and 25 °C on June 21 at noon in Paris with a clear sky; photolytic reactions 2, 4, 16, and 17 depend on the zenith angle, cloudiness, and atmospheric particulate matter concentration; all reactions, except the photolytic reactions, depend on temperature. Sources of the rate constants: Jacobson (2005) except for Reaction 9, which uses the kinetics of propene as the generic hydrocarbon (Calvert et al., 2000), Reaction 11 (Finlayson-Pitts and Pitts, 2000) and Reaction 18 (Mollner et al., 2010). The kinetics of peroxyacetyl nitrate was used for the generic peroxyacyl nitrate (called PAN here), that of acetaldehyde for the generic aldehyde, and that of the methylperoxyl radical for Reaction 20.

$$\frac{d[O(^1D)]}{dt} = k_4[O_3] - k_5[O(^1D)][M] - k_6[O(^1D)][H_2O] = 0 \quad (8.23)$$

That is:
$$[O(^1D)] = \frac{k_4[O_3]}{k_5[M] + k_6[H_2O]}$$

The stationary-state solution for [OH] is as follows:

$$\frac{d[OH]}{dt} = 2\ k_6[O(^1D)][H_2O] + k_7[NO][HO_2] + k_8[O_3][HO_2] \\ - k_9[RH][OH] - k_{12}[R'CHO][OH] - k_{18}[NO_2][OH] = 0 \quad (8.24)$$

That is:
$$[OH] = \frac{2\ k_6[O(^1D)][H_2O] + k_7[NO][HO_2] + k_8[O_3][HO_2]}{k_9[RH] + k_{12}[R'CHO] + k_{18}[NO_2]}$$

The stationary-state solution for the concentration of the hydroperoxyl radical (HO_2) requires the solution of a quadratic algebraic equation:

$$\frac{d[HO_2]}{dt} = k_{10}[NO][RO_2] + k_{17}[R'CHO] \\ - 2k_{19}[HO_2]^2 - k_{20}[RO_2][HO_2] - k_7[NO][HO_2] - k_8[O_3][HO_2] = 0 \quad (8.25)$$

That is:

$$[HO_2] = \frac{-b_{HO_2} + \sqrt{(b_{HO_2})^2 - 4\,a_{HO_2}\,c_{HO_2}}}{2a_{HO_2}}$$

$$a_{HO_2} = 2k_{19}; \quad b_{HO_2} = k_{20}[RO_2] + k_7[NO] + k_8[O_3];$$

$$c_{HO_2} = -(k_{10}[NO][RO_2] + k_{17}[R'CHO])$$

The concentrations of the organic peroxyl radicals, RO_2 and $R'O_2$, are lumped as $[RO_2]$. Since their self-reaction is neglected compared to their reaction with HO_2 (which is present in greater concentration), the solution is simpler than that for HO_2:

$$\frac{d[RO_2]}{dt} = k_9[RH][OH] + k_{13}[R'C(O)O_2][NO] + k_{17}[R'CHO] \\ - k_{10}[NO][RO_2] - k_{20}[HO_2][RO_2] = 0 \quad (8.26)$$

That is:
$$[RO_2] = \frac{k_9[RH][OH] + k_{13}[R'C(O)O_2][NO] + k_{17}[R'CHO]}{k_{10}[NO] + k_{20}[HO_2]}$$

The algebraic equation for the peroxyacyl radical is as follows:

$$\frac{d[R'C(O)O_2]}{dt} = k_{12}[R'CHO][OH] + k_{15}[PAN] - k_{13}[NO][R'C(O)O_2] \\ - k_{14}[NO_2][R'C(O)O_2] = 0 \quad (8.27)$$

That is: $[R'C(O)O_2] = \dfrac{k_{12}[R'CHO][OH] + k_{15}[PAN]}{k_{13}[NO] + k_{14}[NO_2]}$

For the other species, the ordinary differential equations are as follows:

$$\frac{d[O_3]}{dt} = k_2[NO_2] - k_1[NO][O_3] - k_4[O_3] - k_8[HO_2][O_3] \quad (8.28)$$

$$\frac{d[NO_2]}{dt} = k_1[NO][O_3] + k_7[NO][HO_2] + k_{10}[NO][RO_2] + k_{13}[NO][R'C(O)O_2] \\ + k_{15}[PAN] - k_2[NO_2] - k_{14}[R'C(O)O_2][NO_2] - k_{18}[OH][NO_2] \quad (8.29)$$

$$\frac{d[NO]}{dt} = k_2[NO_2] - k_1[NO][O_3] - k_7[NO][HO_2] - k_{10}[NO][RO_2] - k_{13}[NO][R'C(O)O_2]$$
$$(8.30)$$

$$\frac{d[RH]}{dt} = -k_9[RH][OH] \quad (8.31)$$

$$\frac{d[R'CHO]}{dt} = k_{10}[NO][RO_2] - k_{12}[R'CHO][OH] - k_{16}[R'CHO] - k_{17}[R'CHO] \quad (8.32)$$

$$\frac{d[PAN]}{dt} = k_{14}[NO_2][R'C(O)O_2] - k_{15}[PAN] \quad (8.33)$$

$$\frac{d[HNO_3]}{dt} = k_{18}[NO_2][OH] \quad (8.34)$$

$$\frac{d[H_2O_2]}{dt} = k_{19}[HO_2]^2 \quad (8.35)$$

$$\frac{d[R'H]}{dt} = k_{16}[R'CHO] \quad (8.36)$$

$$\frac{d[CO]}{dt} = k_{16}[R'CHO] + k_{17}[R'CHO] \quad (8.37)$$

The numerical solution of these ordinary differential equations provides the temporal evolution of the concentrations of these chemical species given their initial concentrations. The numerical methods that can be used to solve such a set of equations are described in Chapter 7.

8.4.3 Spatio-temporal Variability of Photochemical Air Pollution

The diurnal profile of ozone concentrations shows a peak in midday, since ozone formation results from the photolysis of NO_2 and the photolysis kinetics is maximum at noon. Ozone formation continues during the afternoon until the ozone destruction reactions (e.g., the NO titration reaction) begin to overcome the formation reaction, i.e., the NO_2 photolysis, as radical (OH and HO_2) production starts to slow down (i.e., when the Sun comes down toward the horizon). Figure 8.8 depicts temporal profiles of measured and simulated ozone concentrations corresponding to a photochemical air pollution episode in the Los Angeles Basin in 1974, that is, at a time when air pollution levels were still very high (maximum hourly concentrations >300 ppb, i.e., about 600 μg m^{-3}).

Figure 8.8 shows temporal profiles of measured and simulated NO_2 concentrations for the same Los Angeles air pollution episode. The diurnal profile of the nitrogen dioxide (NO_2) concentrations shows peaks during the traffic rush hour (i.e., in the morning and in late afternoon) and lower concentrations during the day when emissions are lower and photolysis of this compound is more important. The maximum NO_2 concentration reaches 300 ppb (about 600 μg m^{-3}) during the early-morning peak of the second day.

The spatial distribution of the ozone concentrations typically shows lower concentrations near nitrogen oxides (NO_x) sources because of the titration of O_3 by NO. Thus, the maximum O_3 concentrations occur downwind of the source regions, since

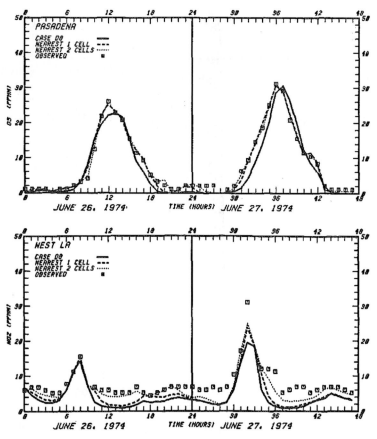

Figure 8.8. Temporal profiles of ozone (O_3, top figure) and nitrogen dioxide (NO_2, bottom figure) concentrations on June 26 and 27, 1974 in the Los Angeles Basin, California (in pphm, "parts per hundred million"; 1 pphm = 10 ppb). The small squares correspond to the measurements and the lines correspond to the results of a numerical simulation. The solid lines correspond to the simulation at the location of the measurement; the dotted lines correspond to optimal solutions obtained at one or two model grid cells of the measurement location. O_3 concentrations are in Pasadena, downwind of Los Angeles; NO_2 concentrations are in downtown Los Angeles. Source: Tesche et al. (1984).

it takes some time for the kinetics of the photochemical reactions to occur and to lead to ozone formation. On the other hand, the maximum NO_2 concentrations occur mostly near NO_x sources, i.e., in urban areas near major roadways. For example, the O_3 concentrations shown in Figure 8.8 correspond to a location in Pasadena, i.e., downwind of the industrial areas (located on the coast) and of the downtown area. Figure 8.9 depicts maps of simulated concentrations of O_3 and NO_2 for the Paris region in 2009. The high NO_2 concentrations appear in the downtown Paris area, near major roadways, near the airports, and near an industrial site. On the other hand, the O_3 concentrations are negligible in these locations because of the titration reaction with NO and they are high outside of the urban area.

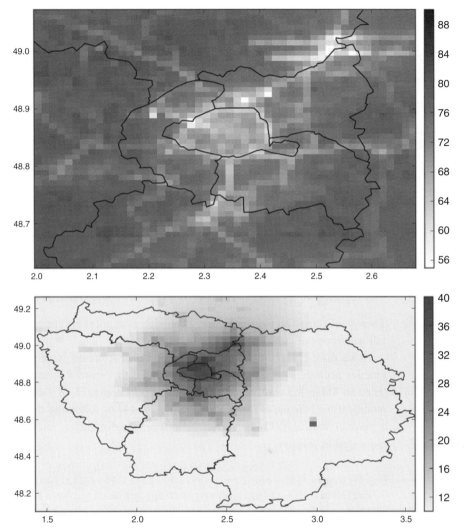

Figure 8.9. Spatial distribution of simulated O_3 and NO_2 concentrations for 2009 in the Paris region. The simulation was performed with the Polyphemus numerical model. Top figure: O_3 concentrations (annual average in µg m^{-3} of the daily maximum 8-hour average concentrations); bottom figure: NO_2 concentrations (annual average in µg m^{-3}). Concentrations are indicated on the right vertical axis. Latitude (°N) and longitude (°E) are indicated on the left vertical axis and horizontal axis, respectively. Source: Ademe (2014).

8.5 Emission Control Strategies for Photochemical Air Pollution

8.5.1 Chemical Regimes of Photochemical Smog

Ozone formation occurs when the Leighton photostationary state undergoes a perturbation due to the addition of VOC or CO, thereby leading to the formation of peroxyl radicals, which can convert NO to NO_2 without consuming O_3. NO_2 photolysis leads to the formation of one molecule of O_3 and regenerates NO. NO can then be converted back to NO_2, either by reaction with O_3 (null O_3 balance) or by reaction with a peroxyl radical (potential formation of an O_3 molecule). Therefore, there is competition between these two NO oxidation pathways and we may distinguish two chemical regimes:

- A regime where the peroxyl radical concentrations are not limiting compared to NO (or more generally NO_x) concentrations
- A regime where the peroxyl radical concentrations are limiting compared to NO (or more generally NO_x) concentrations

The peroxyl radicals are formed by oxidation of VOC or CO. However, CO is significantly less reactive than most VOC and, therefore, one may consider that high peroxyl radical concentrations result from high VOC concentrations. These two regimes may, therefore, be defined in terms of the relative concentrations of NO_x and VOC:

- A regime rich in VOC and poor in NO_x (hereafter referred to as low-NO_x regime)
- A regime poor in VOC and rich in NO_x (hereafter referred to as high-NO_x regime)

There is of course an intermediate regime. However, most atmospheric conditions tend to fall within one of these two extreme regimes.

These different chemical regimes can be characterized in terms of their chemical kinetics according to the mechanism presented in Figure 8.7, using the rate constants listed in Table 8.2. We use here the conceptual approach of Jacob (1999) with some modifications. Ozone is formed by photolysis of NO_2 following the oxidation of NO by a peroxyl radical (RO_2 or HO_2). Therefore, the formation rate of O_3, PO_3, can be expressed as follows:

$$PO_3 = k_7 [NO][HO_2] + k_{10}[NO][RO_2] \tag{8.38}$$

The OH and HO_2 radicals, represented together as HO_x, play a key role in this scheme of O_3 formation. They are formed mostly by the photolysis of O_3 and nitrous acid (HNO_2) in the case of OH and by the photolysis of aldehydes in the case of HO_2. Their formation kinetics are represented here by POH and PHO_2, respectively. Furthermore, we define $PHO_x = POH + PHO_2$, to represent the formation of all HO_x radicals. POH is greater than PHO_2, the former being typically about 50 to 80 % of PHO_x (Mao et al., 2010). The main termination reactions are, for OH, the formation of nitric acid (HNO_3) and, for HO_2, the formation of peroxides by reaction with HO_2 or RO_2. At steady state, their production rate is equal to the rate of their termination reactions and we have the following relationship:

$$PHO_x = POH + PHO_2 = k_{18}[NO_2][OH] + 2k_{19}[HO_2]^2 + k_{20}[RO_2][HO_2] \tag{8.39}$$

Assuming steady state for these radicals:

$$\frac{d[OH]}{dt} = POH + k_7[NO][HO_2] - k_9[RH][OH] - k_{18}[NO_2][OH] = 0 \tag{8.40}$$

$$\frac{d[HO_2]}{dt} = PHO_2 + k_{10}[NO][RO_2] - k_7[NO][HO_2] - 2k_{19}[HO_2]^2$$
$$-k_{20}[RO_2][HO_2] = 0 \qquad (8.41)$$

The rates of the propagation reactions of these radicals are assumed to be significantly faster than their formation and termination reactions (the ratio of these rates is typically on the order of 5 to 10; Mao et al., 2010). Therefore, we obtain the following relationships:

$$k_7[NO][HO_2] \approx k_9[RH][OH] \qquad (8.42)$$

$$k_{10}[NO][RO_2] \approx k_7[NO][HO_2] \qquad (8.43)$$

Then, the production of O_3 can be expressed as follows:

$$PO_3 \approx 2k_7[NO][HO_2] \approx 2k_9[RH][OH] \qquad (8.44)$$

In addition, since the values of the rate constants k_7 and k_{10} are similar, the following simplifying assumption may be introduced:

$$[RO_2] \approx [HO_2] \qquad (8.45)$$

Thus, replacing $[RO_2]$ by $[HO_2]$ in the steady-state relationship between production and termination of the HO_x radicals:

$$PHO_x \approx k_{18}[NO_2][OH] + (2k_{19} + k_{20})[HO_2]^2 \qquad (8.46)$$

Low-NO$_x$ Regime

In a low-NO$_x$ regime, the formation of peroxides governs the termination reactions. Therefore, the first term of PHO_x (Equation 8.46) can be neglected and we have:

$$[HO_2] \approx \left(\frac{PHO_x}{(2k_{19} + k_{20})}\right)^{1/2} \qquad (8.47)$$

Then, we can write the formation of O_3 (Equation 8.44) simply as a function of HO_x production and the concentration of NO:

$$PO_3 \approx 2k_7[NO][HO_2] \approx 2k_7[NO]\left(\frac{PHO_x}{2k_{19} + k_{20}}\right)^{1/2} \qquad (8.48)$$

Thus, in a low-NO$_x$ regime, O_3 formation is proportional to the concentration of NO and to the square root of the production of HO_x radicals.

High-NO$_x$ Regime

In a high-NO$_x$ regime, nitric acid formation governs the termination reactions and the second term of $PHOx$ (Equation 8.46) can be neglected. The OH concentration may then be calculated as follows:

$$[OH] \approx \frac{PHO_x}{k_{18} [NO_2]} \tag{8.49}$$

The production of O_3 can be expressed simply as a function of the HO_x production and the concentrations of RH and NO_2:

$$PO_3 \approx 2 \, k_9 \, [RH] \, [OH] \approx 2 \, PHO_x \, \frac{k_9 \, [RH]}{k_{18} \, [NO_2]} \tag{8.50}$$

This result implies that the rate of propagation of the radicals is much greater than that of their termination (assumption made to obtain Equations 8.42 and 8.43). However, this assumption is not always verified. For example, the ratio of these rates is only about 2 in New York (Mao et al., 2010) and could be close to 1 in Paris (based on reactivity and ambient concentrations measurements; Borbon et al., 2013; Airparif, 2015). Then, the steady-state relationship for [OH] leads to the more general relationship:

$$k_7 \, [NO] \, [HO_2] = k_9 \, [RH] \, [OH] + k_{18} \, [NO_2] \, [OH] - POH \tag{8.51}$$

The steady-state assumption made for $[HO_2]$ leads to the following relationship (neglecting the termination reactions of the peroxyl radicals, which is appropriate in a high-NO_x regime):

$$k_{10} \, [NO] \, [RO_2] \approx k_7 \, [NO] \, [HO_2] - PHO_2 \tag{8.52}$$

Therefore, substituting these two terms in the equation for the production of O_3 (Equation 8.38), leads to the following solution:

$$PO_3 \approx 2 \, (k_9 \, [RH] \, [OH] + k_{18} \, [NO_2] \, [OH] - POH) - PHO_2 \tag{8.53}$$

Substituting [OH] by its value as a function of PHO_x:

$$PO_3 \approx 2 \, (k_9 \, [RH] + k_{18} \, [NO_2]) \, \frac{PHO_x}{k_{18} \, [NO_2]} - 2 \, POH - PHO_2 \tag{8.54}$$

Then, writing POH as a function of PHO_x and PHO_2, the general equation for the formation of O_3 in a high-NO_x regime is obtained:

$$PO_3 \approx 2 \, PHO_x \left(\frac{k_9 \, [RH]}{k_{18} \, [NO_2]} \right) + PHO_2 \tag{8.55}$$

The previous equation for PO_3 in the high-NO_x regime is obtained when $k_{18} \, [NO_2] \ll k_9 \, [RH]$. In a high-$NO_x$ regime, O_3 formation is proportional to the production of HO_x radicals and to the $[RH]/[NO_2]$ ratio. If the regime is very high NO_x, then, the O_3 formation rate tends toward the production rate of HO_2 radicals, which corresponds mostly to the photolysis rate of aldehydes.

Nitrogen oxides have a relatively short lifetime (a few hours for NO_2 by reaction with OH); therefore, in the absence of important NO_x emissions, the atmospheric concentrations of NO and NO_2 are low (on the order of a few ppb).

VOC have variable lifetimes ranging from a few hours (e.g., aldehydes) to a few days (e.g., benzene). In addition, the oxidation of a VOC with several carbon atoms will occur over several oxidation steps and will, therefore, last for some time. Also, VOC emissions are both anthropogenic (urban areas, industrial sites ...) and biogenic (rural areas, forests). Therefore, although VOC concentrations can be high near major sources (e.g., in urban areas because of on-road traffic), they are rarely negligible given the diverse lifetimes of VOC and their oxidation products as well as the spatial distribution of their emissions. Given these general patterns on the atmospheric concentrations of NO_x and VOC, one may assume that the atmosphere is generally high-NO_x in urban areas, near major traffic routes (shipping, on-road, and air traffic), and near some industrial sites and is generally low-NO_x in rural areas. However, some urban areas and industrial sites may have important VOC concentrations, for example, if the surrounding biogenic emissions are important. Thus, we may have specific cases with high VOC regimes (relatively to NO_x) in some urban areas and industrial sites.

8.5.2 Strategies to Reduce Ozone Precursor Emissions

It is important to identify the chemical regime of ozone formation, because it affects the efficiency of various strategies available for reducing the emissions of ozone precursors. Let us consider the two cases described in Section 8.5.1: low-NO_x regime and high-NO_x regime.

Low-NO_x Regime

NO_x emissions consist mostly of NO and include little NO_2 (<10 % except for some diesel vehicles with catalytic particle filters, see Chapter 2). If one reduces NO_x emissions (i.e., mostly NO concentrations), one reduces the formation of NO_2 by oxidation of NO by peroxyl radicals and, therefore, O_3 formation by photolysis of NO_2. Indeed, Equation 8.48 shows that O_3 formation is proportional to the NO concentration. In a low-NO_x regime, a reduction of the NO_x emissions is efficient to reduce O_3 concentrations.

If one reduces VOC emissions (i.e., VOC concentrations), one reduces the production of peroxyl radicals and, therefore, O_3 formation. However, although the production of peroxyl radicals decreases, their concentration does not decrease proportionately because they are consumed via two different reactions pathways: (1) their reaction with NO (which is proportional to their concentration) and (2) their self-reaction, which leads to the formation of peroxides (see Figure 8.10) and is proportional to the square of their concentration. The kinetics of the second consumption pathway decreases, therefore, faster than their production rate and, as a result, their concentration decreases proportionately less than the VOC concentration. The O_3 formation rate given by Equation 8.48 depends on VOC emissions only via the production of HO_x radicals (PHO_x), i.e., aldehyde photolysis, which implies that VOC emissions have little effect on ozone production. In addition, alkenes react with O_3 (see Section 8.3.6) and if their concentrations are important, a decrease of their concentration may lead to less O_3 consumption and, therefore, a small increase in O_3 concentrations. A VOC reduction strategy in a high-VOC regime (i.e., low-NOx regime) is not very efficient to reduce O_3 concentrations.

High-NO$_x$ Regime

If NO$_x$ concentrations are high compared to VOC concentrations, the production of peroxyl radicals by VOC oxidation will be relatively low and will, therefore, have only a small effect on the Leighton photostationary state. Therefore, if NO$_x$ emissions are reduced (i.e., mostly NO concentrations), the consumption of O$_3$ by NO will be reduced (since the NO/NO$_2$/O$_3$ cycle is not influenced much by peroxyl radicals in that case) and there will be an increase in the O$_3$ concentration. In addition, in a high-NO$_x$ regime, the termination of the O$_3$ formation cycle occurs mostly by reaction of OH radicals with NO$_2$ (since the NO$_2$ concentration is high compared to the VOC concentration, see Figure 8.10). A decrease in NO$_x$ concentrations (i.e., a decrease in the NO$_2$ concentration), implies that there will be a greater availability of OH radicals to react with VOC and produce peroxyl radicals, thereby favoring the oxidation of NO into NO$_2$ and O$_3$ production. This analysis is consistent with the inverse relationship between the O$_3$ formation rate (PO_3) and the NO$_2$ concentration of Equation 8.55. In a high-NO$_x$ regime, a reduction of NO$_x$ emissions has an antagonistic effect leading to an increase of the O$_3$ concentrations.

On the other hand, a reduction of VOC emissions (i.e., a reduction of VOC concentrations), leads to a reduction of peroxyl radicals and, therefore, a reduction in O$_3$ formation. This is consistent with the proportional relationship between the O$_3$ formation rate (PO_3) and the VOC concentration of Equation 8.55. Therefore, an efficient strategy to reduce O$_3$ concentrations in a high-NO$_x$ regime (i.e., a low-VOC regime) is to reduce VOC emissions.

Design of Emission Control Strategies

In summary, an efficient strategy to reduce ozone concentrations must focus on the reduction of the precursor (NO$_x$ or VOC) that has the lower atmospheric concentration and is, therefore, limiting for O$_3$ formation. A reduction of the emissions of the precursor that has the higher atmospheric concentrations may not be efficient and may even be counterproductive because it may lead to an increase of the O$_3$ concentration. Figure 8.10 depicts these two regimes with the main termination reactions: peroxide formation in a low-NO$_x$ regime and nitric acid formation in a high-NO$_x$ regime. These mechanisms imply that it is possible to determine experimentally whether the O$_3$ concentrations measured in the atmosphere have been formed under a low-NO$_x$ or high-NO$_x$ regime (Sillman, 1995). For example, one only needs to measure the HNO$_3$ and H$_2$O$_2$ concentrations. If the [HNO$_3$]/[H$_2$O$_2$] ratio is high, then O$_3$ has been formed under a high-NO$_x$ regime. If this ratio is low, O$_3$ has been formed under a low-NO$_x$ regime. It is then possible to determine which emission reduction strategy will be the most efficient to reduce O$_3$ concentrations.

EKMA (« Empirical Kinetic Modeling Approach ») diagrams have been developed in the U.S. to provide simple information to help develop efficient emission reduction strategies to reduce O$_3$ concentrations. They are developed based on simulations of the atmospheric chemistry of O$_3$. These simulations may use simple or advanced atmospheric chemistry

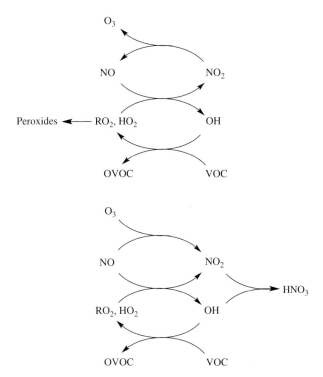

Figure 8.10. Schematic representations of ozone chemistry in a low-NO_x regime (top figure) and a high-NO_x regime (bottom figure). VOC: volatile organic compound, OVOC: oxidized volatile organic compound. NO, NO_2, and O_3 are involved in the Leighton photostationary-state reactions, and the directions of the arrows indicate the main effect of the change in NO_x emissions on the O_3 concentration for each regime.

models. Initially, they were based on trajectory model simulations, i.e., a box-model that follows a mean wind trajectory. An ensemble of simulations was conducted with different VOC and NO_x emission levels, to construct a diagram such as the one presented in Figure 8.11. This diagram highlights the two chemical regimes: high-NO_x in the top left part of the figure and low-NO_x in the bottom right part. In the high-NO_x case, a reduction of NO_x emissions leads toward isopleths corresponding to higher O_3 concentrations and a reduction of VOC emissions is required to reduce O_3 concentrations. In the low-NO_x case, a reduction of VOC emissions is not very efficient, because the change in O_3 concentrations is nearly parallel to the O_3 isopleths. On the other hand, a reduction of NO_x emissions leads toward isopleths corresponding to lower O_3 concentrations. A [VOC]/[NO_x] ratio, expressed as (ppb C)/(ppb NO_x), of about 8 is typically considered to be the limit between the high-NO_x and low-NO_x regimes.

Figure 8.12 shows this type of information obtained for the Paris region with a more comprehensive model than those typically used for EKMA diagrams. The Polyphemus numerical model used here is a three-dimensional model that includes all relevant atmospheric emissions, transport, dispersion, transformation, and deposition processes.

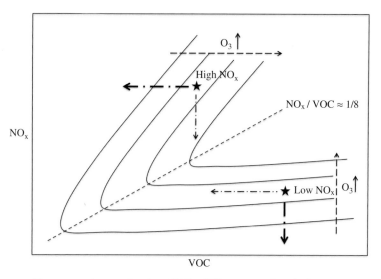

Figure 8.11. Isopleth diagram of O_3 concentrations as a function of VOC and NO_x emissions. It is also known as an EKMA diagram, where EKMA stands for "Empirical kinetic modeling approach." A reduction of VOC emissions leads to lower O_3 concentrations in a high-NO_x regime (i.e., low-VOC regime), whereas a NO_x emission reduction is more efficient in a low-NO_x regime. Reducing NO_x emissions in a high-NO_x regime leads to an increase in O_3 concentrations.

The limit between the high-NO_x and low-NO_x regimes is located beyond the downtown Paris area. A simulation conducted with a 15 % reduction in NO_x emissions shows an increase in O_3 concentrations in the Paris urban area, which corresponds to the high-NO_x regime. On the other hand, a simulation conducted with a 15 % reduction in VOC emissions leads to a significant reduction of the O_3 concentrations within the Paris urban and suburban areas, but negligible reductions in the surrounding rural areas where the low-NO_x regime does not favor VOC emission reductions.

8.5.3 Ozone Formation Potentials of VOC

VOC differ significantly in terms of their chemical reactivity and, therefore, in terms of their potential to form ozone. The lifetimes presented in Table 8.1 provide useful information concerning the oxidation kinetics of selected VOC. Such information is relevant to their potential to form ozone in the presence of sunlight and NO_x. However, the products of these oxidation reactions also affect ozone formation and the kinetics of the first oxidation step (called kinetic reactivity) is not sufficient to estimate the ozone formation potential of a VOC. The number of ozone molecules formed per VOC molecule that has been oxidized gives some information on the potential of that VOC to form O_3. However, this measure of the ozone formation potential (called mechanistic reactivity) does not account for the time needed to form O_3, which is important in the atmosphere where transport and deposition processes must also be taken into account. Note that this mechanistic reactivity differs from the theoretical ozone yields mentioned in Section 8.3 for some VOC (for example, two O_3 molecules per molecule of formaldehyde being oxidized or photolyzed), because the

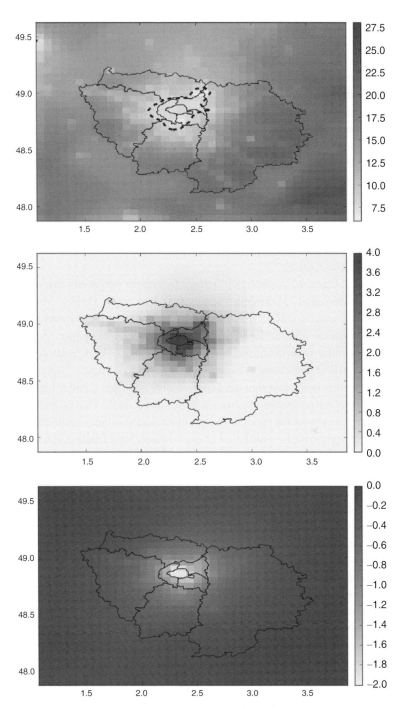

Figure 8.12. Influence of NO_x and VOC emissions on O_3 in the Paris region. Top figure: VOC/NO_x atmospheric concentration ratio for the Paris region (the dotted line corresponds to [VOC]/[NO_x] = 8 mole C/mole NO_x). Changes in O_3 concentrations in µg m^{-3} (average from May to September 2005 of the maximum daily 8-h average O_3 concentrations) due to emission reductions over the Paris region: 15 % NO_x reduction (middle figure) and 15 % VOC reduction (bottom figure). The VOC/NO_x ratio and changes in O_3 concentrations are indicated on the right vertical axis. Latitude (°N) and longitude (°E) are indicated on the left vertical axis and horizontal axis, respectively. All simulations were performed with the Polyphemus numerical model. Source: Cerea (2011).

chemistry of a VOC depends on other chemical species present in the atmosphere, which in turn are influenced by the VOC being studied.

Therefore, one should see these two approaches (mechanistic reactivity and kinetic reactivity) as complementary, one providing valuable information on the amount of O_3 that will eventually be formed and the other indicating the time needed to form O_3. Thus, it is appropriate to use a metric of the ozone formation potential that combines these two reactivity concepts. To that end, one may estimate the amount of O_3 formed per molecule (or mass) of VOC added to a mixture of VOC and NO_x. This metric is called the incremental reactivity, because one measures the VOC reactivity in terms of its increment of ozone formation. There is no unique definition for this type of ozone formation potential, because (1) it depends on the conditions under which it has been estimated (e.g., mixture of VOC being used, relative amounts of VOC and NO_x, and duration of the simulation or experiment) and (2) the amount of O_3 formed can be defined in several ways (maximum concentration, concentration averaged over a certain period, etc.).

Three main metrics of ozone formation potentials that are commonly used may be mentioned: Maximum Incremental Reactivity (MIR), Maximum Ozone Incremental Reactivity (MOIR), and Equal Benefits Incremental Reactivity (EBIR). These different metrics give different values for the ozone formation potential, but they are rather similar in terms of the reactivity of a specific VOC relative to those of the other VOC. MIR is used here to illustrate this concept of ozone formation potential. "Maximum" refers to the fact that the experiment is conducted under high-NO_x conditions for which the formation of O_3 from a VOC is optimal. The estimation of the MIR of a VOC depends on the mixture composition, the sunlight conditions (photolysis rates), and the duration of the simulation or experiment. Nevertheless, once those conditions have been set, it is possible to estimate in a consistent manner the MIR of a large number of VOC under conditions that approximate rather well those observed during air pollution episodes. The estimation of MIR can be conducted either experimentally (in smog chambers) or via numerical simulations using a chemical kinetic mechanism, which has been previously evaluated satisfactorily against smog chamber data (see Section 8.6).

Figure 8.13 shows some results obtained with MIR estimated using simulations conducted with the SAPRC-07 mechanism (Carter, 2010). A simulation time of a few hours and a high-NO_x mixture ([VOC]/[NO_x] = 3.7) including VOC representative of an urban area were used for these simulations. There is good agreement between the MIR scale and the rate constants of the OH reaction for some species (for example, for the biogenic compounds). However, there are significant differences for most of the other species. For example, the alkane MIR values are similar, whereas their OH reaction kinetics are nearly proportional to their number of carbon atoms (see Figure 8.2). Furthermore, toluene and xylene have MIR values similar to those of the biogenic compounds, despite having much lower OH reaction rate constants. These results highlight the importance of mechanistic reactivity as a complement to kinetic reactivity. According to the results presented in Figure 8.13, the most reactive VOC in terms of O_3 formation are the alkenes (including biogenic compounds), the aromatic compounds (except for benzene), and the aldehydes.

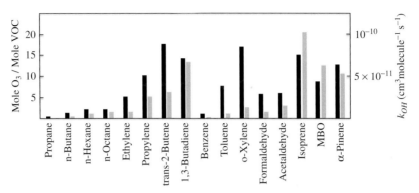

Figure 8.13. Maximum incremental reactivity (MIR) and oxidation kinetics of selected VOC. MIR, here in number of moles of O_3 formed per mole of VOC added to the initial mixture, obtained by numerical simulation using the SAPRC-07 chemical kinetic mechanism (in black), left scale, and rate constant of the OH reaction (in gray), right scale. Sources: Carter (2010) for MIR and references from Table 8.1 for k_{OH}. See text for the interpretation of these values.

We will see in Chapter 9 that one obtains different results when VOC reactivity is considered in terms of secondary organic particulate matter formation.

8.6 Numerical Modeling of the Gas-phase Chemistry of Photochemical Air Pollution

The simulation of ozone formation is performed with a chemical kinetic mechanism that represents the major chemical reactions involved. The number of inorganic reactions among different nitrogenous species (NO_y), CO, and oxidants (O_3, OH, HO_2 ...) is limited and it is, therefore, possible to include all these reactions in an exhaustive manner. On the other hand, that is not the case for VOC, because (1) they are emitted in large numbers as primary species and (2) they lead to even larger numbers of secondary species via successive oxidation steps. Chemical kinetic mechanisms that represent explicitly the chemical reactions of very large numbers of VOC have been developed, such as MCM (Saunders et al., 2003; Bloss et al., 2005) and GECKO-A (Aumont et al., 2005). However, the use of such mechanisms, which include thousands of chemical species and reactions, in operational air pollution models (which must include, in addition to chemical reactions, emissions, transport, dispersion, solar radiation, and deposition) has been limited because of the overwhelming computational times involved. It is, therefore, necessary to reduce the size of the organic chemical mechanism in terms of numbers of species and reactions. To that end, two major approaches have been developed and applied.

- Use of surrogate organic molecules
- Decomposition of the organic molecules into functional groups

The first approach consists of assuming that organic molecules that have similar formulas and properties will behave chemically similarly. For example, alkanes containing a few carbon atoms will have similar kinetics and similar oxidation products. Of course, the oxidation kinetics of propane, butane, and pentane are different. Nevertheless, one may define a surrogate molecule that is representative of these three species and has a kinetics obtained by weighing the kinetics of these three species (the weighing can be performed using the ambient concentrations or the emission rates of these three species for the area being studied). The two mechanisms of this kind that are the most widely used are RACM (Regional Atmospheric Chemistry Mechanism) and SAPRC (Statewide Air Pollution Research Center). The most recent versions are RACM 2 (Goliff et al., 2013) and SAPRC 07 (Carter, 2010). The advantage of this approach is that it represents the evolution of chemical molecules well. The disadvantage is that it does not conserve mass, since molecules with different numbers of carbon atoms are lumped into a single surrogate molecule. In addition, the rate constants of the reactions of this surrogate molecule are defined by weighing relative amounts of the molecules being represented by this surrogate molecule. Since these relative amounts depend on the local emission inventory, they are likely to vary among different areas, as well as in time (for example, a strategy for VOC emission reductions may not affect all VOC in the same way).

The second approach consists of breaking a molecule into its functional groups and making the assumption that its chemical behavior is equivalent to the chemical behaviors of its functional groups. For example, an alkene with four carbon atoms will be broken down into a group of two carbon atoms with a double bond (which will follow alkene chemistry) and two groups containing each one carbon atom with single bonds (which will follow alkane chemistry). These mechanisms are called "carbon-bond" mechanisms. Some recent versions include CB05 (Yarwood et al., 2005) and CB05-TU (with a modified toluene chemistry; Whitten et al., 2010). The advantage of this approach is that it conserves carbon mass, since carbon mass is defined exactly in the various functional groups. In addition, it is independent of the emission inventory and a carbon-bond mechanism is applicable to any region or period as is. The disadvantage is that breaking down molecules into functional groups loses the molecular structure and, therefore, the evolution of the VOC molecule cannot be followed clearly over several oxidation steps. In particular, the formation of semi-volatile organic compounds (SVOC), which is important for the formation of secondary organic aerosols (SOA, see Chapter 9), cannot be represented by a carbon-bond mechanism, because the number of carbon atoms of the initial molecules is lost during the decomposition into functional groups. Then, some chemical reactions specific to SOA formation must be added to a carbon-bond mechanism.

These chemical kinetic mechanisms are developed based on available mechanistic and kinetic data. In addition, they are evaluated with experimental data obtained in smog chambers and, if needed, they are modified to improve the agreement with the experimental data. The two approaches presented include simplifying assumptions and offer advantages and shortcomings (Dodge, 2000). Comparisons of different chemical kinetic mechanisms that are commonly used have been conducted. For O_3 formation, these two distinct approaches lead to similar results, with differences that are small when compared to other sources of uncertainties (e.g., Kim et al., 2009). The set of ordinary differential equations that represents the chemical kinetic mechanism is numerically stiff, and they must be solved using an appropriate numerical algorithm (see Chapter 7).

The simulation of ozone formation requires to take into account emissions, transport by the mean wind, dispersion, and deposition of air pollutants, in addition to their chemistry. Such simulations are conducted using three-dimensional (3D) chemical-transport models based on the atmospheric diffusion equation (see Chapter 6). The first 3D chemical-transport model for photochemical smog was developed for the Los Angeles Basin in California by Reynolds et al. (1973). Some of the current chemical-transport models are mentioned in Chapter 15.

Problems

Problem 8.1 Chemical kinetics
a. The rate constant of the oxidation of heptane (alkane with seven carbon atoms) by OH radicals is 6.5×10^{-12} cm^3 molecule^{-1} s^{-1} (or cm^3 radical^{-1} s^{-1}). If $[OH] = 2 \times 10^6$ radicals cm^{-3}, what is the half-life of heptane?
b. What is the lifetime (also called residence time) of heptane for this same reaction?
c. Will the half-life of cetane (an alkane with 16 carbon atoms) be longer, shorter or similar? Explain why.

Problem 8.2 Reduction of tropospheric ozone of photochemical smog
a. Measurements conducted in an urban area give atmospheric concentrations of volatile organic compounds (VOC) of 120 ppb C and concentrations of nitrogen oxides (NO$_x$) of 30 ppb. The emission ratio of VOC and NO$_x$ is (in moles of C and moles of NO$_x$) VOC/NO$_x$ = 3. One assumes that the limit between a low-NOx regime and a high-NO$_x$ regime is VOC/NO$_x$ = 8. Which precursor category (NO$_x$ or VOC) should be reduced to decrease ozone concentrations in an optimal manner?
b. Why is the VOC/NO$_x$ ratio of the emissions different from that of the concentrations?
c. In a different urban area, measurements of hydrogen peroxide (H$_2$O$_2$) and nitric acid (HNO$_3$) are available. There is more of the former (H$_2$O$_2$) than of the latter (HNO$_3$), i.e., [H$_2$O$_2$]/[HNO$_3$] $\gg$ 1. To decrease the ozone concentration, which precursor emissions should be reduced?

Problem 8.3 Nitrogen dioxide concentrations
Concentrations of nitrogen dioxide (NO$_2$) near a roadway have remained nearly constant in the Paris region from 2009 to 2015, while the emissions of nitrogen (NO$_x$) have decreased significantly during the same period. Explain the reason(s) for this phenomenon.

Problem 8.4 Chemical mechanism
In a CB05 carbon-bond mechanism, two carbon atoms with a double bond (C=C) are represented by "OLE," one carbon atom with single bond (–CH$_2$- or –CH$_3$) is represented

by "PAR," and an aldehyde group (-CHO) is represented by 'FORM." Write the CB05 formulation for 1-hexene (H_2 C=CH-CH_2-CH_2-CH_2-CH_3).

References

Ademe, 2014. *Évaluation des impacts sur la qualité de l'air d'action et de mesures orientées « villes et territoires durables »*, Final report, Contract 1162c0014, Project performed by CEREA and LISA for *Agence de l'Environnement et de la Maitrise de l'Énergie* (Ademe), Paris.

Aiparif, 2015. *Rapports d'activités et bilan de la qualité de l'air 2014*, Airparif, Paris.

Aumont, B., S. Szopa, and S. Madronich, 2005. Modelling the evolution of organic carbon during its gas-phase tropospheric oxidation: Development of an explicit model based on a self generating approach, *Atmos. Chem. Phys.*, **5**, 2497–2517.

Bloss, C., V. Wagner, A. Bonzazini, M.E. Jenkin, K. Wirtz, M. Martin-Reviejo, and M.J. Pilling, 2005. Evaluation of detailed aromatic mechanisms (MCMv3 and MCMv3.1) against environmental chamber data, *Atmos. Chem. Phys.*, **5**, 623–639.

Borbon, A., J.B. Gilman, W.C. Kuster, N. Grand, S. Chevaillier, A. Colomb, C. Dolgorouky, V. Gros, M. Lopez, R. Sarda-Esteve, J. Holloway, J. Stutz, H. Petetin, S. McKeen, M. Beekmann, C. Warneke, D.D. Parrish, and J.A. de Gouw, 2013. Emission ratios of anthropogenic volatile organic compounds in northern mid-latitude megacities: Observations versus emission inventories in Los Angeles and Paris, *J. Geophys. Res.*, **118**, 2041–2057.

Calvert, J.G., R. Atkinson, J.A. Kerr, S. Madronich, G.K. Moortgat, T.J. Wallington, and G. Yarwood, 2000. *The Mechanisms of Atmospheric Oxidation of the Alkenes*, Oxford University Press, Oxford, UK.

Calvert, J.G., R. Atkinson, K.H. Becker, R.M. Kamens, J.H. Seinfeld, T.J. Wallington, and G. Yarwood, 2002. *The Mechanisms of Atmospheric Oxidation of Aromatic Hydrocarbons*, Oxford University Press, Oxford, UK.

Calvert, J.G., R.G. Derwent, J.J. Orlando, G.S. Tyndall, and T.J. Wallington, 2008. *The Mechanisms of Atmospheric Oxidation of the Alkanes*, Oxford University Press, Oxford, UK.

Calvert, J.G., A. Mellouki, J.J. Orlando, M.J. Pilling, and T.J. Wallington, 2011. *The Mechanisms of Atmospheric Oxidation of the Oxygenates*, Oxford University Press, Oxford, UK.

Carter, W.P.L., 2010. *Development of the SAPRC-07 Chemical Mechanism and Updated Ozone Reactivity Scales*, Final report, Contracts N° 03–318, 06–408, and 07–730, California Air Resources Board, Sacramento, CA; available at: www.engr.ucr.edu/~carter/SAPRC/saprc07.

Cerea, 2011. *Évaluation de l'influence des émissions anthropiques de polluants en Europe sur les niveaux d'ozone en Île-de-France*, Engineer Laboratory Internship Report by E. Foessel, Y. Roustan, advisor, *Centre d'Enseignement et de Recherche en Environnement Atmosphérique*, École des Ponts ParisTech, Marne-la-Vallée, France.

Dodge, M.C., 2000. Chemical oxidant mechanisms for air quality modeling: Critical review, *Atmos. Environ.*, **34**, 2103–2130.

Finlayson-Pitts, B.J. and J.N. Pitts, Jr., 2000. *Chemistry of the Upper and Lower Atmosphere: Theory, Experiments, and Applications*, Academic Press, New York.

Goliff, W.S., W.R. Stockwell, and C.W. Lawson, 2013. The regional atmospheric chemistry mechanism, version 2, *Atmos. Environ.*, **68**, 174–185.

Jacob, D.J., 1999. *Introduction to Atmospheric Chemistry*, Princeton University Press, Princeton, NJ.

Jacobson, M.Z., 2005. *Fundamentals of Atmospheric Modeling*, Cambridge University Press, Cambridge, UK.

Kim, Y., K. Sartelet, and C. Seigneur, 2009. Comparison of two gas-phase chemical kinetic mechanisms of ozone formation over Europe, *J. Atmos. Chem.*, **62**, 89–119.

Mao, J., X. Ren, S. Chen, W.H. Brune, Z. Chen, M. Martinez, H. Harder, B. Lefer, B. Rappenglück, J. Flynn, and M. Leuchner, 2010. Atmospheric oxidation capacity in the summer of Houston 2006: Comparison with summer measurements in other metropolitan studies, *Atmos. Environ.*, **44**, 4107–4115.

Mollner, A.K., S. Valluvadasan, L. Feng, M. K. Sprague, M. Okumura, D. B. Milligan, W. J. Bloss, S. P. Sander, P. T. Martien, R. A. Harley, A. B. McCoy, and W. P. L. Carter, 2010. Rate of gas phase association of hydroxyl radical and nitrogen dioxide, *Science*, **330**, 646–649.

Reynolds, S.D., P.M. Roth, and J.H. Seinfeld, 1973. Mathematical modeling of air pollution – 1: Formulation of the model, *Atmos. Environ.*, **7**, 1033–1061.

Roustan, Y., M. Pausader, and C. Seigneur, 2011. Estimating the effect of on-road vehicle emission controls on future air quality in Paris, France, *Atmos. Environ.*, **45**, 6828–6836.

Saunders, S.M., M.E. Jenkin, R.G. Derwent, and M.J. Pilling, 2003. Protocol for the development of the Master Chemical Mechanism MCM v3 (Part A): Tropospheric degradation of non-aromatic volatile organic compounds, *Atmos. Chem. Phys.*, **3**, 161–180.

Sillman, S., 1995. The use of NO_y, H_2O_2, and HNO_3 as indicators for ozone-NO_x-hydrocarbon sensitivity in urban locations, *J. Geophys. Res*, **100**, 14175–14188.

Tesche, T.W., C. Seigneur, W.R. Oliver, and J.L. Haney, 1984. Modeling ozone control strategies in Los Angeles, *J. Environ. Eng.*, **110**, 208–225.

Whitten, G.Z., G. Heo, Y. Kimura, E., McDonald-Bullet, D.T. Allen, W.P.L. Carter, and G. Yarwood, 2010. A new condensed toluene mechanism for Carbon Bond CB05-TU, *Atmos. Environ.*, **44**, 5346–5355.

Yarwood, G., S. Rao, M. Yocke, and G.Z. Whitten, 2005. *Updates to the carbon bond mechanism: CB05. Report to the U.S. Environmental Protection Agency*, Research Triangle Park, NC (www.camx.com/publ/pdfs/CB05_Final_Report_120805.pdf).

9 Atmospheric Particles

Atmospheric particles and, in particular, fine particles are one of the major components of air pollution. They lead to significant adverse health effects, degrade atmospheric visibility, are involved in cloud formation and precipitation, and play a role in climate change. Particles have various sizes, ranging from ultrafine and fine to coarse, and different chemical compositions, since they may contain a large number of different inorganic and organic species. In addition, particles typically include a primary fraction, which has been emitted from various sources directly into the atmosphere, and a secondary fraction, which has been formed in the atmosphere via chemical reactions from precursor gases. The secondary fraction generally dominates the mass of fine particles. Therefore, the development of efficient emission control strategies to decrease the ambient concentrations of atmospheric particles is a challenging task, because it requires identifying the numerous sources of atmospheric particles, including those of the gaseous precursors of the secondary fraction, in order to properly characterize the processes that govern particulate matter (PM) formation and understand the complex relationships that link gaseous precursors and the secondary PM fraction. This chapter describes the processes that lead to the formation of atmospheric PM and, in particular, those that govern the particle size distribution and chemical composition. Chapter 10 addresses the aqueous chemical reactions taking place in clouds and fogs and the formation of particulate matter following the evaporation of cloud and fog droplets. Examples based on actual case studies are presented to illustrate the complex relationships between emissions and PM concentrations and the difficulties associated with the development of optimal emission control strategies to reduce ambient levels of PM. Finally, the main approaches used to simulate atmospheric PM are summarized.

9.1 General Considerations on Atmospheric Particles

By definition, an ensemble of atmospheric particles in suspension in a gas is called aerosol. Therefore, an aerosol includes the condensed phase and its surrounding gas phase. However, the term "aerosol" is also used occasionally to represent only the particle in suspension in a gas (i.e., without the surrounding gas phase).

Particles display a large variety in terms of sizes. There are several possible definitions of particle size and the two main ones used for atmospheric particles are defined here. The Stokes diameter of a particle is the diameter of a spherical particle that has the same density and sedimentation velocity as the particle of interest. The aerodynamic diameter of a particle is the diameter of a spherical particle with a density of 1 g cm^{-3} that has the same sedimentation velocity as the particle of interest. The aerodynamic diameter, $d_{p,a}$, is related to the Stokes diameter, $d_{p,St}$, via the particle density, ρ_p:

$$d_{p,a} = \sqrt{\rho_p} \, d_{p,St} \tag{9.1}$$

Particle diameters may range from a few nanometers (nm), in the case of a new particle formed by nucleation of gaseous molecules, to several micrometers (μm) in the case of some primary particles, such as those emitted from industrial sources, sea salt particles, soil particles, etc. By definition, a nanoparticle is a particle that has one dimension that is between a few nanometers and 100 nm. An ultrafine particle has a representative diameter that is less than 100 nm. Therefore, a nanotube with a length much greater than 100 nm is not an ultrafine particle, but an ultrafine particle is a nanoparticle. A fine particle is a particle with an aerodynamic diameter equal to or less than 2.5 μm. Fine particles are typically referred to as $PM_{2.5}$. Therefore, $PM_{2.5}$ includes fine and ultrafine particles, although in terms of mass, that of ultrafine particles is negligible compared to that of fine particles (see Section 9.2.1). PM_{10} corresponds to particles that have an aerodynamic diameter that is equal to or less than 10 μm. PM_{10} particles are sometimes referred to as inhalable particles. Coarse particles have an aerodynamic diameter greater than 2.5 μm. Therefore, PM_{10} includes ultrafine particles, fine particles, and a fraction of coarse particles.

Chemical species in atmospheric particles include inorganic and organic species. They can be primary (emitted directly in the atmosphere as particles) or secondary (formed in the atmosphere). They may originate from anthropogenic sources (traffic, industry, biomass burning, etc.) or natural sources (volcanic eruptions, vegetation, oceans, soils, wildfires, etc.).

Soot refers to carbonaceous particulate matter emitted from combustion processes (for example, particles emitted from diesel engines or biomass burning). It contains strongly light-absorbing carbon and organic compounds. The strongly light-absorbing carbon gives soot its black color (Bond and Bergstrom, 2006). It can be measured by combining a thermal method (increasing temperature using several stages to eliminate organic carbon from the sample; seven stages are used in the NIOSH and IMPROVE methods) and an optical method (measuring transmittance or reflectance) (Chow et al., 2001). It can also be measured continuously with an optical method (aethalometer), which may use several wavelengths to differentiate between different sources of light-absorbing carbon (e.g., diesel engines and biomass burning) (Petzold et al., 2005). These different methods may lead to different results and the measurement method must, therefore, be mentioned when presenting the results. Typically, strongly light-absorbing carbon is called either elemental carbon (EC) or black carbon (BC). Here, EC is used to refer to concentrations obtained with thermal-optical methods and BC is used to refer to concentrations obtained with optical methods (i.e., aethalometers). Soot is used here to refer to primary PM from a combustion process, which includes a combination of strongly light-absorbing carbon (EC or BC) and the associated organic compounds.

Secondary inorganic compounds that are present in atmospheric PM are mostly sulfate, nitrate, and ammonium. Primary inorganic compounds include a large number of chemical species such as metals, sea salt compounds (sodium, chloride, sulfate, etc.), and sulfate emitted from other sources (a fraction of sulfur emissions from combustion processes is sulfate).

There is a large number of secondary organic compounds present in atmospheric PM. This particulate fraction is typically referred to as secondary organic aerosols (SOA). One refers to primary organic aerosols (POA) for the organic fraction of PM that is directly emitted in the atmosphere as particles.

Photochemical air pollution (also called photochemical smog) leads to the formation of gaseous air pollutants such as ozone (O_3, see Chapter 8), as well as the formation of particulate matter. Photochemical smog particles typically consist of a core corresponding to a primary particle on which chemical species have condensed as atmospheric chemical reactions occurred. This condensed fraction is the secondary PM fraction of photochemical air pollution. These particles have diameters that range mostly from about 0.1 to 1 µm, because the physico-chemical processes involved in the formation of those particles favor the accumulation of particles in that size range (see Section 9.2). These are fine particles. They may penetrate deeply into the human respiratory system where they may lead to significant adverse health effects (see Chapter 12). The poor visibility associated with photochemical air pollution episodes is due to those fine particles, because they scatter light and reduce the amount of light transmitted from an observed target toward the eye of the observer (see Chapter 5). In addition, the presence of soot in the core of some of those particles contributes to visibility degradation via light absorption.

9.2 Dynamics of Atmospheric Particles

9.2.1 Size Distribution of Atmospheric Particles

The dynamics of particles covers by definition the processes that affect directly the size distribution of a population of aerosols, i.e., a population of particles in suspension in a gas. Therefore, emission and deposition processes are not included in aerosol dynamics; they are addressed separately in Chapters 2 and 11, respectively.

The size distribution of a population of particles may be represented in several ways. For example, the concentration of particles may be represented in terms of their number, N_p, their surface, S_p, their volume, V_p, or their mass, M_p. These different variables are interrelated via the Stokes diameter and the particle density. For example, for a population of particles with a same size, $d_{p,St}$ (monodispersed population), and density, ρ_p:

$$M_p = V_p \, \rho_p = N_p \, \pi \frac{d_{p,St}^3}{6} \rho_p \tag{9.2}$$

However, atmospheric particles cover a large range of sizes. Therefore, the particle concentration must be represented by a concentration distribution as a function of size. Such distributions are typically presented as a function of the particle volume or diameter, and the particles are assumed to be spherical. It is a reasonable assumption for aged particles, which have undergone condensation of semi-volatile chemical species. It is, however, a poor assumption for freshly emitted primary particles, which may have fractal shapes. For the sake of simplicity, one uses here d_p rather than $d_{p,St}$ to represent the particle diameter. The two particle size distributions that are the most commonly used are the number distribution and the mass distribution. The mass distribution is equivalent to the volume distribution if the particle density is uniform for all particles (then, ρ_p is the factor converting volume to mass). Let $n_p(v_p)$ be the distribution of the number concentration of particles as a function of their volume, v_p, and $n_{p,d}(d_p)$ be the distribution of the number concentration of particles as a function of their diameter d_p. The main equations for those number concentration distributions are as follows:

$$N_p = \int_0^\infty n_p(v_p)\, dv_p = \int_0^\infty n_{p,d}(d_p)\, dd_p; \quad n_{p,d}(d_p) = \frac{\pi}{2} d_p^2 n_p(v_p)$$

$$V_p = \int_0^\infty n_p(v_p)\, v_p\, dv_p = \int_0^\infty n_{p,d}(d_p) \frac{\pi}{6} d_p^3\, dd_p \quad (9.3)$$

$$M_p = \int_0^\infty n_p(v_p)\, v_p\, \rho_p(v_p)\, dv_p = \int_0^\infty n_{p,d}(d_p) \frac{\pi}{6} d_p^3\, \rho_p(d_p)\, dd_p$$

Here, N_p, V_p, and M_p represent the total concentration in terms of number, volume, and mass, respectively, of the polydispersed particle population. Typically, one distinguishes three modes in the particle size distribution (Whitby, 1978):

– A nucleation mode
– An accumulation mode
– A coarse mode

The nucleation mode corresponds to particles that have been formed from gaseous molecules and have later grown via the condensation of other gaseous molecules and coagulation with other nucleated particles. This mode is located within the ultrafine fraction of PM.

The accumulation mode results from the emissions of fine particles and from dynamic processes such as condensation and coagulation. It is called the accumulation mode because these dynamic processes lead to the accumulation of particles in that size range (see Sections 9.2.3 and 9.2.4).

The coarse mode consists mostly of particles emitted via mechanical processes (abrasion, wind erosion, etc.). Condensation and coagulation have little effect on these particles, because their number concentration is low (although their mass concentration can be significant) and particle dynamics favors fine particles, which are typically present in greater numbers and have greater available surface area for condensation and coagulation processes to occur.

Figure 9.1 shows idealized volume and number concentration distributions of a typical urban particle population. Lognormal distributions are used here for the volume and number concentrations (see Section 9.7 for a description of the lognormal

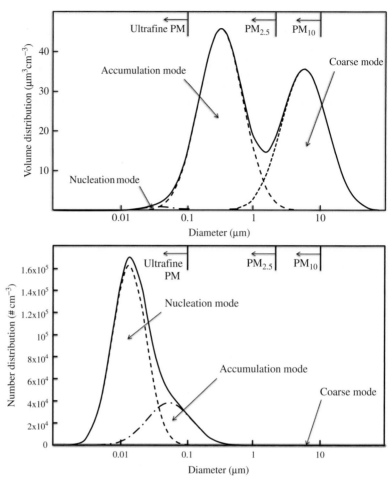

Figure 9.1. Schematic representation of the size distribution of the volume concentration (top) and number concentration (bottom) of atmospheric particles typical of a polluted urban area. Source of the data: Whitby (1978).

representation of particle concentration distributions). It appears that the volume (and, therefore, the mass) of the particles is present mostly in the accumulation and coarse modes and that the number concentration is dominated by the ultrafine particles of the nucleation mode.

One should note that this classification into three modes is too simplistic. For example, several accumulation modes can be found in the atmosphere, because an accumulation mode resulting from the condensation of gaseous pollutants will have a different size distribution than an accumulation mode resulting from the evaporation of cloud and fog droplets where aqueous-phase chemical reactions have taken place (Hering and Friedlander, 1982).

Three processes are considered to govern particle dynamics and affect the size distribution of a population of particles:

- Nucleation
- Condensation (and the reverse process, which is evaporation)
- Coagulation

The mathematical representation of these three processes is called the general dynamic equation, GDE (Friedlander, 2000). The different terms of this equation are presented in detail below.

9.2.2 Nucleation

Nucleation is the formation of a new particle from gaseous molecules. This nucleation process may involve two or three different chemical species. Nucleation involving two species (for example, sulfuric acid and water) is referred to as binary nucleation. Nucleation involving three species (for example, sulfuric acid, ammonia, and water) is referred to as ternary nucleation. Nucleation may occur spontaneously in the gas phase; then, one refers to homogeneous nucleation. In some cases, nucleation may be favored by the interaction of gaseous chemical species with a surface; then, one refers to heterogeneous nucleation. Nucleation rates are difficult to estimate theoretically and uncertainties may be of several orders of magnitude. The most robust algorithms have been developed for the homogeneous binary nucleation of sulfuric acid and water. A detailed review of nucleation algorithms has been conducted by Zhang et al. (2010), who recommended the algorithm of Kuang et al. (2008) for the nucleation rate, J_n (particles cm^{-3} s^{-1}), of sulfuric acid particles:

$$J_n = 1.6 \times 10^{-14} \left(N_{H_2SO_4}\right)^2 \tag{9.4}$$

where N_{H2SO4} is the number of gaseous sulfuric acid molecules (or other molecules containing sulfate, such as ammonium bisulfate) (molec cm^{-3}).

The nucleation process is actually in competition with condensation (described in Section 9.2.3), which transfers gaseous molecules toward existing particles. If the particle concentration is important, then condensation tends to dominate. Therefore, nucleation is mostly observed in the atmosphere when the concentration of gaseous molecules of a chemical species with a low saturation vapor pressure is high and when the concentration of existing particles is low or moderate. Thus, one observes nucleation of new particles near emission sources (e.g., near vehicle exhaust) or under atmospheric conditions where the formation of a chemical species with a low saturation vapor pressure is important. This is the case, for example, in forested areas where the atmosphere is pristine (i.e., with low particle concentrations) and where emissions of biogenic VOC are important. These biogenic VOC can be oxidized rapidly to form compounds with low volatility, which may then nucleate. In addition, episodes of sulfate particle nucleation have been observed under conditions where the oxidation of sulfur dioxide (SO_2) is rapid and important.

Nucleation increases the number of particles, as well as the total mass of particles. However, the particle mass created by nucleation is typically very low compared to the mass of existing fine and coarse particles.

9.2.3 Condensation and Evaporation

Condensation consists of the transfer of gaseous molecules toward an existing particle. This particle can be liquid or solid and the new condensed mass can also be liquid or solid. The reverse process is evaporation; it consists of the transfer of molecules from the particle toward the gas phase. Condensation (or evaporation) occurs when the gas phase and the condensed phase are not in thermodynamic equilibrium and transfer between these phases must occur to establish this thermodynamic equilibrium. The two phases may be out of equilibrium, for example, because of a change of concentration in the gas phase (formation of a semi-volatile or non-volatile species) or because of a change of temperature, pressure or relative humidity.

The evolution of the distribution of the particle number concentration as a function of volume, $n_p(v_p)$, where v_p is the particle volume, is given by the following equation:

$$\frac{\partial n_p(v_p)}{\partial t} = -\frac{\partial I_v n_p(v_p)}{\partial v_p}$$

$$I_v = \frac{dv_p}{dt}$$

(9.5)

where I_v is called the growth law. It represents the rate of growth by condensation of a particle of volume v_p. This growth law corresponds to the flux of molecules from the gas phase to the particle and depends on (1) the mass transfer flux by diffusion of the molecules from the gas phase toward the particle surface and (2) the particle surface area available for condensation of these gaseous molecules. For spherical coarse particles, continuum mechanics implies that the mass transfer flux is proportional to the molecular diffusion coefficient and inversely proportional to a characteristic distance, which is the particle radius. Thus, the growth law is given as the product of three terms, which are the particle surface area, the mass transfer coefficient (diffusion coefficient/particle radius), and the condensing species concentration difference between the bulk gas phase and the particle surface (expressed in terms of molecular volume per unit volume of air):

$$\frac{dv_p}{dt} = \pi d_p^2 \frac{2D_m}{d_p} v_m (C_g - C_{g,e}) = 2\pi d_p D_m v_m (C_g - C_{g,e})$$

(9.6)

where v_p (cm^3) is the particle volume, d_p is the particle diameter (cm), D_m is the diffusion coefficient of the condensing molecules in the air (cm^2 s^{-1}), v_m is the molecular volume of the condensing molecule (cm^3 molec^{-1}), C_g is the gas-phase concentration of the condensing molecules (expressed as number of molecules per unit volume of air; here, molec cm^{-3}), and $C_{g,e}$ is the gas-phase concentration of the condensing molecules in thermodynamic equilibrium at the particle surface (molec cm^{-3}). The molecular volume is obtained by dividing the molecular mass of the condensing species by its density in the condensed phase. Since the particle surface area is proportional to the square of the particle diameter and the mass transfer flux is inversely proportional to the particle diameter, the growth law is proportional to the particle diameter.

For particles that have a diameter significantly less than the mean free path in the air (which is about 70 nm at an atmospheric pressure of 1 atm), particle dynamics can no longer be represented by continuum mechanics, but must be treated as a free molecular flow. As a result, the mass flux of molecules condensing on a particle is governed by the kinetic theory of gases and does not depend on the particle size. Therefore, the particle diameter only appears in the particle surface area available for condensation, which is proportional to the square of the particle diameter:

$$\frac{dv_p}{dt} = \frac{\pi}{4} d_p^2 \left(\frac{8 k_B T}{\pi m_g}\right)^{\frac{1}{2}} v_m (C_g - C_{g,e}) \tag{9.7}$$

where m_g is the mass of the condensing gaseous molecule (g), k_B is the Boltzmann constant (1.381×10^{-23} J K^{-1} or, here, 1.381×10^{-16} erg K^{-1}) and T is the temperature (K). The term in parentheses to the power 1/2 is the mean thermal velocity of the condensing molecule, $\overline{c_t}$.

The Fuchs-Sutugin equation (Fuchs and Sutugin, 1971) is commonly used to calculate the growth law over the full range of particle sizes (there are other similar formulas to represent the growth law):

$$\frac{dv_p}{dt} = \frac{2\pi d_p D_m}{1 + \left(\frac{1.33 \text{Kn} + 0.71}{1 + \text{Kn}}\right) \text{Kn}} v_m \left(C_g - C_{g,e} \exp\left(\frac{4\sigma_p v_m}{d_p k_B T}\right)\right) \tag{9.8}$$

where Kn is the Knudsen number, which is equal to ($2 \lambda_{m,a}/d_p$), $\lambda_{m,a}$ is the mean free path of a molecule in the air (cm), and σ_p is the surface tension of the particle (erg cm^{-2}). The term in the denominator (which is a function of Kn) accounts for the transition between the molecular regime for condensation on ultrafine particles and the continuum regime for condensation on coarse particles. Condensation on fine particles (i.e., d_p between about 0.1 and 2.5 µm) occurs in this intermediate regime. For ultrafine particles, Kn becomes important and the growth law becomes proportional to the square of the particle diameter. For coarse particles, Kn tends toward 0 and the growth law becomes proportional to the particle diameter.

The exponential term represents the Kelvin effect, which results in the possible evaporation of molecules from some ultrafine particles because of their strong curvature. For a sphere, the concentration above its surface will be greater than the thermodynamic equilibrium concentration obtained over a flat surface. Conceptually, this phenomenon is due to the fact that the curvature of the spherical surface implies that a molecule in the particle will be farther statistically from molecules in the gas phase, thereby allowing more space for molecules in the gas phase near the particle surface. This phenomenon depends on the surface tension of the particle. It can be significant: for example, it leads to a 10 % increase in the saturation vapor pressure of water over a 10 nm water droplet. It implies that condensation may not take place on the smallest ultrafine particles and that instead these particles may undergo evaporation of some species present in solution.

Figure 9.2 shows the growth law for the condensation of sulfuric acid, H_2SO_4, immediately neutralized into ammonium sulfate, $(NH_4)_2SO_4$. One notes that the growth law

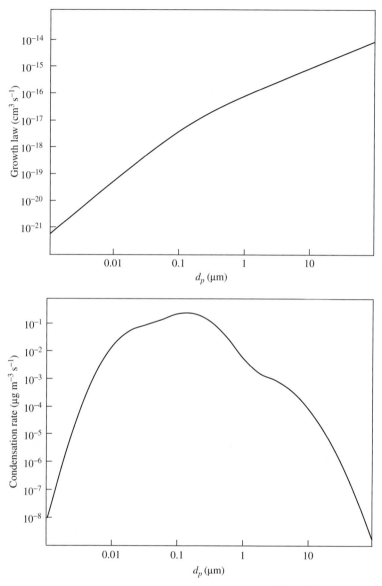

Figure 9.2. Condensation. Top figure: growth law due to gas/particle conversion by condensation of 1 ppb of sulfuric acid (immediately neutralized as ammonium sulfate) as a function of the particle diameter; bottom figure: distribution of the condensation rate of 1 ppb of sulfuric acid (immediately neutralized as ammonium sulfate) as a function of particle size for a particle population typical of a polluted urban area (corresponding here to that of Figure 9.1).

increases with the particle diameter, but that the slope decreases as the diameter increases, since the growth law is proportional to d_p^2 for ultrafine particles and proportional to d_p for coarse particles. However, in air pollution, one is interested in the overall condensation rate, i.e., the growth of a particle population due to condensation of gaseous molecules on all

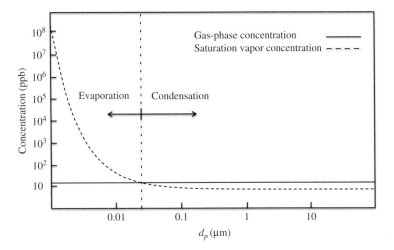

Figure 9.3. Kelvin effect. Comparison of the gas-phase concentration and saturation vapor concentration of nonadecane as a function of particle diameter. The gas-phase concentration was taken equal to twice the saturation vapor concentration over a flat surface. Source: Adapted from Devilliers et al. (2013); reprinted with authorization, © 2012 Elsevier Ltd.

those particles. Figure 9.2 also shows the overall condensation rate, i.e., the growth law for a single particle of a given size multiplied by the number of particles of that size, for a particle population typical of a polluted urban area (see Figure 9.1). This condensation rate is maximum here for particles with diameters ranging from about 0.03 to 0.3 µm, which corresponds mostly to the accumulation mode. We will see in Section 9.2.4 that coagulation leads to a similar result, which explains why atmospheric particles accumulate in this size range.

Figure 9.3 depicts the Kelvin effect for an organic compound, nonadecane (an alkane with 19 carbon atoms), which is present in diesel engine exhaust. The saturation vapor pressure of nonadecane above a flat surface is 6.1×10^{-9} atm, i.e., 6.1 ppb at $P = 1$ atm. One investigates here the case where its gas-phase concentration is twice its saturation vapor pressure, i.e., 12.2 ppb. One notes that the Kelvin effect increases the saturation vapor pressure (or concentration) significantly below 0.025 µm. Thus, evaporation of nonadecane occurs for particles with a diameter less than about 0.025 µm. These particles shrink, whereas particles with a diameter greater than about 0.025 µm grow by nonadecane condensation. This process may occur in the exhaust of a diesel vehicle, where alkanes are initially very concentrated in the gas phase and condense on soot particles until the exhaust dilution leads to lower gas-phase concentrations, which may then become less than the saturation vapor pressure for the ultrafine particles. The decrease of the diameter of these ultrafine particles will favor their transfer toward the accumulation mode via coagulation (see Section 9.2.4).

Condensation increases the particle mass, but does not change the number of particles. Conversely, evaporation decreases the particle mass, but does not change the number of particles.

9.2.4 Coagulation

Coagulation corresponds to the collision of two particles and the formation of a single particle from the two original particles. One typically assumes that the collision of two particles results in the formation of a single particle and that, therefore, every collision leads to coagulation. Although this is likely to be the case for liquid or liquid-coated particles, it may not always be the case for solid particles. For the sake of simplicity, one will assume here that every collision leads to coagulation. The coagulation rate is minimum for two particles of a same size and increases as the difference between the sizes of the two coagulating particles increases. Therefore, freshly nucleated ultrafine particles will coagulate more readily with a fine or coarse particle than with an ultrafine particle. However, the number concentration of particles must be taken into account as well and, if the number of ultrafine particles is very high compared to those of fine and coarse particles, then coagulation among ultrafine particles becomes important. The change as a function of time of the distribution of the particle number concentration as a function of the particle volume, $n_p(v_p)$, consists of two terms: one represents the increase of the number of particles of volume v_p due to coagulation of particles of smaller volumes (i.e., a particle of volume v_p' coagulating with a particle of volume $(v_p - v_p')$ to create a particle of volume v_p) and the other represents the decrease of the number of particles of volume v_p due to coagulation of those particles with other particles. This overall change is given by the following equation, where the terms $\beta(v_p, v_p')$ are the coagulation coefficients between a particle of volume v_p and a particle of volume v_p':

$$\frac{dn_p(v_p)}{dt} = \frac{1}{2}\int_0^{v_p} \beta(v_p', v_p - v_p')n_p(v_p')n_p(v_p - v_p')dv_p' - \int_0^{\infty} \beta(v_p, v_p')n_p(v_p)n_p(v_p')dv_p' \tag{9.9}$$

As it was done for the growth law representing the transfer of gaseous molecules toward a particle, the collision between two particles can be addressed by considering the extreme cases of the continuous regime for coarse particles and the kinetic theory of gases for ultrafine particles in the free molecular regime. Between these two extremes, an intermediate regime must be parameterized for the size range that corresponds approximately to fine particles. For coarse particles, the coefficient of coagulation between two particles, $\beta(v_{p,i}, v_{p,j})$ (expressed here in units of cm^3 per particle per second), is given by the following equation:

$$\beta(v_{p,i}, v_{p,j}) = 2\pi(D_{p,i} + D_{p,j})(d_{p,i} + d_{p,j}) \tag{9.10}$$

where, for particle i, $v_{p,i}$ is the particle volume (cm^3), $D_{p,i}$ is the brownian diffusion coefficient in the air (cm^2 s^{-1}), and $d_{p,i}$ is the particle diameter (cm). The coagulation coefficient, $\beta(v_{p,i}, v_{p,j})$, is minimum when $d_{p,i} = d_{p,j}$ and increases when the difference in size of the two particles increases.

For particles of diameter less than the mean free path in the air, the kinetic theory of gases applies:

$$\beta(v_{p,i}, v_{p,j}) = \frac{\pi}{4}(\bar{c}_{t,i}^2 + \bar{c}_{t,j}^2)^{1/2}(d_{p,i} + d_{p,j})^2 \qquad (9.11)$$

In the intermediate regime (which concerns fine particles), the Fuchs equation is commonly used (Fuchs, 1964):

$$\beta(v_{p,i}, v_{p,j}) = \frac{2\pi (D_{p,i} + D_{p,j})(d_{p,i} + d_{p,j})}{\left[\frac{d_{p,i} + d_{p,j}}{d_{p,i} + d_{p,j} + 2(g_i^2 + g_j^2)^{1/2}} + \frac{8(D_{p,i} + D_{p,j})}{(\bar{c}_{t,i}^2 + \bar{c}_{t,j}^2)^{1/2}(d_{p,i} + d_{p,j})}\right]}$$

$$D_{p,i} = \frac{k_B T c_c}{3\pi \mu_{v,a} d_{p,i}}; \quad c_c = 1 + \frac{2\lambda_{m,a}}{d_{p,i}}\left(1.257 + 0.400 \exp(-\frac{0.55 d_{p,i}}{\lambda_{m,a}})\right)$$

$$\bar{c}_{t,i} = 1.6\left(\frac{k_B T}{m_{p,i}}\right)^{\frac{1}{2}}; \quad g_i = \frac{0.47}{d_{p,i} \lambda_{m,i}}\left[(d_{p,i} + \lambda_{m,i})^3 - (d_{p,i}^2 + \lambda_{m,i}^2)^{\frac{3}{2}}\right] - d_{p,i}; \quad \lambda_{m,i} = \frac{8 D_{p,i}}{\pi \bar{c}_{t,i}}$$

$$(9.12)$$

where c_c is the Cunningham coefficient and, for particle i, $v_{p,i}$ is the volume (cm³), $D_{p,i}$ is the brownian diffusion coefficient in the air (cm² s⁻¹), $d_{p,i}$ is the diameter (cm), $\bar{c}_{t,i}$ is the mean thermal velocity (cm s⁻¹), $m_{p,i}$ is the particle mass, and g_i is a coefficient (cm) that is a function of the particle diameter i and its mean free path in the air $\lambda_{m,i}$. This formula tends toward the solution of the continuous regime when one of the diameters becomes large and toward the kinetic theory of gases when both diameters become less than the mean free path. This brownian coagulation coefficient is shown in Figure 9.4 for particles of diameters 0.001, 0.01, 0.1, 1, 10, and 100 μm coagulating with particles with diameters ranging from 0.001 to 100 μm in diameter. For a particle of a given diameter, the coagulation coefficient is minimum for coagulation with a particle of the same diameter.

This coagulation coefficient includes the assumption that all collisions lead to coagulation of the two particles to form a single particle. If that is not the case, one must introduce a correction term to account for the fact that the coagulation probability is less than 1. In addition, this equation corresponds to the brownian motion of particles. For ultrafine particles, the van der Waals forces are no longer negligible and increase the coagulation rate. For electrically charged particles, one must account for the Coulomb forces. Equations are available to take into account these processes if they are relevant (Friedlander, 2000).

The coagulation rate between particles depends not only on the value of the coagulation coefficient, but also on their number concentrations. Therefore, coagulation is more important for ultrafine particles since they are typically present in greater number than fine and coarse particles (see Figure 9.1). For fine particles, coagulation will be important only near sources, because in the ambient background, their number concentrations are typically too low to lead to significant coagulation rates and coagulation can then be neglected. Figure 9.4 shows the distribution of the coagulation rate of a particle of a given diameter (0.001, 0.01, 0.1, 1, and 10 μm) coagulating with particles of a greater diameter using a particle population typical of a polluted urban area. This coagulation rate is equivalent to a pseudo-first-order coagulation rate coefficient, which is defined as

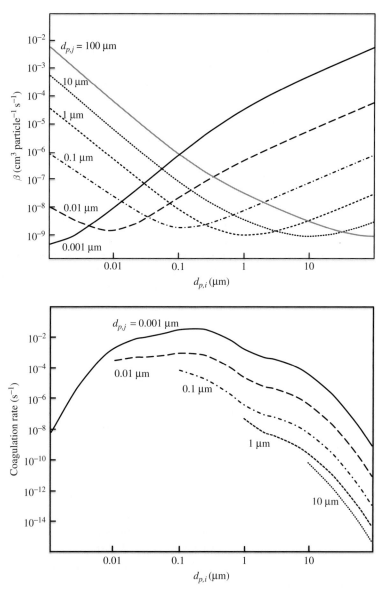

Figure 9.4. Brownian coagulation. Top figure: brownian coagulation coefficient between particles of diameters $d_{p,i}$ and $d_{p,j}$; bottom figure: distribution of the coagulation rate of particle j coagulating with particles i of greater size using a particle population typical of a polluted urban area corresponding to that of Figure 9.1.

the product of the coagulation coefficient and number concentration of the other particles involved in the coagulation process. The particle number concentrations presented in Figure 9.1 were used here. One notes that those coagulation rates are greater for ultrafine particles and that those particles coagulate preferentially with fine particles of diameters ranging from about 0.05 to 0.5 μm. The coagulation rates become negligible for coagulation with particles of diameter greater than 2.5 µm because their number concentrations are very low. Consequently, coagulation increases particulate mass in the accumulation mode (fine particles) and has almost no effect on coarse particles. As for condensation, particles tend to accumulate by coagulation in the size range corresponding to the accumulation mode. We will see in Chapter 11 that atmospheric deposition is less efficient for particles in that size range. Therefore, fine particles, i.e., those in the accumulation mode, have an atmospheric lifetime that is longer than those of ultrafine and coarse particles.

Coagulation decreases the number of particles, but does not affect their total mass.

9.3 Equilibrium Thermodynamics

Thermodynamic equilibrium between a gas phase and a condensed phase must be established for all chemical species present in particles, otherwise condensation or evaporation will take place to establish the equilibrium.

9.3.1 Saturation Vapor Pressure

The saturation vapor pressure, $P_{s,i}$, is the pressure that a gas cannot exceed. If the partial pressure, P_i, of a gaseous species exceeds its saturation vapor pressure, then nucleation and/or condensation occur to decrease the partial pressure and bring the species back to thermodynamic equilibrium. One must note, however, that condensation on liquid particles may occur in cases where the partial pressure is less than the saturation vapor pressure (see Sections 9.3.2 and 9.3.3).

9.3.2 Henry's Law

Henry's law applies to dilute aqueous solutions. It relates the concentration of a chemical species in the gas phase (represented by its partial pressure, P_i) to its activity in the aqueous phase as follows:

$$\gamma_i C_i = H_i P_i \tag{9.13}$$

where C_i is the concentration in the particle in moles per liter (M), γ_i is the activity coefficient of the species in the aqueous phase ($\gamma_i C_i$ is the activity of the species), and H_i is the Henry's law constant (which depends on temperature). The activity coefficient reflects the fact that the solution is not ideal and that interactions among species in solution (molecules and ions) affect their ability to be present in solution. By definition, the activity coefficients are equal to 1 in an ideal solution. If interactions with other species present in solution are repulsive, the species will be less present in solution than if the solution were ideal (i.e., without any interactions with other species); on the other hand, if the interactions are attractions, then the species will be more present in solution than if the solution were ideal. An activity coefficient greater than 1 corresponds to the case where the interactions are repulsive and an activity coefficient less than 1 corresponds to the case where interactions correspond to attractions. A dilute solution tends toward ideality; then, the activity coefficient tends toward 1, and the activity tends toward the concentration (Henry's law was originally formulated for an ideal solution). The partial pressure is generally expressed in atm; then, the Henry's law constant is expressed in M atm^{-1}.

9.3.3 Raoult's Law

Raoult's law applies to concentrated solutions. It relates the concentration of a chemical species in the gas phase (represented by its partial pressure, P_i) to its activity in the liquid phase as follows:

$$P_{s,i}\gamma_i x_i = P_i \tag{9.14}$$

where x_i is the molar fraction of the chemical species in solution. If the solution is ideal, then the molar fraction of the species in solution is equal to the ratio of its partial pressure and its saturation vapor pressure. For a pure solution of the species, its molar fraction is equal to 1 and one obtains the definition of the saturation vapor pressure as being the maximum value of the partial pressure.

9.4 Secondary Inorganic Fraction of Particulate Matter

The three major inorganic species that constitute the secondary inorganic fraction of atmospheric particulate matter are sulfate, nitrate, and ammonium. Sulfate results from the oxidation of sulfur dioxide (SO_2) into sulfuric acid (H_2SO_4) and nitrate results from the oxidation of nitrogen dioxide (NO_2) into nitric acid (HNO_3) (see Chapters 8 and 10). Ammonium originates directly from the association of ammonia (which is emitted mostly from agriculture) with sulfate and nitrate.

9.4.1 Sulfate and Ammonium

Sulfuric acid has a very low saturation vapor pressure and is not likely to remain in the gas phase. Therefore, it undergoes nucleation (either binary with water or ternary with water and ammonia) or condensation on existing particles. In the aqueous phase, sulfuric acid dissociates to form bisulfate (HSO_4^-) and sulfate (SO_4^{2-}) ions:

$$H_2SO_4 \leftrightarrow H^+ + HSO_4^- \tag{R9.1}$$

$$HSO_4^- \leftrightarrow H^+ + SO_4^{2-} \tag{R9.2}$$

H_2SO_4 is a strong acid and the equilibria are strongly displaced toward the formation of the sulfate ions. H_2SO_4 and its ions are partially or totally neutralized by ammonia (a base) present in the atmosphere. Depending on the ammonia and sulfate concentrations present in the atmosphere, there may be formation of ammonium bisulfate (NH$_4$HSO$_4$), ammonium sulfate ((NH$_4$)$_2$SO$_4$) or letovicite:

$$H_2SO_4 + NH_3 \leftrightarrow NH_4HSO_4 \quad (R9.3)$$

$$NH_4HSO_4 + NH_3 \leftrightarrow (NH_4)_2SO_4 \quad (R9.4)$$

Ammonium bisulfate is formed when there are as many moles of ammonia as there are moles of sulfate. Ammonium sulfate is formed when there are at least two moles of ammonia available for each mole of sulfate (sulfuric acid being a diacid, it takes two moles of ammonia to neutralize one mole of sulfuric acid totally). Letovicite corresponds to the intermediate case when there are 1.5 moles of ammonia available per mole of sulfate and its chemical formula is (NH$_4$)$_3$H(SO$_4$)$_2$.

9.4.2 Nitrate and Ammonium

Nitric acid has a rather high saturation vapor pressure and, therefore, it remains preferentially in the gas phase. However, it can react with ammonia to form a semi-volatile species: ammonium nitrate (NH$_4$NO$_3$):

$$HNO_3 + NH_3 \leftrightarrow NH_4NO_3 \quad (R9.5)$$

This species is subject to the following thermodynamic equilibrium:

$$K_{eq} = \frac{[HNO_3(g)][NH_3(g)]}{[NH_4NO_3(p)]} \quad (9.15)$$

where the notations (g) and (p) indicate a gas-phase and particulate-phase species, respectively. In the case of a solid phase, the ammonium nitrate concentration is taken to be 1 by definition and we have the following equilibrium relationship:

$$K_{eq} = [HNO_3(g)][NH_3(g)] \quad (9.16)$$

The first comparison of this theoretical equilibrium relationship with experimental data in the Los Angeles basin, California, showed satisfactory results (Stelson et al., 1979). In the case of an aqueous phase, ammonium nitrate dissociates into an ammonium cation and a nitrate anion:

$$NH_4NO_3 \leftrightarrow NH_4^+ + NO_3^- \quad (R9.6)$$

Then, the activities of the nitrate and ammonium ions must be taken into account and the equilibrium relationship is as follows:

$$K_{eq} = \frac{[HNO_3(g)][NH_3(g)]}{\gamma_{NH_4^+}[NH_4^+(aq)]\,\gamma_{NO_3^-}[NO_3^-(aq)]} \quad (9.17)$$

where the notation (aq) indicates an aqueous phase (this notation is generally ignored for ions, because they are typically present in the aqueous phase in the lower atmosphere). If the product of the concentrations of nitric acid and ammonia is low, then there is no formation of solid particulate ammonium nitrate (at low humidity) and negligible formation of ammonium nitrate in solution (at high humidity). A small fraction of nitric acid may dissolve in aqueous particles (e.g., in sulfate particles), but the liquid water content of particles is very low (<1 mg m^{-3}) and the fraction of nitrate formed via dissolution in particles is negligible.

The equilibrium constant of ammonium nitrate decreases when the temperature decreases and when relative humidity increases (e.g., Stelson and Seinfeld, 1982; Mozurkewich, 1993). Therefore, cold and humid atmospheric conditions (for example, in wintertime) are conducive to the formation of particles containing ammonium nitrate. However, solar radiation is limited during winter. Therefore, the formation of HNO_3, which requires OH radicals or ozone, both of which are produced by photolysis, is low then. Thus, it is typically during springtime or in fall that the concentrations of ammonium nitrate will be highest, because there is sufficient production of HNO_3 and a moderate temperature favors the formation of particulate ammonium nitrate.

Ammonia will preferentially neutralize sulfuric acid before reacting with nitric acid. Therefore, the formation of ammonium nitrate will occur only if (1) there is sufficient ammonia to neutralize sulfuric acid first and (2) the product of the nitric acid and remaining ammonia concentrations is sufficiently high for ammonium nitrate formation.

Nitrate may occasionally be found in coarse particles, because nitric acid can react with sea salt and alkaline soil particles. The thermodynamics is favorable to the formation of nitrate salts and the following reactions may occur:

$$\text{Sea salt}: HNO_3(g) + NaCl(p) \rightarrow NaNO_3(p) + HCl(g) \quad \text{(R9.7)}$$

$$\text{Soil particles}: 2\,HNO_3(g) + CaCO_3(p) \rightarrow Ca(NO_3)_2(p) + CO_2(g) + H_2O(g) \quad \text{(R9.8)}$$

Similar reactions are possible with gaseous sulfuric acid. However, given its very low volatility, sulfuric acid tends to be in the particulate phase directly via nucleation or condensation, rather than via chemical reactions. Note that sea salt contains a fraction of primary sulfate. Sea salt and soil particle emissions are addressed in Chapter 11.

9.4.3 Deliquescence

When humidity increases, a solid salt (ammonium sulfate, ammonium nitrate ...) will become liquid at a given humidity. This humidity is called the humidity of deliquescence. It depends on temperature. At 25 °C for example, it is 40 % for ammonium bisulfate, 62 % for ammonium nitrate, and 80 % for ammonium sulfate.

When the relative humidity decreases, a salt present in solution will become solid at a given humidity. This humidity of crystallization is less than the humidity of deliquescence and the solution is metastable between these two values of the relative humidity, being in a state of supersaturation for this salt. This phenomenon is called hysteresis.

Thermodynamic data pertaining to the inorganic equilibria of atmospheric aerosols are available in several books on atmospheric chemistry (e.g., Finlayson-Pitts and Pitts, 2000; Jacobson, 2005a; Seinfeld and Pandis, 2016).

9.5 Organic Fraction of Particulate Matter

9.5.1 General Considerations

The organic fraction of particulate matter consists of primary particulate matter, which has been emitted directly into the atmosphere as particles (for example, the organic fraction of particles emitted from biomass burning and diesel vehicles without particle filters), and secondary particulate matter, which was formed in the atmosphere. One typically refers to primary organic aerosols (POA) and secondary organic aerosols (SOA) to differentiate these two categories of particulate organic species. Regarding SOA, the semi-volatile organic compounds (SVOC) that condense on existing particles (or in some cases lead to new ultrafine particles via nucleation) may have been emitted into the atmosphere directly and have subsequently condensed because of a decrease in temperature that favors their partitioning toward the particulate phase or they may have been formed in the atmosphere via the oxidation of volatile or semi-volatile organic compounds (VOC and SVOC). A SVOC is a compound that partitions between the gas and particulate phases, whereas a VOC is entirely present in the gas phase. The European Union (EU) defines a VOC as an organic compound that has a boiling point equal to or less than 250 °C at a standard atmospheric pressure (1 atm) (the boiling point is anti-correlated with the saturation vapor pressure; the saturation vapor pressure increases as the boiling point decreases; Perry's Chemical Engineers' Handbook, 2008). There is no official definition of SVOC, but one may consider that they have saturation vapor pressures between 10^{-14} and 10^{-4} atm (Weschler and Nazaroff, 2008).

A potentially important source of SOA is the semi-volatile fraction of organic emissions from combustion processes (vehicle engines, biomass burning, etc.). This fraction may condense on soot particles, for example, during the cooling of the exhaust plume. It may also be oxidized, leading then to SVOC that are heavier (in terms of molar mass) and, therefore, more likely to condense. This SVOC source is currently uncertain, because only the particulate fraction of SVOC is typically included in emission inventories. The gas-phase fraction must, therefore, be estimated, which is difficult given the small number of available experimental data sets.

9.5.2 Chemistry of SOA Formation

Since the atmosphere is an oxidizing medium, VOC and SVOC undergo oxidation reactions, which lead to the formation of new compounds, which are oxygenated (e.g., Ziemann and Atkinson, 2012; Ng et al., 2017). The main oxidants are the hydroxyl radical (OH), the nitrate radical (NO_3), and for alkenes ozone (O_3). The oxidation step that consists of the addition of a functional group to the initial organic molecule and results in the addition of oxygen to that organic molecule is called functionalization. Once the organic molecule has been oxidized, other reactions may occur leading to successive functionalization steps with addition not only of oxygen, but also nitrogen (organic nitrates) and sulfur

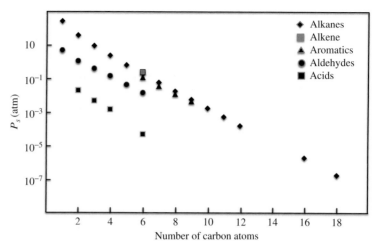

Figure 9.5. Saturation vapor pressures of selected organic compounds (atm). The saturation vapor pressures are shown for n-alkanes (methane to octadecane), n-hexene, benzene and its mono-substituted alkyl derivatives (toluene, ethyl-, and propyl-benzene), aldehydes (formaldehyde to hexanal), and monocarboxylic acids (acetic acid to hexanoic acid). Source of the data: Schwarzenbach et al. (2003).

(organic sulfates). It is also possible that an oxidation reaction leads to the fragmentation of the organic molecule and the formation of two (or more) molecules that are smaller (in terms of the number of carbon atoms) than the original molecule. Then, there is a loss of carbon atoms from the original molecule.

The addition of oxygen atoms to an organic molecule will increase its molar mass because the oxygen atom replaces a hydrogen atom (for example, in the case of aldehydes, ketones, and acids) or is added to the organic molecule leaving the number of hydrogen atoms unchanged (for example, in the case of alcohols and peroxides). In general, given a VOC or SVOC with a given number of carbon atoms, the augmentation of the molar mass via an oxidation reaction leads to a less volatile compound, i.e., a compound with a lower saturation vapor pressure. Figure 9.5 shows the saturation vapor pressures of some organic compounds. There is little difference in terms of volatility among an alkane, an alkene, and an aromatic compound with the same number of carbon atoms. On the other hand, for a given number of carbon atoms, an aldehyde is less volatile than an alkane and a carboxylic acid is less volatile than an aldehyde. The oxidation products are, therefore, more likely to be present in the particulate phase than the original compound. As the number of oxidation steps increases, the number of functionalizations can increase and the saturation vapor pressure will decrease, favoring the partitioning of the compound toward the particulate phase. Similarly, the formation of an organic nitrate or sulfate will lead to a decrease in volatility compared to that of the original compound.

On the other hand, a reaction leading to the fragmentation of an organic compound will lead to compounds having lower molar masses and, therefore, higher saturation vapor pressures. The products of such a reaction are, therefore, more likely to be present in the gas

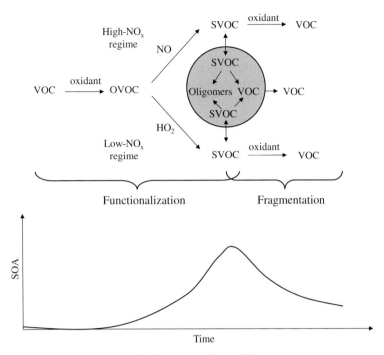

Figure 9.6. Schematic representation of the evolution of SOA formation from a VOC.

phase than in the particulate phase, as compared to the original compound. The probability of fragmentation is significantly greater for compounds that are more functionalized, thereby limiting the number of functionalizations that are possible.

Figure 9.6 summarizes schematically the main processes involved in SOA formation. After a period leading to the formation of SVOC by functionalization of organic molecules, fragmentation reactions begin to occur and lead to more volatile products. These processes occur over several hours in the atmosphere. SVOC partitioning between the gas phase and the particulate phase may occur in an organic liquid phase or an aqueous phase depending on the hydrophobic or hydrophilic characteristics of the SVOC and those of the existing atmospheric particles (Saxena et al., 1995). The chemical reactions and products corresponding to these processes are described in detail below.

Gas-phase Reactions

SOA formation in the gas phase results from the oxidation of a VOC or SVOC after one or more oxidation steps (one may, however, consider more generally that the condensation of a SVOC following a decrease in temperature corresponds also to SOA formation). The main characteristics of the mechanisms leading to the formation of different semi-volatile organic molecules are discussed here for the main VOC categories.

The oxidation of alkanes was shown for n-pentane in Figure 8.1. The first oxidation step of alkanes by OH consists of the abstraction of a hydrogen atom, which results in the formation of a water molecule (H_2O) and an alkyl radical. This radical reacts rapidly with molecular oxygen (O_2) to form an organic peroxyl radical (R_1-HC(O_2)-R_2). Then, this radical reacts with NO according to two possible reaction pathways: formation of an organic nitrate or formation of NO_2 and an alkoxy radical (R_1-HC(O)-R_2). The organic nitrate is semi-volatile and may, therefore, contribute to SOA formation. The alkoxy radical may decompose (leading to a fragmentation into an aldehyde and an alkyl radical, i.e., volatile products) or isomerize. Isomerization leads to the formation of an alcohol group and, after reaction with O_2, a peroxyl group. A series of reaction schemes similar to those just mentioned follows, leading to either formation of an organic nitrate or formation of an alkoxy radical. The products of these successive oxidation steps include, for example, hydroxynitrates, hydroxycarbonyls, hemiacetals, carbonyl-esters, and furans. In the absence of NO_x, the peroxyl radicals react preferentially with other peroxyl radicals (organic peroxyl, RO_2, or hydroperoxyl, HO_2) to form, for example, organic peroxides. The oxidation of alkanes by the NO_3 radical is similar for the first oxidation step, leading to the formation of the same alkyl radical and a molecule of nitric acid (HNO_3), instead of the water molecule. The following oxidation steps should lead to products in part similar to those obtained by OH oxidation under high-NO_x conditions (but without the NO reactions); however, no experimental data on SOA formation from the oxidation of alkanes by NO_3 was found.

The oxidation of linear alkenes by OH occurs preferentially on the double bond (see Figure 8.3). An organic radical with an alcohol group in the alpha position is formed. The following steps are similar to those described for the alkanes. The organic radical reacts with O_2 to form a peroxyl radical. Next, the reaction with NO leads either to a hydroxynitrate (R_1-HC(OH)-HC(NO_3)-R_2) or an alkoxy radical. In the latter case, formation of organic dihydroxynitrates follows. Other semi-volatile oxidation products include trihydroxynitrates, dihydroxycarbonyls, trihydroxycarbonyls, and dihydrofurans. The oxidation of the NO_3 radical is similar to that of OH for the first oxidation step, but it leads mostly to the formation of organic nitrates via subsequent steps. The oxidation by addition of ozone to the double bond leads to the addition of functional groups after the breakup of the original molecule at the double bond (see Figure 8.5).

The oxidation of aromatic compounds by OH leads either to a cleavage of the phenyl ring (the main chemical pathway) or to oxidation retaining the phenyl ring (see Figure 8.6). In the former case, the resulting organic molecules may lead to compounds with two or three functional groups, which have low volatility. However, the products may also undergo decomposition leading to smaller compounds that are then more volatile. In the latter case, the addition of oxygen and nitrate groups to the phenyl ring leads rapidly to the formation of carboxylic acids, carbonyls, and oxocarboxylic acids (pyruvic acid, glyoxylic acid . . .). SOA formation from the oxidation of aromatic compounds by NO_3 has not been studied experimentally, because this oxidation pathway is slow (see Chapter 8).

The oxidation of isoprene by OH leads to compounds such as methacrolein, methyl vinyl ketone, and hydroxy hydroperoxides. The latter compounds, which are formed under low-NO_x conditions, lead to the formation of epoxydiols, commonly called IEPOX. IEPOX react in the particulate phase to lead to the formation of various compounds including tetrols (which have been identified in the ambient atmosphere). The oxidation of isoprene by NO_3 leads to the formation of a nitrooxy hydroxyepoxide, commonly called INHE, which undergoes in the particulate phase similar reactions to those undergone by IEPOX. The main oxidation products of the oxidation of isoprene by NO_3 are organic nitrates. The NO_3 radicals are present mostly at night (since they are photolyzed during the day) when isoprene emissions are less important (since they depend on sunlight). However, the SOA formation rate by reaction with NO_3 is much greater than that of the reaction with OH. Therefore, the oxidation of isoprene by NO_3 can be an important reaction pathway for SOA formation, despite its nocturnal pattern. The oxidation of 2-methyl-3-buten-2-ol (MBO) by OH radicals leads to less SOA formation than that of isoprene, but the oxidation products are similar, since one finds tetrols, dicarboxylic acids, and multifunctional compounds with alcohol and aldehyde groups.

The oxidation of monoterpenes by OH leads to a large number of compounds, including aldehydes, carboxylic acids, and multifunctional compounds. If the chemical composition of SOA is similar for the two chemical regimes (low-NO_x and high-NO_x), i. e., mostly organic acids, it seems that some acids (pinonic acid, pinic acid, and hydroxy-pinonic acid) are more abundant under low-NO_x conditions, whereas organic nitrates are present under high-NO_x conditions. The oxidation of monoterpenes by NO_3 leads to the formation of organonitrates. These organonitrates can be hydrolyzed in the aqueous phase leading to the formation of HNO_3 and an organic compound. The oxidation of monoterpenes by ozone has shown the presence of compounds with very low volatility (a few % for monoterpenes with an internal double bond such as limonene and α-pinene).

The oxidation of sesquiterpenes by OH leads readily to SOA formation given the low saturation vapor pressure of terpenes with 15 carbon atoms. The oxidation products are mostly hydrophobic and contain functional groups identical to the oxidation products of monoterpenes.

Particulate-phase Reactions

Chemical reactions in the particulate phase occur and lead to further oxidation of SOA, their fragmentation or the formation of other species (sulfates, nitrates, etc.). In particular, some reactions lead to the combination of several species to form new chemical species with a greater molar mass and, therefore, lower volatility. Such reactions include the formation of hemiacetals (R_1R_2-C(OH)-O-R_3), acetals (R_1R_2-C(OR_3)-O-R_4), aldol condensation products (R_1-C(O)-CH=CHR_2), and esters (R_1-C(O)-O-R_2):

$$\text{aldehyde} + \text{alcohol} \rightarrow \text{hemiacetal} \quad (R9.9)$$

$$\text{hemiacetal} + \text{alcohol} \rightarrow \text{acetal} \quad (R9.10)$$

$$\text{enol} + \text{aldehyde} \rightarrow \text{aldol condensation product} \quad (R9.11)$$

$$\text{acid} + \text{alcohol} \rightarrow \text{ester} \quad (R9.12)$$

The first three types of reactions involve organic species. The last type of reaction can occur by reaction between an alcohol and an inorganic acid, such as sulfuric acid. Then, organic sulfates are formed. The reaction of alcohols with nitric acid is not efficient and organic nitrates are not formed via this chemical pathway, but are formed instead in the gas phase via the reaction of organic radicals and nitrogen oxides (mostly by reaction between peroxyl radicals and nitric oxide, NO).

Oligomerization means the formation of organic compounds by the combination of two or more organic molecules (oligomers are small polymers). Some of the reactions listed in this section such as the formation of hemiacetals and aldol condensation products may lead to the formation of oligomers. For example, oligomers may be formed from glyoxal and methylglyoxal under acidic conditions. An oligomer has a very low saturation vapor pressure and, therefore, is non-volatile. Thus, an oligomerization reaction favors the transfer of monomers from the gas phase toward the particulate phase and increases SOA formation. However, some oligomerization reactions could be reversible, which would then lead to a decrease in the SOA concentrations if the equilibrium would evolve toward a decomposition of the oligomer toward its original monomers. The contribution of oligomerization to SOA formation still needs to be better quantified.

Aqueous-phase Reactions

Some organic compounds are water-soluble and can, therefore, be involved in aqueous-phase reactions, for example, in clouds and fogs (the liquid water content of atmospheric particles is too small for these reactions to have much impact in the particulate phase). Simulations conducted with aqueous-phase chemical kinetic mechanisms suggest as notable products oxalic acid, oligomers of glyoxal and methylglyoxal, glycolic, glyoxylic, and pyruvic acids, tetrols, and organosulfates (Ervens, 2015). The contribution of aqueous chemistry to the total formation of SOA is low (<10 % on average), according to numerical simulation results (e.g., Couvidat et al., 2013), except perhaps for specific conditions. As expected, SOA formation in the aqueous phase is strongly correlated with the presence of clouds and reactive organic compounds.

9.5.3 Experimental Estimates of SOA Formation Yields

Smog Chamber Experiments

The oxidation of VOC (and SVOC) has been studied experimentally in smog chambers, which allow one to follow the evolution of the concentrations of the main chemical species (reactants and products) over several hours. These experiments can be conducted in the presence of natural sunlight, artificial lighting or in the dark. The oxidants used are those present in the atmosphere and include the hydroxyl radical (OH), ozone (O_3), and the nitrate radical (NO_3). Most of these experiments have been conducted with OH as the oxidant. One may distinguish two main categories of experiments conducted with OH: (1) those where the OH radicals are produced by reaction between NO_x and the VOC being studied or by photolysis of nitrous acid (HONO) and (2) those where the OH radicals are produced in the absence of NO_x. In the former case, one will typically be in a high-NO_x regime and nitrogen oxides will then be involved in some reactions of the VOC oxidation products. In addition, the presence of NO_x is generally sufficient to titrate the ozone formed by reactions between NO_x and VOC, which allows one to study solely the OH oxidation pathway (in the case of alkenes). However, in some experiments, the NO_x concentrations are kept sufficiently low that one may consider those experiments to be typical of a low-NO_x regime. In the latter case, one is in a no-NO_x regime, which is similar to a low-NO_x regime in the atmosphere. The OH radicals can be produced by photolysis of hydrogen peroxide (H_2O_2) or by ozone (O_3) photolysis in the presence of water vapor (H_2O). The first oxidation step by OH will be identical in both cases, but the products of the subsequent reactions could be very different depending on the NO_x regime. Typically, the formation of organic nitrates prevails in a high-NO_x regime and the formation of peroxides prevails in a low-NO_x regime. Since the saturation vapor pressure of the peroxides and nitrates may differ significantly, different SOA formation yields may be obtained depending on whether the experiment was conducted under high- or low-NO_x conditions. Clearly, these two types of experiments are relevant, since VOC oxidation may occur in the atmosphere under high-NO_x (e.g., urban or industrial areas) or low-NO_x (e.g., rural areas) conditions.

Experiments conducted in smog chambers have initially targeted anthropogenic compounds (mostly aromatic compounds), but have rapidly been extended to include biogenic compounds. Today, experiments that have been conducted in smog chambers include, for example, biogenic compounds such as hemiterpenes and their products, monoterpenes, and sesquiterpenes, and anthropogenic compounds such as aromatic compounds, alkanes, alkenes, and polycyclic aromatic hydrocarbons (PAH). The terms terpenes and terpenoids have been defined in Chapter 8 and the term terpenes is used here to cover both terms.

It is challenging to summarize all the results obtained in a large number of smog chamber experiments, because those studies have been conducted under conditions that are sometimes very different (for example, in terms of chemical regime, lighting, duration, temperature, humidity, presence of seed particles, wall effects ...). Some smog chambers are located indoor and use artificial lighting, which must then approximate sunlight; others are located outdoor, thereby using natural sunlight, but being affected by the possible presence of clouds. The size of the chamber impacts directly the duration of the experiments that can be conducted, because deposition to the walls is less in a large chamber (the surface/volume ratio is inversely proportional to the characteristic dimension of the chamber). The maximum duration of the experiments conducted in a large smog chamber is typically of several hours (but less than a day). In addition, the measurement of the SOA yield (the ratio of the particulate mass formed and that of the VOC reacted, expressed as a fraction or percentage) is subject to interpretation. First, this yield depends on the duration of the experiment. The longer the experiment, the more the yield tends to increase, because few experiments last long enough for fragmentation processes to prevail over functionalization reactions. Second, as the particulate mass increases, it favors the gas/particulate equilibrium toward the particulate phase, because a larger particulate volume becomes available for absorption of the semi-volatile compounds. Therefore, an experiment conducted with a large initial concentration of a precursor (for example, 1 ppm instead of 10 ppb) will lead to a greater yield, because the particulate mass will be more important (for example, ~1 mg m^{-3} instead of ~10 µg m^{-3}) and will favor the absorption of semi-volatile compounds. If the results are not very sensitive to this initial concentration, it implies that the oxidation products have low volatility and are, therefore, mostly in the particulate phase (i.e., their particulate concentrations do not depend much on the particulate mass available for their absorption).

The temperature used for the experiments is important because, according to the van't Hoff equation (see Chapter 10), the thermodynamic equilibrium favors the semi-volatile compounds toward the gas phase as the temperature increases. In addition, a low relative humidity does not allow hydrophilic compounds to condense on an aqueous phase. Such dissolution in aqueous particles may occur for some SVOC in the atmosphere (Saxena et al., 1995), in which case the smog chamber experiments conducted at low relative humidity may not be representative of atmospheric conditions.

Concentrations of the organic particulate mass present in the atmosphere are generally lower than those used in smog chamber experiments (higher concentrations make chemical measurements easier). Therefore, the maximum yields obtained in smog chamber experiments will typically overestimate the actual yields observed in the atmosphere. Normalizing those yields with respect to a particulate mass concentration of 10 µg m^{-3}, for example, should provide more realistic SOA yields. One must also take into account the fact that gases and particles will deposit on the chamber walls. Therefore, some assumptions must be made regarding the absorption (or not) of semi-volatile compounds on the particles deposited on the walls; depending on those assumptions, yields may differ by a factor of two. Finally, measurement methods for SOA have uncertainties, which can affect the estimation of the SOA yield.

A summary of selected smog chamber experiments is presented in Figure 9.7 as an illustration of typical SOA yields for a range of VOC.

SOA Yields for Anthropogenic Compounds

The main aromatic compounds that have been studied include benzene, toluene, and xylenes. The SOA yields are greater without NO_x, which suggests that the formation of peroxides leads to greater SOA formation (>30 % in terms of mass yield) than the formation of organic nitrates (<30 %).

Alkanes with a few carbon atoms do not lead to any significant formation of semi-volatile products and, therefore, most experiments have targeted alkanes with at least eight carbon atoms. For a given number of carbon atoms, the most important SOA mass yields have been obtained with cycloalkanes; the lowest yields have been obtained with alkylalkanes, because they are more subject to fragmentation. As expected, the SOA yields increase with the number of carbon atoms and range, for example, from <1 % for a linear C8 alkane to >50 % for a linear C15 alkane. SOA formation seems to be greater in the presence of NO_x (ranging from 11 to 98 %) than without NO_x (ranging from 3 to 86 % for the same linear alkanes).

Alkenes have not been studied much compared to alkanes and aromatic compounds. SOA formation is low for alkenes with fewer than eight carbon atoms. OH oxidation

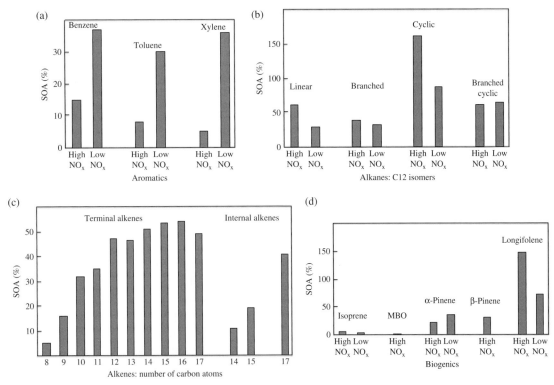

Figure 9.7. Examples of SOA yields obtained by OH oxidation under high-NO$_x$ and low-NO$_x$ regimes for several VOC categories. Other experiments have been conducted that can lead to significantly different yields for some of these compounds. Therefore, these yields must be seen in terms of comparison between high-NO$_x$ and low-NO$_x$ regimes or among VOC of a same category. Figures (a) to (d) from top to bottom:

(a) SOA mass yields for aromatic compounds for an organic particulate mass of 10 µg m^{-3} (Ng et al., 2007a).
(b) Maximum SOA mass yields for alkanes with 12 carbon atoms, which include a linear alkane (dodecane), a branched alkane (methylundecane), a cyclic alkane (cyclododecane), and a branched cyclic alkane (hexylcycloalkane) (Loza et al., 2014). These yields were obtained with high organic particulate mass concentrations (>100 µg m^{-3} in most cases) and, therefore, may not be representative of atmospheric yields.
(c) SOA molar yields for terminal and internal alkenes (Matsunaga et al., 2009) calculated from measurements of molecular compounds in a high-NO$_x$ regime. It was assumed that the dihydroxycarbonyls are semi-volatile and partition between the gas and particulate phases.
(d) Maximum SOA mass yields for biogenic compounds (Kroll et al. (2006) for isoprene; Jaoui et al. (2012) for 2-methyl-3-buten-2-ol (MBO); Eddingsaas et al. (2012) for α-pinene; Lee et al. (2006) for β-pinene; Ng et al. (2007b) for longifolene).

experiments conducted with linear alkenes ranging from C8 to C15 have led to SOA molar yields ranging from 0 for C8 to >50 % for C15. Alkenes with a terminal double bond have a greater SOA yield than the corresponding internal alkenes. The oxidation of alkenes by NO$_3$ radicals can also lead to SOA formation, including organic nitrates as first-generation products, while nitrated furans appear as second-generation products.

SOA Yields for Biogenic Compounds

Biogenic compounds with five carbon atoms (hemiterpene and hemiterpenoid) include isoprene (emitted mainly from deciduous trees) and methylbutenol (MBO, emitted mainly from coniferous trees). Experiments conducted with MBO using OH as the oxidant have not led to any measurable SOA formation. Those conducted with isoprene with OH as the oxidant have led to SOA mass yields ranging from 0 to 7 %. The differences could be due to the NO_x regime, lighting (natural solar radiation, artificial lighting), relative humidity, etc. In particular, the oxidation products of isoprene are water-soluble and are, therefore, more likely to condense on aqueous particles than on organic particles. Smog chamber experiments often use a low relative humidity and could, therefore, underestimate SOA formation under humid conditions. If the yields measured in smog chambers seem low, the contribution to PM concentrations in the atmosphere could nevertheless be notable, because isoprene is the main biogenic VOC emitted in the atmosphere. The aqueous chemistry of isoprene has been studied and the experimental results suggest greater SOA yields than under dry conditions (e.g., Brégonzio et al., 2016); however, the experimental data are currently insufficient to obtain quantitative yields of SOA formation in clouds.

A large number of monoterpenes (biogenic compounds with 10 carbon atoms) have been studied in smog chamber experiments. Among those that have been studied the most, one may mention α-pinene, β-pinene, Δ^3-carene, and limonene. The oxidants used include OH, O_3, and NO_3, but OH is the one that has been used the most and for which the most data are available. The SOA mass yields vary greatly, ranging from 1 to 40 %. Experiments conducted with and without NO_x suggest greater SOA formation in a low-NO_x regime.

Biogenic compounds with 15 carbon atoms are clearly less volatile than those with 10 or 5 carbon atoms. Therefore, one may expect that their oxidation leads rapidly to condensable products and significant SOA formation. Humulene, longifolene, and caryophyllene have been studied the most. OH is the oxidant that has been used the most. The mass yields vary greatly ranging from 10 to 150 % (values greater than 100 % are possible because the oxidation products are heavier than the original sesquiterpene).

In summary, sesquiterpenes have SOA yields that are on average greater than those of monoterpenes, which in turn have yields that are greater than those of isoprene and MBO (see Figure 9.7). However, these SOA yields must be weighted by the amounts of the emitted precursor to evaluate the importance of SOA formation in the atmosphere. In terms of global emissions of biogenic organic compounds, isoprene represents about 50 %, monoterpenes about 15 %, and sesquiterpenes only about 3 % (other biogenic VOC include aldehydes, alcohols, and other compounds, see Chapter 2). Using approximate yields of 4 % for isoprene, 40 % for monoterpenes, and 90 % for sesquiterpenes, the SOA yields weighted by the fraction of their emissions in the global inventory are 2 %, 6 %, and 3 %, respectively. Therefore, all those biogenic compound categories may contribute in commensurate amounts to SOA formation.

9.6 Emission Control Strategies for Atmospheric Particulate Matter

PM concentrations (PM_{10} and $PM_{2.5}$) exceed the national ambient regulatory standards and the guideline concentrations of the World Health Organization (WHO) in many countries. The development of control strategies for PM and gaseous precursor emissions is challenging, because the secondary fraction of fine PM is significant (typically more than half of $PM_{2.5}$ mass) and this fraction involves complex nonlinear relationships between the precursor emissions and the ambient concentrations of the secondary fraction. In Paris, France, the secondary fraction of fine PM exceeds 50 % in terms of annual-average mass concentration, in the urban background as well as near roadways (see Figure 9.8). In Beijing, China, and in Delhi, India (see Figure 9.9), the $PM_{2.5}$ annual-averaged concentrations are significantly greater than in Paris (>100 µg m^{-3} versus about 15 µg m^{-3}). Nevertheless, the same major chemical species are present in the fine PM size range. The elemental carbon fraction contributes a larger fraction in Paris, possibly because of the importance of diesel vehicles without particle filters in France in 2009–2010. The organic matter fraction dominates the urban background in all areas with annual contributions ranging from 24 to 36 %. Among the inorganic ions, sulfate dominates in Beijing and Delhi, whereas nitrate dominates in Paris. The regulations on sulfur content in gasoline and diesel in Europe and

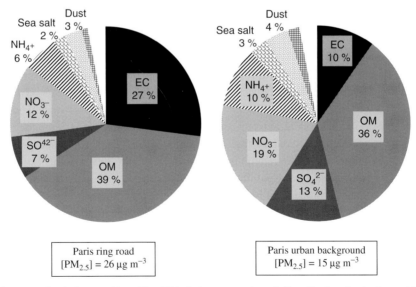

Figure 9.8. Annual-average chemical composition of fine PM in Paris near a roadway (left) and in the urban background (right). Measurements were conducted over one year (from September 11, 2009 to September 10, 2010). EC: elemental carbon measured by a thermal-optical method, OM: organic matter, SO_4^{2-}: non-sea-salt sulfate, NO_3^-: nitrate, NH_4^+: ammonium; the unspecified fractions (gray with fine hatching and white) represent other ions and non-identified chemical species, respectively. Source: Airparif (2011).

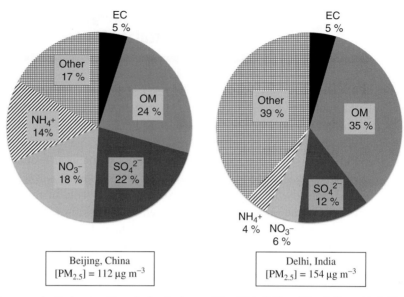

Figure 9.9. Annual-average chemical composition of urban background fine PM in Beijing, China (left), and in Delhi, India (right). Measurements were conducted during four periods from June 2012 to April 2013 amounting to about three months in Beijing and from November 2011 to December 2012 in Delhi. EC: elemental carbon measured by a thermal-optical method, OM: organic matter, SO_4^{2-}: non-sea-salt sulfate, NO_3^-: nitrate, NH_4^+: ammonium; the unspecified fraction (gray with fine hatching) represents other ions and non-identified chemical species. Percentages may not add to 100 %, because of round-off. Data sources: Wang et al. (2015) for Beijing and Dumka et al. (2017) for Delhi.

the small fraction of fossil-fueled power plants for electricity production in France explain the lower sulfate concentrations in Paris.

In Paris, the primary fraction typically dominates during winter because of low photochemical activity and an increase in residential biomass burning. On the other hand, the secondary fraction dominates during spring episodes. Similar seasonal variations are observed in Beijing and Delhi. In Beijing, the highest concentrations are observed during winter (about 170 µg m^{-3}), followed by spring (about 120 µg m^{-3}). The carbonaceous fraction is important in winter due to fossil-fuel and biomass combustion and to meteorological conditions that are conducive to air pollution (little dilution and low precipitation). On the other hand, the nitrate fraction is important in spring due to traffic, photochemical activity producing secondary pollutants such as nitric acid, and moderate temperatures favoring the particulate partitioning of semi-volatile compounds such as ammonium nitrate (Wang et al., 2015). In Delhi, $PM_{2.5}$ concentrations are highest (>200 µg m^{-3}) during post-monsoon (October and November) and winter (December to February) seasons due to fossil-fuel and biomass combustion and to unfavorable meteorology, whereas they are lowest (<100 µg m^{-3}) during the monsoon season (July to September) due to significant particle scavenging by precipitation (Dumka et al., 2017). Although the $PM_{2.5}$ concentrations and chemical composition differ among these three cities, similar seasonal trends are observed that result from variations in emissions and meteorology. These seasonal variations and their impact on emission control strategies are analyzed in Sections 9.6.4 and 9.6.5 for Paris via the analysis of two case studies, which include a winter episode dominated by primary PM and a spring episode dominated by secondary PM.

9.6.1 Primary PM

Primary particles can be captured before being emitted in the atmosphere (see Chapter 2). In the case of diesel vehicle particulate emissions, the recent European regulatory emission standards require that particle filters be installed on all new on-road vehicles. There are also incentives provided by the French government for the replacement of old wood stoves (which are large sources of primary PM) by modern and more efficient stoves. The development of efficient control strategies for primary PM requires the identification of the major sources. A case study of a wintertime PM pollution in the Paris region is presented in Section 9.6.4 to illustrate this point.

9.6.2 Secondary Inorganic PM

The reduction of PM containing sulfate, nitrate, and ammonium can be challenging if one does not account for the relationships among these different species. The reduction of sulfate concentrations is obtained by decreasing the emissions of its precursor, SO_2. The strategy to reduce SO_2 emissions from coal-fired power plants in the U.S. has shown that this approach was efficient, in particular once SO_2 ambient concentrations became less than those of hydrogen peroxide (see Chapters 10, 13, and 15). However, the reduction of sulfate leads to a transfer of secondary particulate ammonium toward the gas phase as ammonia. Then, this gaseous ammonia becomes available for reaction with nitric acid to form ammonium nitrate. Therefore, a decrease in sulfate concentrations may lead to an increase in nitrate concentrations, thereby limiting the benefits of SO_2 emission controls. These antagonistic effects must be taken into account when selecting the emissions to be controlled, in order to avoid setting up an emission control strategy that could be inefficient.

The reduction of nitrate concentrations can occur either via the reduction of nitric acid or via the reduction of ammonia. Therefore, one must determine which of these two precursors will lead to the best result in terms of reducing the ammonium nitrate concentration. The amount of ammonia available for reaction with nitric acid, $[NH_3]_a$, is defined as follows:

$$[NH_3]_a = [NH_3]_t - 2\,[H_2SO_4] \qquad (9.18)$$

where $[NH_3]_t$ represents the total ammonia concentration, i.e., the concentration before any reaction or partitioning between the gas and particulate phases. This equation, expressed in moles or in ppb, reflects the fact that each mole of sulfuric acid requires two moles of ammonia to be neutralized and that, therefore, these moles of ammonia will not be available for reaction with nitric acid.

Next, one defines the excess ammonia concentration, $[NH_3]_e$, which represents the available ammonia concentration that is in excess of the total concentration of nitric acid:

$$[NH_3]_e = [NH_3]_a - [HNO_3]_t \qquad (9.19)$$

One can show that if $[NH_3]_e < 0$, then, it is more efficient to reduce ammonia emissions rather than nitric acid emissions (ammonia-poor regime). On the other hand, if $[NH_3]_e > 0$, then, it is more efficient to reduce the nitric acid concentrations rather than those of ammonia (ammonia-rich regime).

Another way to express this relationship is to use the ratio of the gaseous precursors of ammonium nitrate, i.e., the available ammonia and total nitric acid (Ansari and Pandis, 1998):

$$GR = \frac{[NH_3]_a}{[HNO_3]_t} \qquad (9.20)$$

Thus, GR < 1 for the ammonia-poor regime and GR > 1 for the ammonia-rich regime.

The reduction of nitric acid concentrations is not straightforward. Nitric acid is formed by the oxidation of NO_2. A decrease in NO_x emissions may lead to an increase in oxidant concentrations in a high-NO_x regime (see Chapter 8); therefore, nitric acid formation, which is proportional to the concentrations of NO_2 and the oxidant (OH or NO_3), may not necessarily decrease and, in some cases, could increase.

Example: Calculation of PM concentrations for a sulfate/nitrate/ammonium mixture and different control strategies of the gaseous precursors.

Atmospheric conditions: 1 atm, 15 °C, and low relative humidity (<40 %).

Sulfuric acid has a very low saturation vapor pressure and is, therefore, present entirely in the particulate phase as sulfate. On the other hand, nitric acid is volatile and is converted to particulate nitrate via its reaction with ammonia. At 15 °C and a low humidity, the gas/particle equilibrium constant of ammonium nitrate is as follows:

$$K_{eq} = [HNO_3][NH_3] = 2 \text{ ppb}^2$$

Case 1: Reference

The following conditions are provided: sulfate = 5 µg m^{-3}
At 1 atm and 15 °C : $n/V = P/RT = 1/(8.205 \times 10^{-5} \times 288) = 42.3$ moles m^{-3}
The molar mass of sulfate is 96 g mole^{-1}, thus:
1 µg m^{-3} = 10^{-6} / (96 g mole^{-1}) / (42.3 moles m^{-3} atm^{-1}) × (10^9 ppb atm^{-1}) = 0.25 ppb
Therefore: sulfate = 1.23 ppb

Nitrate = 2 μg m^{-3}
1 μg m^{-3} = 10^{-6} / (62 g mole^{-1}) / (42.3 moles m^{-3} atm^{-1}) × (10^9 ppb atm^{-1}) = 0.38 ppb
Therefore: nitrate = 0.76 ppb
To form ammonium nitrate, sulfate must first be entirely neutralized as ammonium sulfate, $(NH_4)_2SO_4$: ammonium = ammonium associated with sulfate + ammonium associated with nitrate;
that is: ammonium = (1.23 × 2) + 0.76 = 3.22 ppb
1 ppb = (10^{-9} atm ppb^{-1}) × (42.3 moles m^{-3} atm^{-1}) × 18 ×10^6 μg mole^{-1} = 0.76 μg m^{-3};
therefore: ammonium = 3.22 × 0.76 = 2.45 μg m^{-3}, including 1.87 μg m^{-3} associated with sulfate and 0.58 μg m^{-3} associated with nitrate

The concentration of gaseous HNO_3 is 4 ppb. Since we are at equilibrium for ammonium nitrate: $[NH_3] = K_{eq} / 4 = 2 / 4 = 0.5$ ppb. The concentration of total ammonia (gaseous ammonia + particulate ammonium) is: 3.22 + 0.5 = 3.72 ppb

The PM concentration is here the sum of sulfate, nitrate, and ammonium:

$$[PM] = 5 + 2 + 2.45 = 9.45 \text{ μg m}^{-3}$$

Case 2: The sulfate concentration is reduced by 50 %

Thus, we have: sulfate = 2.5 μg m^{-3}
Ammonium associated with sulfate, as ammonium sulfate, is now: 1.87 / 2 μg m^{-3} = 0.94 μg m^{-3}. The ammonia concentration available to form ammonium nitrate is, therefore, the total ammonia concentration less that associated with sulfate, i.e.: $[NH_3]_a$ = 3.72 − 1.23 = 2.49 ppb
The nitric acid concentration available is the sum of gaseous nitric acid and particulate nitrate, i.e., total nitrate: $[HNO_3]_t$ = 0.76 + 4 = 4.76 ppb. To calculate the new concentration of ammonium nitrate, the following equilibrium relationship must be verified:
$[HNO_3] [NH_3]$ = 2 ppb^2, i.e., $([HNO_3]_t - [NH_4NO_3]) ([NH_3]_a - [NH_4NO_3]) = 2$ ppb^2
Solving for $[NH_4NO_3]$, one obtains a quadratic equation for ammonium nitrate:
$[NH_4NO_3]^2 - [NH_4NO_3] ([HNO_3]_t + [NH_3]_a) + ([NH_3]_a × [HNO_3]_t - 2) = 0$
The solution is: $[NH_4NO_3]$ = 1.81 ppb
1 ppb of NH_4NO_3 = (10^{-9} atm ppb^{-1}) × (42.3 moles m^{-3} atm^{-1}) × 80 ×10^6 μg mole^{-1} = 3.38 μg m^{-3}; thus, there is 6.12 μg m^{-3} of ammonium nitrate. Therefore, the PM concentration is:

$$[PM] = 2.5 + 0.94 + 6.12 = 9.56 \text{ μg m}^{-3}$$

A reduction in sulfate (due, for example, to a reduction in SO_2 emissions) leads to a displacement of particulate ammonium toward the gas phase, which leads to an increase of ammonium nitrate. In this specific case, the increase of ammonium nitrate is slightly greater than the decrease of ammonium sulfate and, as a result, there is a slight increase by 1 % of the total PM mass concentration.

Case 3: Sulfate remains at its original value of 5 µg m^{-3} and the total concentration of nitrate (gaseous nitric acid + particulate nitrate) is reduced by 50 %

Sulfate = 5 µg m^{-3}
Therefore, ammonium associated with sulfate as ammonium sulfate is: 1.87 µg m^{-3}. Ammonia available to form ammonium nitrate is the total ammonia concentration less that associated with sulfate, i.e.: $[NH_3]_a = 3.72 - 2.46 = 1.26$ ppb
Nitric acid available is half the previous concentration of total nitrate, i.e.: $[HNO3]_t = 4.76 / 2 = 2.38$ ppb
Solving for [NH$_4$NO$_3$] as in Case 2 leads to the following solution:
[NH$_4$NO$_3$] = 0.30 ppb
1 ppb of NH$_4$NO$_3$ = 3.38 µg m^{-3}; therefore, there is 1.01 µg m^{-3} of ammonium nitrate. Thus, the PM concentration is:

$$[PM] = 5 + 1.87 + 1.01 = 7.88 \text{ µg m}^{-3}$$

If the total nitric acid concentration is reduced by 50 % (for example, by reduction of the NO$_x$ emissions, without an increase in oxidant concentrations), the ammonium nitrate concentration is reduced by 60 %, which leads to a reduction in the PM concentration of 17 % (ammonium sulfate remains constant).

Case 4: Sulfate and total nitric acid remain at their original values and the total ammonia concentration (i.e., gaseous ammonia + particulate ammonium) is reduced by 50 %

Therefore, the total concentration of ammonia is: 3.72 / 2 = 1.86 ppb
Sulfate = 5 µg m^{-3}, i.e., 1.23 ppb.
The amount of ammonia needed to neutralized sulfate is: 2 × 1.23 = 2.46 ppb. Therefore, there is not enough ammonia to totally neutralize sulfate and there will be a mixture of ammonium sulfate and ammonium bisulfate. All ammonia will be present as particulate ammonium and there will not be any formation of ammonium nitrate, because there is no gaseous ammonia available to react with nitric acid. The ammonium concentration is: 1.86 × 0.76 = 1.41 µg m^{-3}. Therefore, the PM concentration is:

$$[PM] = 5 + 1.41 = 6.41 \text{ µg m}^{-3}$$

The PM concentration has been reduced by 32 %.
Here, a decrease of the ammonia concentration (for example, a reduction in agricultural emissions) is the most efficient method to reduce the PM concentration. This example highlights the complex nature of the relationships between ambient concentrations of inorganic PM and its gaseous precursor emissions. The case study of a spring air pollution episode in Paris will illustrate in Section 9.6.5 the challenge of developing emission control strategies that are efficient.

9.6.3 Secondary Organic PM

A decrease in the anthropogenic precursors of particulate organic species can be an efficient emission control strategy for reducing $PM_{2.5}$ concentrations. However, secondary organic PM contains an important biogenic fraction. VOC emissions from vegetation cannot be reduced (one could consider the use of tree species that emit lower amounts of VOC that are precursors of SOA, but it is more likely that economic considerations rather than environmental considerations govern the selection of the tree species used in commercial forests). Nevertheless, it is possible to affect the PM mass available for SVOC absorption. SOA formation consists mostly of the partitioning between the gas phase and the particulate phase (organic and, to a lower extent, aqueous). Therefore, the SVOC fraction that is present in the particulate phase depends on the volume of the organic particulate phase available to absorb SVOC: The greater this particulate organic phase, the greater the SVOC absorption. Conversely, a decrease of the volume of this organic particulate phase leads, for a given total concentration (i.e., gas- + particulate-phase) of a SVOC, to a decrease of its particulate fraction.

Accordingly, the strategies implemented to control primary PM emissions (for example, reduction of soot particles from diesel engines and wood burning) and secondary inorganic PM (i.e., the liquid phase content of those deliquescent particles) are beneficial and contribute to a decrease in particulate organic species concentrations, including those corresponding to SVOC that have unchanged total concentrations. In addition, a decrease in oxidant concentrations (such as ozone and the OH and NO_3 radicals produced by ozone reactions) will lead to a slower oxidation of VOC into SVOC and, therefore, a reduction in SVOC concentrations and a decrease of their contribution to organic PM.

9.6.4 Winter Air Pollution Episodes in Paris

We describe here the characteristics of the chemical composition and origin of PM leading to typical winter air pollution episodes in the Paris region, using a case study from December 2016, which provides a comprehensive data set. This episode lasted more than a week and was due to particularly strong anticyclonic conditions, i.e., the presence of a high-pressure system over the Paris region. For example, on December 5 and 6, the wind speed at 10 m was less than 2 m s^{-1} and the planetary boundary layer (PBL) height was about 100 m at night and 150 m during the day according to measurements available at Sirta (*Site instrumental de recherche par télédétection atmosphérique*), an experimental meteorological station located in Saclay, about 15 km southwest of Paris. Therefore, primary PM emissions were not dispersed much due to the low wind speeds and the limited volume of the PBL in which they were diluted, thereby leading to high concentrations. Figure 9.10 depicts the temporal evolution of the fine PM concentration (here PM_1, i.e., particles with an aerodynamic diameter ≤1 μm) and their chemical composition measured at Sirta. One notes the importance of carbonaceous PM, including black carbon and organic compounds. Although a fraction of organic carbon may be secondary, the low photochemical activity during winter suggests that primary organic carbon dominates the organic PM fraction. It is possible, however, that SVOC emissions may have led to condensation of a fraction of SVOC on primary particles because of the very low

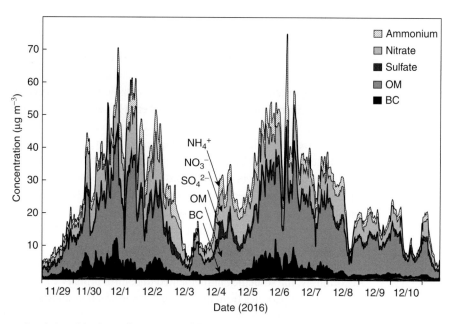

Figure 9.10. Temporal evolution of the chemical composition of fine PM (PM$_1$) during the air pollution episode of December 2016 in the Paris region. BC: black carbon measured with a light absorption method, OM: organic matter, SO$_4^{2-}$: sulfate, NO$_3^-$: nitrate, NH$_4^+$: ammonium. Data sources: SIRTA / IPSL − LSCE / INERIS.

ambient temperature (near 0 °C). The main sources of primary carbonaceous PM are diesel vehicles without particle filters and residential wood burning.

Inorganic species are less important, but are nevertheless present in notable amounts. Although ammonium sulfate contributes little to fine PM mass, ammonium nitrate contributes about 1/3 of fine PM mass on December 5 and 6. This ammonium nitrate fraction results from the oxidation of NO$_x$ into HNO$_3$ and the subsequent reaction of HNO$_3$ with NH$_3$, which is favored here by the very low temperature. The inorganic fraction is less than 50 % of fine PM mass during winter episodes; however, it becomes dominant during spring episodes as discussed in Section 9.6.5.

Numerical modeling of this air pollution episode with the Polyphemus/Polair3D chemical-transport model confirmed that local emissions were the main source of air pollutants (>90 %) and that the imported fraction was a minor contribution.

9.6.5 Spring Air Pollution Episodes in Paris

The meteorology of a spring air pollution episode in the Paris region is described in Chapter 3. We are interested here in investigating which emission control strategies can be developed to reduce the concentrations of fine PM during such episodes. Figure 9.11 depicts the chemical composition of fine PM measured at Sirta, southwest of Paris, during the first half of March 2014 (see also Fritz et al., 2015). PM$_{10}$ hourly concentrations exceeded the daily regulatory value of 50 µg m^{-3}, since the hourly PM$_{2.5}$ concentrations exceeded that concentration several times from March 11 to 15 (there is no daily regulatory concentration

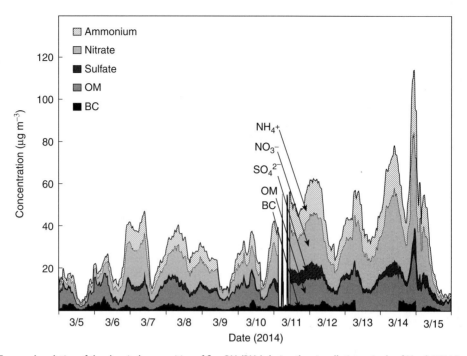

Figure 9.11. Temporal evolution of the chemical composition of fine PM (PM$_1$) during the air pollution episode of March 2014 in the Paris region. BC: black carbon measured with a light absorption method, OM: organic matter, SO$_4^{2-}$: sulfate, NO$_3^-$: nitrate, NH$_4^+$: ammonium. Some periods have missing data: part of March 11 for inorganic and organic compounds and some periods from March 13 to 15 for black carbon. Data sources: SIRTA / IPSL – LSCE / INERIS.

for PM$_{2.5}$ in Europe, see Chapter 15). One notes that the secondary fraction of fine PM dominates and that the contributions of black carbon and primary organic matter are low, particularly during the peak events. Therefore, this PM episode is mostly secondary and dominated by ammonium nitrate and SOA. This episode is interesting to investigate because it involves relationships between secondary PM and its gaseous precursors, as well as relationships between oxidants and precursors. Table 9.1 summarizes simulation results obtained for March 13, 2014 with the Polyphemus/Polair3D numerical chemical-transport model. These simulations used the EMEP emission inventory, which has a relatively coarse spatial resolution. Nevertheless, these simulations are useful to illustrate the main aspects of the relationships between pollutant sources and the ambient levels of air pollution.

The first emission control scenario targeted primary PM emissions within the Paris region with a 43 % reduction in those emissions. The decrease of the PM$_{10}$ concentration is only 11 %, which is consistent with the primary fraction of PM shown in Figure 9.11. It appears clearly that it is necessary to target emission controls of gaseous precursors of the secondary fraction of PM.

The second scenario targeted the reduction of gaseous precursors corresponding to the main chemical species present in the particles during the air pollution episode (reductions of the Paris region emissions are indicated in parentheses), i.e., NO$_x$ (−33 %) for nitrate,

Table 9.1. Emission control scenarios for the fine PM pollution episode of March 2014 in the Paris region (simulations conducted with the Polyphemus/Polair3D numerical chemical-transport model using the EMEP emission inventory).

Scenario*	Simulated PM_{10} concentration	Decrease in $[PM_{10}]$
Reference scenario	105 µg m^{-3}	–
Reduction in emissions of primary PM	93 µg m^{-3}	−11 %
Reduction in emissions of primary PM and gaseous precursors	92 µg m^{-3}	−12 %
Reduction in emissions of primary PM, gaseous precursors, and oxidant levels	85 µg m^{-3}	−19 %
Reduction in local emissions and imported air pollution	70 µg m^{-3}	−33 %

* See text for details.

NH_3 (−76 %) for ammonium, SO_2 (−7 %) for sulfate, and VOC (−23 %) for the organic fraction. The effect of these reductions is negligible, which seems surprising at first. However, a reduction in NO_x emissions leads to an increase in the oxidant levels in a high-NO_x regime (see Chapter 8) and, therefore, the daytime reaction leading to nitric acid formation shows little change in its kinetics ([NO_2] decreases, but [OH] increases). The important reduction in NH_3 emissions shows little effect (−6 % only for ammonium), because (1) NH_3 is not the limiting species for ammonium nitrate formation in the Paris region and (2) NH_3 has an atmospheric lifetime on the order of a week, which implies that the NH_3 concentrations observed in the Paris region are due in great part to long-range atmospheric transport originating from other regions and/or countries. The emissions of SO_2, the precursor of sulfate, are low in the Paris region and sulfate is transported over long distances from other regions. The VOC emission reductions only target anthropogenic VOC (which decrease by 7 %), but has only a small effect on biogenic SOA, which are major contributors to organic PM in spring.

The third scenario takes into account the high-NO_x regime of the Paris region and, therefore, targets a stronger reduction of VOC emissions (−85 %) in order to decrease the oxidant levels, and a more limited reduction (−15 %) of NO_x emissions. This strategy is efficient (−19 % for [PM_{10}]), but still fails to decrease PM_{10} concentrations down to their daily regulatory value. The reason is that a significant fraction of the atmospheric PM concentrations in the Paris region originates from other regions via long-range atmospheric transport.

Therefore, a fourth scenario targeted a larger domain for reductions of PM and gaseous precursor emissions. A reduction by one third of the PM_{10} concentration is then simulated. Although this decrease is still not sufficient to reach the regulatory value, it suggests that a strategy that targets a reduction of some selected gaseous precursors combined with a decrease (or at a minimum a status quo) of the oxidant concentrations over a domain that is much larger than the Paris region is needed to reduce PM_{10} levels significantly during spring air pollution episodes.

9.7 Numerical Modeling of Atmospheric Particulate Matter

9.7.1 Particle Size Distribution

The particle size distribution can be represented by a continuous distribution with a range of atmospheric particle sizes ranging from a few nanometers to several micrometers. However, such a representation is not applicable to air quality modeling because of the very large computational times that would be required to calculate its evolution due to nucleation, coagulation, condensation, and evaporation. Therefore, simpler representations are used. They use the logarithm (typically with base 10, i.e., decimal logarithm) of the particle diameter in order to cover the full spectrum of particle sizes in an optimal manner. There are two main approaches used in air quality simulation models: the modal approach and the sectional approach.

The Modal Approach

This approach is based on the assumption that atmospheric particles can be represented by distinct modes, which include mainly the nucleation mode (ultrafine particles), the accumulation mode (mostly fine particles), and the coarse mode (mostly primary particles produced by mechanical processes) (Whitby, 1978). The observation of atmospheric particle concentrations suggests the presence of such modes, although in some cases the use of only three modes is too simplistic, because there may be more than three modes present in the atmosphere (see Section 9.2.1). Measurements suggest that the number or volume distribution of particles in a mode can be approximately represented by a lognormal distribution of the particle concentration as a function of the logarithm of the particle diameter. Therefore, the distribution of the particle concentrations can be represented by three lognormal functions, each representing a mode and being characterized by a median diameter (or mean diameter since these two diameters are equivalent for a lognormal function), a standard deviation, and a total (or maximum) concentration of the mode. The standard mathematical representation of a lognormal distribution typically uses the natural logarithm. However, in practice, the decimal logarithm is used for the particle diameters. Therefore, the decimal logarithm is used here, except for the relationships between properties of the number and volume concentration distributions, for which the natural logarithm is more appropriate.

For example, the representation of the particle number concentration distribution as a function of the diameter, $\eta_{p,a}(d_p)$, is expressed as follows for one of the modes (the accumulation mode is used here; it is indicated by the subscript a):

$$\eta_{p,a}(d_p) = \frac{N_a}{(2\pi)^{1/2}\log(\sigma_a)} \exp\left(-\frac{\left(\log(d_p) - \log(d_{p,a,n})\right)^2}{2\left(\log(\sigma_a)\right)^2}\right) \qquad (9.21)$$

where $d_{p,a,n}$ is the median (also mean) diameter of the mode, σ_a is the standard deviation and N_a is the total number concentration of particles in that mode. If the distribution of the particle number concentration is lognormal, then the distributions of the surface and volume concentrations (and, assuming uniform particle density, the distribution of the mass concentration) are also lognormal. For example, the particle volume concentration is given as follows:

$$v_{p,a}(d_p) = \frac{V_a}{(2\pi)^{1/2}\log(\sigma_a)} \exp\left(-\frac{\left(\log(d_p) - \log(d_{p,a,v})\right)^2}{2\left(\log(\sigma_a)\right)^2}\right) \quad (9.22)$$

where $v_{p,a}(d_p)$ is the log-normal distribution of the volume concentration as function of the particle diameter, d_p, $d_{p,a,v}$ is the median (also mean) diameter of the mode, σ_a is the standard deviation, and V_a is the total volume concentration of particles in that mode. The standard deviation is identical in all distributions (number, surface, and volume). However, the median diameter differs depending on the variable considered in the distribution. The relationship between the median diameters of the number and volume concentration distributions is as follows:

$$\ln(d_{p,a,v}) = \ln(d_{p,a,n}) + 3\left(\ln(\sigma_a)\right)^2 \quad (9.23)$$

The relationship between the total number concentration and the total volume concentration is as follows:

$$V_a = \frac{\pi}{6} N_a \exp\left(3\ln(d_{p,a,n}) + 4.5\left(\ln(\sigma_a)\right)^2\right) \quad (9.24)$$

The advantage of the modal representation is that it is consistent with the conceptual model of atmospheric aerosols (three modes representing three distinct groups of particles) and it is mathematically simple (lognormal functions allow one to convert from number to surface or volume concentrations analytically). However, it is essential to use variable median diameters and standard deviations when solving the general dynamic equation.

There are a few disadvantages. There is no analytical solution to the general dynamic equation, because the concentration distribution is no longer lognormal following coagulation and condensation/evaporation processes. Therefore, some approximations are necessary to constrain the solution and maintain lognormal distributions for all modes. Such approximations are typically appropriate for the accumulation and coarse modes, but they can lead to erroneous results when simulating ultrafine particles. Indeed, (1) the effect of coagulation on the nucleation mode is poorly represented with lognormal distributions and (2) the Kelvin effect cannot be simulated with a single nucleation mode. Taking into account a larger number of modes could help minimize such problems, but the model formulation then becomes more difficult to develop because of significant overlaps between modes.

Despite these shortcomings (which affect mostly ultrafine particles), the modal approach is used in a large number of air quality models, including that of the U.S. Environmental Protection Agency (EPA) (CMAQ model), and many global models, due to its efficient computational times.

The Sectional Approach

The sectional approach consists of discretizing the distribution (number, volume, mass concentration ...) as a function of the particle diameter or its logarithm (Gelbard and Seinfeld, 1980). Therefore, the distribution corresponds to a histogram of the concentrations by particle size sections. The particle number and volume concentrations for particle size section i are expressed as follows:

$$N_i = \int_{d_{p,i-1}}^{d_{p,i}} n_{p,d}(d_p) \, \mathrm{d}d_p \qquad (9.25)$$

$$V_i = \int_{d_{p,i-1}}^{d_{p,i}} n_{p,d}(d_p) \, \frac{\pi}{6} d_p^3 \, \mathrm{d}d_p \qquad (9.26)$$

where $n_{p,d}(d_p)$ is the distribution of the number concentration as a function of the diameter, d_p, and the subscript i corresponds to the section bounded by $d_{p,i-1}$ and $d_{p,i}$. Generally, the number and volume (or mass) distributions are represented as a function of the decimal logarithm of the particle diameter, $\log(d_p)$. If the sections are selected to be of equal range on a logarithmic scale, then $(d_{p,i} / d_{p,i-1})$ = constant. However, it is not necessary that all sections be of equal range.

The advantage of this approach is that it is mathematically simple to manage, since it consists of a discretization of the particle diameter spectrum into sections. Then, to solve the general dynamic equation, it suffices to discretize the equations for coagulation and condensation/evaporation (see Section 9.7.4). Nucleation is treated simply as the creation of new particles in the section with the smallest particle diameter.

The disadvantages of this approach are as follows. If a small number of sections is used, (1) the representation of the particle distribution is not very accurate (for example, nucleated particles may appear in a section that covers a range of particles that is too large) and (2) the numerical solution may be inaccurate (for example, due to numerical diffusion when solving for condensation). If a large number of sections is used, then the representation of the particle distribution should be sufficiently fine and the solution of the aerosol dynamics should be fairly accurate. However, the computational costs may become large. Therefore, one must optimize between accuracy of the solution and computational burden.

Another disadvantage of the sectional approach is that the solution is obtained for only one variable of the particle concentration (number or mass, in most cases). Then, the calculation of the other variables (for example, the number concentration from the mass concentration) leads to a change of that variable with time that is not consistent with the general dynamic equation (except if the number of sections is infinitely large). A solution to this problem involves solving the dynamic equation for the number concentration for the sections with small diameters (since the mass of those sections is negligible) and the dynamic equation for the volume (or mass) for the sections with large diameters (since the number concentration in those sections is negligible). The transition diameter between these two solutions is typically about 100 nm. Another approach is to let the diameter representative of a given section vary, which allows one to maintain the relationship between the number and mass concentrations of each section with this moving diameter and to jointly simulate the number and mass concentrations (see "Numerical modeling" in Section 9.7.4).

An advantage of the sectional approach is that it is possible to use a large number of sections, unlike the modal approach. Then, a good resolution of the chemical composition of particles as a function of their size may be obtained. In addition, it is possible to discretize also the chemical composition of the particles, which presents the advantage of simulating particles of a same size, but of different chemical compositions (i.e., an external aerosol mixture).

The sectional approach is currently used in most air quality models that are applied at regional and urban scales, as well as in some global models. The number of sections being used varies among models and also depending on the application. Typically, a number of sections on the order of six to eight provides a good compromise between computational time and accuracy.

Comparative evaluation studies for lognormal and sectional approaches have been performed, which illustrate the advantages and shortcomings of these two approaches (Zhang et al., 1999; Sartelet et al., 2006; Devilliers et al., 2013).

9.7.2 Inorganic Aerosols

The inorganic chemical composition of particles may vary when there is condensation of secondary chemical species (sulfate, nitrate, ammonium) or evaporation of semi-volatile species (for example, ammonium nitrate). A particle will tend toward equilibrium with the surrounding gas phase to minimize its energy. Thus, it suffices to calculate the Gibbs energy of that particle, which should be minimum at thermodynamic equilibrium. This calculation requires calculating the chemical potential of each species present in the particle and minimizing the Gibbs energy of the particulate system. When a large number of species must be taken into account, such a minimization can be difficult to perform (see "Numerical modeling" in Section 9.7.4). Therefore, the solution tends to be costly in terms of computational time.

Another approach consists of simplifying the chemical system by using conceptual models. For example, if the system is ammonia-poor, one assumes that there will not be any ammonium nitrate formation. One may also estimate, based on the relative molar amounts of ammonia and sulfate, whether there will be formation of ammonium bisulfate or ammonium sulfate. Thus, it is possible to eliminate some species from the system, thereby simplifying the system of equations to be solved. Next, one writes the system thermodynamics in terms of thermodynamic equilibria, which are then solved numerically as a system of algebraic equations. This approach is computationally more efficient than the former one. Clearly, it is not as accurate since some approximations are involved when simplifying the system. However, a comparison of these two distinct approaches has shown that the latter approach can be sufficiently accurate for most cases. Several models have been developed using this latter approach, such as MARS, EQUISOLV, and ISORROPIA.

ISORROPIA (Fountoukis and Nenes, 2007) is currently the most widely used inorganic aerosol model in air quality simulation models.

9.7.3 Organic Aerosols

Organic aerosols contain a very large number of species, most of which have not yet been identified experimentally. Therefore, organic aerosol models must necessarily involve a larger number of assumptions than used in inorganic aerosol models.

The first studies of SOA formation in smog chambers have been analyzed using parameterizations of the SOA yield as a function of the particulate organic matter formed. If one assumes that there is partitioning of semi-volatile compounds between a gas phase and an organic particulate phase, then Raoult's law can be applied to quantify this partitioning. When the particulate organic mass increases, the partitioning of the semi-volatile compounds is displaced toward the particulate phase, since there is more organic phase volume available for SVOC absorption. It is possible to parameterize this relationship using only two surrogate species in order to obtain quantitative relationships for the SOA yield as a function of the particulate organic mass. The yield increases with the particulate organic mass until it tends toward an asymptotic value, usually for high particulate mass concentrations. One should note that these high concentrations are typically not representative of atmospheric conditions and that yields can be significantly lower in the ambient atmosphere than those obtained from smog chamber experiments. Nevertheless, this type of parameterization, generally referred to as two-compound Odum parameterization, allows one to represent SOA formation from gaseous precursors taking into account kinetic aspects, which depend on the first oxidation step of the VOC studied, and thermodynamic aspects, which depend on the partitioning estimated from the statistical regression performed on the experimental data (Odum et al., 1996). The four parameters obtained from the statistical regression are the stoichiometric coefficients and the partitioning coefficients of the two SVOC. The SOA yield, Y, resulting from the oxidation of one (or several) VOC is defined as the ratio of the particulate organic mass formed, ΔM_o in g m^{-3}, and the VOC mass that has reacted, $\Delta[\text{VOC}]$ in g m^{-3}:

$$Y = \frac{\Delta M_o}{\Delta[\text{VOC}]} \tag{9.27}$$

One assumes that the particulate organic mass is composed of a limited number (typically two) of SVOC and that their partitioning between the gas phase and the particulate phase is governed by Raoult's law, which is expressed as follows for SVOC$_i$ (Pankow, 1994):

$$K_{om,i} = \frac{RT}{MM_{om}\gamma_i P_{s,i}} \tag{9.28}$$

where $K_{om,i}$ is the gas/particle partitioning coefficient in m^3 g^{-1}, R is the ideal gas law constant in m^3 atm mol^{-1} K^{-1}, T is the temperature in K, MM_{om} is the molar mass of the particulate organic phase in g mol^{-1}, γ_i is the activity coefficient of SVOC$_i$ in the particulate phase, and $P_{s,i}$ is the saturation vapor pressure of SVOC$_i$ in atm.

The concentration of SVOC$_i$ formed is (in g m^{-3}):

$$[\text{SVOC}_i] = \alpha_i \, \Delta[\text{VOC}] \tag{9.29}$$

where α_i is the stoichiometric coefficient of the reaction leading to the oxidation of VOC into SVOC_i. Since SVOC_i partitions between the gas and particulate phases according to Raoult's law, given a particulate organic mass M_o, the amount of SVOC_i can be written in terms of an amount present in the gas phase and an amount present in the particulate phase:

$$[\text{SVOC}_i] = [\text{SVOC}_{i,g}] + [\text{SVOC}_{i,om}] \tag{9.30}$$

where $[\text{SVOC}_{i,g}]$ is the concentration of SVOC_i in the gas phase in g m^{-3} and $[\text{SVOC}_{i,om}]$ is the concentration of SVOC_i present in the particulate phase in g m^{-3}. The partitioning between these two phases is quantified using a partitioning constant that is the ratio of the concentrations of SVOC_i in the particulate phase (i.e., expressed in g of SVOC_i per g of M_o) and the concentration of SVOC_i in the gas phase. Therefore, this partitioning constant, $K_{om,i}$, is expressed in m^3 of air per g of particulate organic mass:

$$K_{om,i} = \frac{[\text{SVOC}_{i,om}]}{M_o \, [\text{SVOC}_{i,g}]} \tag{9.31}$$

Thus:

$$[\text{SVOC}_i] = \alpha_i \, \Delta[\text{VOC}] = [\text{SVOC}_{i,om}] + \frac{[\text{SVOC}_{i,om}]}{M_o \, K_{om,i}} \tag{9.32}$$

$$[\text{SVOC}_{i,om}] = \frac{\alpha_i \, M_o \, K_{om,i} \Delta[\text{VOC}]}{(1 + M_o K_{om,i})} \tag{9.33}$$

In the case where there are several SVOC_i, the particulate organic mass formed is as follows:

$$\Delta M_o = \sum_i [\text{SVOC}_{i,om}] = \Delta[\text{VOC}] \sum_i \frac{\alpha_i \, M_o \, K_{om,i}}{(1 + M_o \, K_{om,i})} \tag{9.34}$$

The yield may then be expressed in terms of the stoichiometric coefficients of formation of the SVOC_i, α_i, their partitioning coefficients, $K_{om,i}$, and the particulate organic mass formed, M_o:

$$Y = \frac{\Delta M_o}{\Delta[\text{VOC}]} = \sum_i \left(\frac{\alpha_i \, M_o \, K_{om,i}}{(1 + M_o \, K_{om,i})} \right) \tag{9.35}$$

If there is no particulate organic mass present at the start of the experiment, $\Delta M_o = M_o$. Y and M_o are measured at different times of the experiment and the coefficients α_i and $K_{om,I}$ are calculated by regression. Therefore, in the case of two SVOC_i, four coefficients must be estimated and, therefore, measurements must be made at four distinct times at the minimum (a greater number of measurement sets provides a regression that is statistically more robust).

One must note that most of those experiments have been conducted at fairly low relative humidity. Therefore, only absorption into an organic phase was taken into account and absorption into an aqueous phase was not considered. For some compounds such as those obtained from isoprene oxidation, aqueous absorption (dissolution) prevails over the hydrophobic absorption into an organic phase in the ambient atmosphere. Then, this type of parameterization would underestimate the yields in the ambient atmosphere under humid conditions (Couvidat and Seigneur, 2011).

Modeling of SOA formation from experimental data obtained in smog chambers can be done with other approaches than the two-compound Odum approach. Two main approaches are currently used: (1) the VBS approach (Donahue et al., 2006) and (2) the surrogate molecule approach (Pun et al., 2006). The VBS approach involves using a fixed number of organic compounds with predefined saturation vapor pressures (or concentrations): it is the volatility basis set (VBS). The regression is then performed with these compounds for a given experimental data set to obtain the stoichiometric coefficients (their partitioning coefficients are predefined based on their saturation vapor concentrations; see Equation 9.28). The surrogate molecule approach involves identifying the main compounds present in SOA and to use a limited number of these products to represent the whole SOA. The partitioning coefficients are estimated using theoretical or semi-empirical methods (functional group methods) and the stoichiometric coefficients are obtained from smog chamber data. One advantage of the surrogate molecule approach is that it can treat both hydrophobic SVOC partitioning into an organic phase and hydrophilic SVOC partitioning into an aqueous phase.

The addition of oxygen atoms tends to increase the water solubility of the compounds. Both methods have been modified to account for additional functionalization of the organic molecules following successive oxidation steps. The VBS method uses a second dimension (the first dimension being the saturation vapor concentrations), which corresponds to the O:C ratio (Donahue et al., 2011). The surrogate molecule approach adds additional oxidation steps and surrogate molecules that are more oxidized than those of the first oxidation step (Couvidat et al., 2012). Other methods have been developed to parameterize the evolution of SOA as a function of their oxidation state and volatility; they have been summarized by Seinfeld and Pandis (2016).

9.7.4 Numerical Modeling

Numerical modeling of atmospheric particles consists of two components: (1) the solution of the general dynamic equation and (2) the solution of the chemical composition of particles.

The general dynamic equation is an integro-differential equation. The term representing the condensation and evaporation processes is hyperbolic. Therefore, the numerical diffusion problem found for the advection term of the atmospheric diffusion equation (see Chapter 6) is also present here. Therefore, the numerical algorithms presented in Chapter 6 to address this problem can also be used to solve the condensation/evaporation term of the particle dynamic equation. The other two terms, i.e., nucleation and coagulation do not present any particular numerical difficulty once the size distribution has been defined. Zhang et al. (1999) conducted a comparative evaluation of several numerical algorithms for the various terms of the general dynamic equation. The moving diameter method of Jacobson and Turco (1995) seems to be the most robust (Zhang et al., 1999;

Devilliers et al., 2013) and is, therefore, recommended for the simulation of the particle size distribution with a sectional representation.

The solution of the chemical composition of particles requires treating the thermodynamic equilibria among the various chemical species present in the particles (in both the liquid and solid phases) and those between the particle phase and the gas phase. In addition, it may be necessary to treat the mass transfer between the gas phase and the particles explicitly, i.e., by accounting for the kinetics of this mass transfer. For fine particles, this mass transfer is sufficiently fast that the assumption of equilibrium between the particles and the gas phase is appropriate. On the other hand, the mass transfer kinetics slows down for coarse particles (see Chapter 11) and mass transfer must be taken into account for coarse particles. It is possible to separate fine and coarse particles in terms of their numerical treatment in order to assume equilibrium for fine particles and explicitly treat the mass transfer kinetics for coarse particles (Capaldo et al., 2000). The numerical solution of the mass transfer equation (see Chapter 11) does not present any particular difficulty, but it may affect the numerical stability of the solution of the equations treating the chemical composition of particles.

The solution of the equations that govern the chemical composition of particles can be obtained according to the second law of thermodynamics by minimizing the Gibbs energy of the chemical system (assuming that pressure and temperature are locally constant). However, this system of equations may be large and difficult to minimize. In particular, it is essential to find the global minimum of the system and not a local minimum, which requires using appropriate numerical algorithms. Therefore, this approach may be computationally demanding and other approaches have been developed that are less accurate, but computationally more efficient. Such approaches, which were mentioned in Section 9.7.2, are based on a system of equations based on thermodynamic equilibrium constants to obtain the particulate-phase concentrations of a reduced set of chemical species. The electroneutrality of the particulate phase and mass conservation between the gas phase and the particulate phase must of course be verified. Zhang et al. (2000) compared different numerical models available to calculate the chemical composition of inorganic particles. Currently, the ISORROPIA model (Fountoukis and Nenes, 2007), which uses simplifying assumptions based on various chemical regimes, is widely used in air quality models. Mass transfer between the gas and particulate phases of several chemical species may lead to numerical oscillations arising from opposite modifications of some chemical concentrations affected by (1) mass transfer and (2) chemical equilibria. The PNG-EQUISOLV II model (Jacobson, 2005b) avoids those numerical oscillations, while maintaining long integration time steps. For the chemical equilibria calculation, PNG-EQUISOLV II is based on the solution of chemical equilibria (rather than a minimization of the Gibbs energy), but it uses fewer simplifying assumptions than ISORROPIA II.

Solving for the chemical composition of the organic fraction of particles follows the same approach as used for inorganic species. However, it may be more complicated if one takes into account the possible separation of several organic phases (for example, distinguishing between hydrophobic and hydrophilic organic species) and/or the viscosity of particles, which may affect the diffusion of chemical species within the particles. Couvidat and Sartelet (2015) have developed a model, which treats these processes; other similar models are referenced in their article.

The joint solution of the particle size distribution and chemical composition is a difficult problem, unless one assumes that all particles of a given size (or size range) have the same chemical composition (internal mixing assumption). If particles of a given size have different chemical compositions, particles are then considered to be externally mixed. Such external mixing has been observed in the atmosphere, in particular near emission sources where freshly emitted particles are present near aged particles (Hughes et al., 2000). A bi-dimensional discretization (for both particle size and chemical composition) can be used to solve this problem (Zhu et al., 2015).

Problems

Problem 9.1 Atmospheric particles: Mass and number
Given 1 µg m^{-3} of particles with a diameter of 0.5 µm and 0.01 µg m^{-3} of particles with a diameter of 0.05 µm, which particles (of diameter of 0.5 or 0.05 µm) have the larger number concentration? One assumes that all these particles have the same density.

Problem 9.2 Aerosol dynamics
On-road traffic emits ultrafine and fine particles into the atmosphere. Particle concentrations are greater near the road and decrease with increasing distance from the road. Between ultrafine and fine particles, which ones have the concentrations that decrease the most rapidly as a function of time? Calculate the concentrations after 1 hour and explain the reason of this difference between the evolution of ultrafine and fine particles. Subscripts u, f, and b are used for ultrafine, fine, and background particles, respectively. The particles are assumed to be monodispersed (all particles in a given mode have the same diameter) and have the following diameters.

- Diameter of ultrafine particles: $d_{p,u} = 0.02$ µm
- Diameter of fine particles: $d_{p,f} = 0.3$ µm
- Diameter of background particles: $d_{p,b} = 0.5$ µm

The density of all particles is 1.6 g cm^{-3}. It is assumed that background particles have a greater concentration than those of particles emitted by traffic and, therefore, it may be assumed that coagulation of particles emitted from traffic occurs preferentially with background particles. The mass concentration of background particles is assumed to be 40 µg m^{-3}. The ambient temperature is 25 °C.

Problem 9.3 Secondary inorganic particles
a. The equilibrium constant for the formation of ammonium nitrate at low humidity is a function of temperature and at 15 °C, $K_{eq,s} = 2$ ppb^2. The initial concentration of sulfate is 1 ppb, that of nitric acid (HNO$_3$) is 2 ppb, and that of ammonia (NH$_3$) is 8 ppb. Is there formation of ammonium nitrate?
b. If there is ammonium nitrate formation, which strategy should be used to reduce its concentration (reduction of nitric acid or ammonia)?

Problem 9.4 Organic particles
An organic particulate mass of 10 µg m^{-3} consists of 4 µg m^{-3} of an anthropogenic semi-volatile organic compound (SVOC) and of 6 µg m^{-3} of a biogenic SVOC. The partitioning coefficients are $K_{om,a} = 0.4$ m^3 µg^{-1} for the anthropogenic compound and $K_{om,b} = 0.2$ m^3 µg^{-1} for the biogenic compound. What will the particulate organic concentration be if the total concentration (i.e., gas + particle) of the anthropogenic compound is reduced by 50 %?

References

Airparif, 2011. *Source Apportionment of Airborne Particles in the Île-de-France Region*, Final report, Airparif, Paris.

Ansari, A.S. and S.N. Pandis, 1998. Response of inorganic PM to precursor concentrations, *Environ. Sci. Technol.*, **32**, 2706–2714.

Bond, T.C. and R.W. Bergstrom, 2006. Light absorption by carbonaceous particles: An investigative review, *Aerosol Sci. Technol.*, **40**, 27–67.

Brégonzio-Rozier, L., F. Siekmann, C. Giorio, E. Pangui, S.B. Morales, B. Temime-Roussel, A. Gratien, V. Michoud, S. Ravier, M. Cazaunau, A. Tapparo, A. Monod, J.-F. Doussin, 2016. Gaseous products and secondary organic aerosol formation during long term oxidation of isoprene and methacrolein, *Atmos. Chem. Phys.*, **15**, 2953–2968.

Capaldo, K.P., C. Pilinis, and S.N. Pandis (2000). A computationally efficient hybrid approach for dynamic gas/aerosol transfer in air quality models, *Atmos. Environ.*, **34**, 3617–3627.

Chow, J.C., J.G. Watson, D. Crow, D.H. Lowenthal, and T. Merrifield, 2001. Comparison of IMPROVE and NIOSH carbon measurements, *Aerosol Sci. Technol.*, **34**, 23–34.

Couvidat, F. and C. Seigneur, 2011. Modeling secondary organic aerosol formation from isoprene under dry and humid conditions, *Atmos. Chem. Phys.*, **11**, 893–909.

Couvidat, F., É. Debry, K. Sartelet, and C. Seigneur, 2012. A hydrophilic/hydrophobic organic (H^{2}O) model: Development, evaluation and sensitivity analysis, *J. Geophys. Res.*, **117**, D10304, doi:10.1029/2011JD017214.

Couvidat, F., K. Sartelet, and C. Seigneur, 2013. Investigating the impact of aqueous-phase chemistry and wet deposition on organic aerosol formation using a molecular surrogate modeling approach, *Environ. Sci. Technol.*, **47**, 914–922.

Couvidat, F. and K. Sartelet, 2015. The Secondary Organic Aerosol Processor (SOAP) model: A unified model with different ranges of complexity based on the molecular surrogate approach. *Geosci. Model Dev.*, **8**, 1111–1138.

Devilliers, M., É. Debry, K. Sartelet, and C. Seigneur, 2013. A new algorithm to solve condensation/evaporation for ultra fine, fine, and coarse particles, *J. Aerosol Sci.*, **55**, 116–136.

Donahue, N.M., A.L. Robinson, C.O. Stanier, and S.N. Pandis, 2006. Coupled partitioning, dilution, and chemical aging of semivolatile organics, *Environ. Sci. Technol.*, **49**, 2635–2642.

Donahue, N.M., S.A. Epstein, S.N. Pandis, and A.L. Robinson, 2011. A two-dimensional volatility basis set: 1. Organic-aerosol mixing thermodynamics, *Atmos. Chem. Phys.*, **11**, 3303–3319.

Dumka, U.C., S. Tiwari, D.G. Kaskaoutis, P.K. Hopke, J. Singh, A.K. Srivastava, D.S. Bisht, S.D. Attri, S. Tyagi, A. Misra, and G.S. Munawar Pasha, 2017. Assessment of PM$_{2.5}$ chemical compositions in Delhi: Primary vs secondary emissions and contribution to light extinction and visibility degradation, *J. Atmos. Chem.*, **74**, 423–450.

Eddingsaas, N.C., C.L. Loza, L.D. Yee, M. Chan, K.A. Schilling, P.S. Chhabra, J.H. Seinfeld, and P.O. Wennberg, 2012. α-Pinene photooxidation under controlled chemical conditions – Part 2: SOA yield and composition in low- and high-NO$_x$ environments, *Atmos. Chem. Phys.*, **12**, 7413–7427.

Ervens, B., 2015. Modeling the processing of aerosol and trace gases in clouds and fogs, *Chem. Rev.*, **115**, 4157–4198.

Finlayson-Pitts, B.J. and J.N. Pitts, Jr., 2000. *Chemistry of the Upper and Lower Atmosphere: Theory, Experiments, and Applications*, Academic Press, New York.

Fountoukis, C. and A. Nenes, 2007. ISORROPIA II: A computationally efficient thermodynamic equilibrium model for K^+-Ca^{2+}-Mg^{2+}-NH_4^+-Na^+-SO_4^{2-}-NO_3^--Cl^--H_2O aerosols, *Atmos. Chem. Phys.*, **7**, 4639–4659.

Friedlander, S.K., 2000. *Smoke, Dust, and Haze – Fundamentals of Aerosol Dynamics*, Oxford University Press, Oxford.

Fritz, A., F. Dugay, C. Honoré, O. Sanchez, V. Ghersi, C. Songeur, P. Pernot, F. Mahé, S. Moukhtar, and J. Sciare, 2015. Bilan de l'épisode de pollution de mars 2014 et évaluation de la mise en place de la circulation alternée le 17 mars 2014 en Île-de-France, Pollution Atmosphérique, **Special issue, March 2015**, 25–34.

Fuchs, N.A., 1964. *Mechanics of Aerosols*, Pergamon, New York.

Fuchs, N.A. and A.G. Sutugin, 1971. High dispersed aerosol, in *Topics in Current Aerosol Research*, G.M. Hidy and J.R. Broch, eds., **2**, 1–60, Pergamon Press, Oxford, UK.

Gelbard, F. and J.H. Seinfeld, 1980. Simulation of multicomponent aerosol dynamics, *J. Colloid Interface Sci.*, **78**, 485–501.

Hering, S.V. and S.K. Friedlander, 1982. Origins of aerosol sulphur size distributions in the Los Angeles basin, *Atmos. Environ.*, **16**, 2647–2656.

Hughes, L.S., J.O. Allen, P. Bhave, M.J. Kleeman, G.R. Cass, D.-Y. Liu, D.P. Fergenson, B.D. Morrical, and K.A. Prather, 2000. Evolution of atmospheric particles along trajectories crossing the Los Angeles basin, *Environ. Sci. Technol.*, **34**, 3058–3068.

Jacobson, M.Z. and R.P. Turco, 1995. Simulating condensational growth, evaporation, and coagulation of aerosols using a combined moving and stationary size grid, *Aerosol Sci. Technol.*, **22**, 73–92.

Jacobson, M.Z., 2005a. *Fundamentals of Atmospheric Modeling*, Cambridge University Press, Cambridge, UK.

Jacobson, M.Z., 2005b. A solution to the problem of nonequilibirum acid/base gas-particle transfer at long time step, *Aerosol Sci. Technol.*, **39**, 92–103.

Jaoui, M., T.E. Kleindienst, J.H. Offenberg, M. Lewandowski, and W.A. Lonneman, 2012. SOA formation from the atmospheric oxidation of 2-methyl-3-buten-2-ol and its implications for $PM_{2.5}$, *Atmos. Chem. Phys.*, **12**, 2173–2188.

Kroll, J.H., N.L. Ng, S.M. Murphy, R.C. Flagan, and J.H. Seinfeld, 2006. Secondary organic aerosol formation from isoprene photooxidation, *Environ. Sci. Technol.*, **40**, 1869–1977.

Kuang, C., P.H. McMurry, A.V. McCormick, and F.L. Eisele, 2008. Dependence of nucleation rates on sulfuric acid vapor concentration in diverse atmospheric locations, *J. Geophys. Res.*, **113**, D10209.

Lee, A., A.H. Goldstein, J.H. Kroll, N.L. Ng, V. Varutbangkut, R.C. Flagan, and J.H. Seinfeld, 2006. Gas-phase products and secondary aerosol yields from the photooxidation of 16 different terpenes, *J. Geophys. Res.*, **111**, D17305.

Loza, C.L., J.S. Craven, L.D. Yee, M.M. Coggon, R.H. Schwantes, M. Shiraiwa, X. Zhang, K.A. Schilling, N.L. Ng, M.R. Canagaratna, P.J. Ziemann, R.C. Flagan, and J.H. Seinfeld, 2014. Secondary organic aerosol yields of 12-carbon alkanes, *Atmos. Chem. Phys.*, **14**, 1423–1439.

Matsunaga, A., K.S. Docherty, Y.B. Lim, and P.J. Ziemann, 2009. Composition and yields of secondary organic aerosol formed from OH radical-initiated reactions of linear alkenes in the presence of NO_x: modeling and measurements, *Atmos. Environ.*, **43**, 1349–1357.

Mozurkewich, M., 1993. The dissociation constant of ammonium nitrate and its dependence on temperature, relative humidity and particle size, *Atmos. Environ. Part A*, **27**, 261–270.

Ng, N.L., J.H. Kroll, A.W.H. Chan, P.S. Chhabra, R.C. Flagan, and J.H. Seinfeld, 2007a. Secondary organic aerosol formation from m-xylene, toluene, and benzene, *Atmos. Chem. Phys.*, **7**, 3909–3922.

Ng, N.L., P.S. Chhabra, A.W.H. Chan, J.D. Surratt, J.H. Kroll, A.J. Kwan, D.C. McCabe, P.O. Wennberg, A. Sorooshian, S.M. Murphy, N.F. Dalleska, R.C. Flagan, and J.H. Seinfeld, 2007b.

Effect of NO$_x$ level on secondary organic aerosol (SOA) formation from the photooxidation of terpenes, *Atmos. Chem. Phys.*, **7**, 5159–5174.

Ng, N.L., S.S. Brown, A.T. Archibald, E. Atlas, R.C. Cohen, J.N. Crowley, D.A. Day, N.M. Donahue, J.L. Fry, H. Fuchs, R.J. Griffin, M.I. Guzman, H. Herrmann, A. Hodzic, Y. Iinuma, J.L. Jimenez, A. Kiendler-Scharr, B.H. Lee, D.J. Luecken, J. Mao, R. McLaren, A. Mutzel, H.D. Osthoff, B. Ouyang, B. Picquet-Varrault, U. Platt, H.O.T. Pye, Y. Rudich, R.H. Schwantes, M. Shiraiwa, J. Stutz, J.A. Thornton, A. Tilgner, B.J. Williams, and R.A. Zaveri, 2017. Nitrate radicals and biogenic volatile organic compounds: Oxidation, mechanisms, and organic aerosol, *Atmos. Chem. Phys.*, **17**, 2103–2162.

Odum, J.R., T. Hoffmann, F. Bowman, D. Collins, R.C. Flagan, and J.H. Seinfeld, 1996. Gas/particle partitioning and secondary organic aerosol yields, *Environ. Sci. Technol.*, **30**, 2580–2585.

Pankow, J.F., 1994 An absorption model of gas/particle partitioning of organic compounds in the atmosphere, *Atmos. Environ.*, **28**, 185–188.

Perry's Chemical Engineers' Handbook, D.W. Green and R.H. Perry, eds., 2008. McGraw Hill, New York.

Petzold, A., H. Schloesser, P.J. Sheridan, W.P. Arnott, J.A. Ogren, and A. Virkkula, 2005. Evaluation of multiangle absorption photometry for measuring aerosol light absorption, *Aerosol Sci. Technol.*, **39**, 40–51.

Pun, B., C. Seigneur, and K. Lohman, 2006. Modeling secondary organic aerosol via multiphase partitioning with molecular data, *Environ. Sci. Technol.*, **40**, 4722–4731.

Sartelet, K., H. Hayami, B. Albriet, and B. Sportisse, 2006. Development and preliminary validation of a modal aerosol model for tropospheric chemistry: MAM, *Aerosol Sci. Technol.*, **40**, 118–127.

Saxena, P., L.M. Hildemann, P.H. McMurry, and J.H. Seinfeld, 1995. Organics alter hygroscopic behavior of atmospheric particles, *J. Geophys. Res.*, **100**, 18755–18770.

Schwarzenbach, R.P., P.M. Gschwend, and D.M. Imboden, 2003. *Environmental Organic Chemistry*, Wiley-Interscience, Hoboken, NJ.

Seinfeld, J.H. and S.N. Pandis, 2016. *Atmospheric Chemistry and Physics – From Air Pollution to Climate Change*, Wiley, New York.

Stelson, A.W., S.K. Friedlander, and J.H. Seinfeld, 1979. A note on the equilibrium relationship between ammonia and nitric acid and particulate ammonium nitrate, *Atmos. Environ.*, **13**, 369–371.

Stelson, A.W. and J.H. Seinfeld, 1982. Relative humidity and temperature dependence of the ammonium nitrate dissociation constant, *Atmos. Environ.*, **16**, 983–992.

Wang, H., M. Tian, X. Li, Q. Chang, J. Cao, F. Yang, Y. Ma, and K. He, 2015. Chemical composition and light extinction contribution of PM$_{2.5}$ in urban Beijing for a 1-year period, *Aerosol Air Quality Res.*, **15**, 2200–2211.

Weschler, C.J. and W.W. Nazaroff, 2008. Semivolatile organic compounds in indoor environments, *Atmos. Environ.*, **42**, 9018–9040.

Whitby, K.T., 1978. The physical characteristics of sulphur aerosols, *Atmos. Environ.*, **12**, 135–159.

Zhang, Y., C. Seigneur, J.H. Seinfeld, M.Z. Jacobson, and F. Binkowski, 1999. Simulation of aerosol dynamics: A comparative review of algorithms used in air quality models, *Aerosol Sci. Technol.*, **31**, 487–514.

Zhang, Y., C. Seigneur, J.H. Seinfeld, M. Jacobson, S.L. Clegg, and F.S. Binkowski, 2000. A comparative review of inorganic aerosol thermodynamic equilibrium modules: Similarities, differences, and their likely causes, *Atmos. Environ.*, **34**, 117–137.

Zhang, Y., P.H. McMurry, F. Yu, and M.Z. Jacobson, 2010. A comparative study of nucleation parameterizations: 1. Examination and evaluation of the formulations, *J. Geophys. Res.*, **115**, D20212, *doi*: 10.1029/2010JD014150.

Zhu, S., K. Sartelet, and C. Seigneur, 2015. A size-composition resolved aerosol model for simulating the dynamics of externally mixed particles: SCRAM (v 1.0), *Geosci. Model Dev.*, **8**, 1595–1612.

Ziemann, P.J. and R. Atkinson., 2012. Kinetics, products, and mechanisms of secondary organic aerosol formation, *Chem. Soc. Rev.*, **41**, 6582–6605.

10 Clouds and Acid Rain

This chapter describes the processes that lead to the formation of atmospheric pollutants in clouds and fogs via aqueous chemical transformations. Although the volume occupied by water droplets in the air is small, important chemical reactions occur in clouds. These reactions modify the atmospheric chemical composition and may lead to an increase of particulate mass when clouds and fogs evaporate or to acid rain when clouds precipitate. First, some general considerations on clouds and fogs are presented. Then, aqueous chemistry is addressed. Chemical equilibria and reactions have been described in other textbooks (e.g., Stumm and Morgan, 1995), and the focus here is on the processes pertaining to air pollution. This chapter treats in particular the transformations leading to the formation of sulfuric acid and nitric acid, two constituents of acid rain (if the cloud precipitates), as well as precursors of fine inorganic particles (if the cloud evaporates). The aqueous chemistry of organic compounds concerns mostly the formation of secondary organic aerosols (SOA) and is treated in Chapter 9. Finally, emission control strategies to reduce acid rain are discussed. Atmospheric deposition processes are treated in Chapter 11, and the impacts of acid rain on ecosystems are summarized in Chapter 13. Public policy programs developed to reduce acid rain in the United States are discussed in Chapter 15.

10.1 General Considerations on Clouds and Fogs

Clouds and fogs are formed in the atmosphere when the water vapor concentration exceeds its saturation vapor pressure (which depends on temperature, see Chapter 4). Water vapor condenses on hygroscopic particles, which grow and form cloud and fog droplets.

For clouds, the exceedance of the water saturation vapor pressure occurs when the air parcel rises and becomes colder, which leads to a lower saturation vapor pressure (see Chapter 4). A weak vertical velocity of the air parcel leads to stratus clouds, whereas a strong vertical velocity (i.e., convection) leads to cumulus clouds. The term nimbus is used to characterize precipitating clouds: nimbostratus and cumulonimbus.

Fog is a cloud that is in contact with the surface of the Earth. There are radiation and advection fogs. A radiation fog is formed when the temperature of the atmosphere decreases due to radiative transfer. For example, surface cooling at night because of infrared radiative transfer leads to a decrease in the temperature of the lower layers of the atmosphere, which are in contact with the surface. An advection fog is formed when an air mass is transported over a colder surface. This occurs, for example, in coastal areas where the ocean is particularly cold (for example, in San Francisco, California); the moist air mass encounters the cold ocean surface, which leads to water vapor condensation.

Cloud precipitation (rain, snow or ice) occurs when the fall velocity of the water droplet (or that of the ice crystal or snow flake) is greater than the vertical air parcel velocity. This fall velocity depends on gravity and the frictional force (see Chapter 11). Since the vertical velocity of an air parcel leading to a stratus cloud is less than that of an air parcel leading to a cumulus cloud, raindrops of a cumulonimbus are larger than those of a nimbostratus.

The liquid water content varies depending on the type of cloud or fog. Stratus clouds have liquid water contents on the order of 0.1 g of water per m^3 of air; cumulus clouds contain more liquid water with water contents up to 1 g m^{-3}. A precipitating cloud has a liquid water content greater than the equivalent non-precipitating cloud (for example, nimbostratus versus stratus), since rain drops are larger than non-precipitating cloud droplets. The liquid water content of a fog is generally low (for example, on the order of 10 mg m^{-3}) and the precipitation of fog droplets is small compared to that of a cloud.

A cloud contains mostly air: for a stratus cloud with a liquid water content of 0.1 g m^{-3}, the mass of water present in one m^3 of air is about 0.01 % of the mass of the air (1 m^3 of air weights about 1.2 kg at 20 °C at sea level, less at higher altitudes). The volume occupied by the cloud droplets is only 0.00001 % of the volume of the air parcel (given that at a pressure of 1 atm, the density of water is about 1,000 times that of the air).

Cloud droplets are larger than atmospheric particles, but sufficiently small that their fall velocity remains less than that of the ascending air. For example, a cloud droplet may have a diameter on the order of 40 μm. A fog droplet may have a diameter on the order of 10 μm and, therefore, it has a very small sedimentation velocity. Raindrops are larger with diameters ranging up to a few mm.

More comprehensive descriptions of clouds and fogs are available in meteorology textbooks (e.g., Ahrens, 2012).

10.2 Aqueous-phase Chemistry

Several processes must be taken into account in aqueous-phase chemistry:

– Mass transfer of chemical species between the gas phase and the liquid phase
– Chemical reactions and equilibria at the interface between the gas phase and the liquid phase
– Chemical reactions and equilibria in the liquid phase
– Non-ideality of concentrated aqueous solutions
– Electroneutrality of the aqueous solution

10.2.1 Mass Transfer of Chemical Species between the Gas Phase and the Liquid Phase

The mass transfer of a chemical species in the gas phase toward the liquid phase can be seen as a sequence of three steps:

- The mass transfer of the gas-phase molecule toward the droplet surface by molecular diffusion
- The thermodynamic equilibrium between the gas-phase and aqueous-phase concentrations at the droplet surface
- The mass transfer of the dissolved molecule in the aqueous phase from the droplet surface toward the inner droplet by molecular diffusion

These processes occur in the reverse direction in the case of a chemical species that volatilizes from the aqueous phase.

In most models of cloud chemistry, the system is assumed to be at equilibrium and the two mass transfer steps are neglected. Nevertheless, it is important to take them into account for the absorption of pollutants by falling raindrops (precipitation scavenging) as well as for heterogeneous reactions, which occur at the droplet surface.

10.2.2 Henry's Law

Thermodynamic equilibrium at the surface of a droplet is governed by Henry's law. Henry's law applies to dilute aqueous solutions. It relates the concentration of a chemical species in the gas phase (characterized by its partial pressure, P_i) to its activity in the aqueous phase as follows:

$$\gamma_i C_i = H_i P_i \qquad (10.1)$$

where C_i is the concentration in the droplet in M (moles per liter), γ_i is the activity coefficient of the chemical species in the aqueous phase, and H_i is the Henry's law constant, which depends on temperature. Partial pressure is generally expressed in atm and the Henry's law constant is, therefore, expressed in M atm^{-1}.

If the solution is very dilute, the solution can be assumed to be ideal and the activity coefficients become 1 (i.e., the chemical species concentrations are equal to their activities). As a matter of fact, the initial formulation of Henry's law used the species concentration, rather than its activity. This assumption is generally appropriate for clouds that have a large liquid water content. However, it may not be applicable to fogs, because they may contain high aqueous-phase concentrations of pollutants, particularly during their formation and evaporation.

The Henry's law constants of selected atmospheric chemical species are provided in Table 10.1.

10.2.3 Ionic Dissociations

In the aqueous phase, some chemical species dissociate into cations (with a positive charge) and anions (with a negative charge). This ionic dissociation leads to a displacement of the

Table 10.1. Henry's law constant, enthalpy of dissolution, effective Henry's law constant (at pH = 5.6), and fraction present in the aqueous phase (for a liquid water content of 1 g m^{-3}) at 25 °C for selected chemical species.

Chemical species	Henry's law constant (M atm^{-1})	Enthalpy of dissolution (kcal mole^{-1})	Effective Henry's law constant[a] (M atm^{-1})	Fraction in the aqueous phase
NO	1.9×10^{-3}	-2.9	1.9×10^{-3}	0.000005 %
NO$_2$	1.2×10^{-2}	-5.0	1.2×10^{-2}	0.00002 %
O$_3$	1.1×10^{-2}	-5.04	1.1×10^{-2}	0.00002 %
CO$_2$	3.4×10^{-2}	-4.85	4×10^{-2}	0.0001 %
SO$_2$	1.23	-6.25	6.5×10^{3}	14 %
NH$_3$	62	-8.17	2.6×10^{5}	87 %
HCHO[b]	6.3×10^{3}	-12.8	6.3×10^{3}	13 %
H$_2$O$_2$	7.45×10^{4}	-14.5	7.45×10^{4}	65 %
HNO$_3$	2.1×10^{5}	–	1.3×10^{12}	100 %

(a) For non-ionic species, the effective Henry's law constant is equal to the Henry's law constant.
(b) The Henry's law constant takes into account the hydrolysis of formaldehyde in solution, which leads to the formation of a diol (methylene glycol).
Source of Henry's law constants and enthalpies of dissolution: Seinfeld and Pandis (2016).

gas/droplet equilibrium if one considers the total aqueous-phase concentration of the chemical species, i.e., the sum of the concentrations of the non-dissociated species and corresponding ions. For example, for a diacid, such as sulfuric acid, H$_2$SO$_4$, or sulfur dioxide (which hydrolyzes into H$_2$SO$_3$):

$$H_2A(g) \leftrightarrow H_2A(aq) \quad H \quad \text{(R10.1)}$$

$$H_2A(aq) \leftrightarrow HA^- + H^+ \quad K_1 \quad \text{(R10.2)}$$

$$HA^- \leftrightarrow A^{2-} + H^+ \quad K_2 \quad \text{(R10.3)}$$

Assuming here an ideal solution, the concentration of the non-dissociated species is obtained according to Henry's law:

$$[H_2A(aq)] = H \, [H_2A(g)] \quad \text{(10.2)}$$

where H is the Henry's law constant in M atm^{-1}, the aqueous-phase concentration of H$_2$A is expressed in moles per liter, i.e., M, and the gas-phase concentration of H$_2$A is expressed in atmosphere (atm).

The dissociation equilibria lead to the formation of HA$^-$ and A^{2-} and their concentrations are related to that of H$_2$A as follows:

$$K_1 = \frac{[HA^-] \, [H^+]}{[H_2A(aq)]} \quad \text{(10.3)}$$

$$K_2 = \frac{[A^{2-}][H^+]}{[HA^-]} \tag{10.4}$$

The ensemble of the concentrations of H_2A in solution, including the ionic species HA^- and A^{2-}, $[H_2A(aq)]_t$, can be defined as follows:

$$[H_2A(aq)]_t = [H_2A(aq)] + [HA^-] + [A^{2-}] \tag{10.5}$$

Thus:
$$[H_2A(aq)]_t = [H_2A(aq)]\left(1 + \frac{K_1}{[H^+]} + \frac{K_1 K_2}{[H^+]^2}\right) \tag{10.6}$$

For example, in the case of sulfur dioxide, H_2SO_3, A is SO_3, which leads to:

$$SO_2(g) \ (+H_2O(l)) \leftrightarrow H_2SO_3(aq) \quad H_{SO2} = 1.23 \ M \ atm^{-1} \tag{R10.4}$$

$$H_2SO_3(aq) \leftrightarrow HSO_3^- + H^+ \quad K_1 = 1.3 \times 10^{-2} \ M \tag{R10.5}$$

$$HSO_3^- \leftrightarrow SO_3^{2-} + H^+ \quad K_2 = 6.6 \times 10^{-8} \ M \tag{R10.6}$$

where the notation (l) indicates liquid water. The values of the Henry's law constant and dissociation equilibrium constants are given at 25 °C. These constants depend on temperature, according to the van't Hoff relationship, as follows:

$$H(T) = H(T_{ref})\exp\left(\frac{\Delta H_A}{R}\left(\frac{1}{T_{ref}} - \frac{1}{T}\right)\right)$$

$$K(T) = K(T_{ref})\exp\left(\frac{\Delta H_R}{R}\left(\frac{1}{T_{ref}} - \frac{1}{T}\right)\right) \tag{10.7}$$

where ΔH_A is the enthalpy of dissolution and ΔH_R is the enthalpy of reaction. The enthalpy of dissolution is negative for most major atmospheric chemical species ($\Delta H_A = -6.25$ kcal mole^{-1} at 25 °C for SO_2). Therefore, if $T < T_{ref}$, the Henry's law constant will be greater than its reference value (generally given at 25 °C). In other words, a lower temperature favors dissolution. The enthalpy of reaction can be positive (for example, for the dissolved species of CO_2, H_2CO_3) or negative (for example, for the dissolved species of SO_2, H_2SO_3). If it is positive, a lower temperature will favor the non-dissociated species (for example, H_2CO_3, rather than HCO_3^- or CO_3^{2-}). If it is negative, a lower temperature will favor the ionic species (for example, SO_3^{2-} and HSO_3^-, rather than H_2SO_3).

The ensemble of species corresponding to SO_2(aq) is generally noted as S(IV), because sulfur is in oxidation state IV. The ensemble of species corresponding to sulfuric acid is noted S(VI), because sulfur is in oxidation state VI. For S(IV):

$$[S(IV)(aq)] = [H_2SO_3(aq)] + [HSO_3^-] + [SO_3^{2-}] \tag{10.8}$$

$$[S(IV)(aq)] = [H_2SO_3(aq)]\left(1 + \frac{K_1}{[H^+]} + \frac{K_1 K_2}{[H^+]^2}\right) \tag{10.9}$$

Therefore, the equilibrium between gas-phase and aqueous-phase SO_2 is expressed as follows:

$$\frac{[S(IV)(aq)]}{[SO_2(g)]} = \frac{[H_2SO_3(aq)]}{[SO_2(g)]}\left(1 + \frac{K_1}{[H^+]} + \frac{K_1 K_2}{[H^+]^2}\right) = H_{SO_2}\left(1 + \frac{K_1}{[H^+]} + \frac{K_1 K_2}{[H^+]^2}\right) \quad (10.10)$$

The relative fractions of the three species of sulfur dioxide in solution can be calculated as a function of pH as follows.

Equation 10.9 provides the fraction of $H_2SO_3(aq)$:

$$\frac{[H_2SO_3(aq)]}{[S(IV)(aq)]} = \left(1 + \frac{K_1}{[H^+]} + \frac{K_1 K_2}{[H^+]^2}\right)^{-1} \quad (10.11)$$

The HSO_3^- and SO_3^{2-} fractions are obtained from the ionic equilibrium relationships:

$$\frac{[HSO_3^-(aq)]}{[S(IV)(aq)]} = \left(1 + \frac{[H^+]}{K_1} + \frac{K_2}{[H^+]}\right)^{-1} \quad (10.12)$$

$$\frac{[SO_3^{2-}(aq)]}{[S(IV)(aq)]} = \left(1 + \frac{[H^+]^2}{K_1 K_2} + \frac{[H^+]}{K_2}\right)^{-1} \quad (10.13)$$

These fractions are illustrated as a function of pH in Figure 10.1. When the solution is acidic, the non-ionic S(IV) fraction dominates. On the other hand, when the solution is basic, the ionic equilibria are displaced toward the formation of H^+ ions and the sulfite ion, SO_3^{2-}, dominates. At pH values typical of clouds (i.e., between about 4 and 5.6), the bisulfite ion, HSO_3^-, dominates.

10.2.4 Effective Henry's Law Constant

An effective Henry's law constant, H_{eff}, includes all forms of a chemical species in solution. For example, for a species with two ionic dissociations:

$$H_{eff} = H\left(1 + \frac{K_1}{[H^+]} + \frac{K_1 K_2}{[H^+]^2}\right) \quad (10.14)$$

This effective Henry's law constant is proportional to the standard Henry's law constant, but it takes into account the dissociation of the molecular species into ionic species (via the dissociation constants, K_1 and K_2) and it depends on the pH of the solution (via $[H^+]$). The effective Henry's law constant is greater than the standard Henry's law constant, which reflects the fact that the dissociation of a molecular species in solution increases its overall solubility. This effective Henry's law constant is illustrated as a function of pH for sulfur dioxide in Figure 10.1. As indicated by Equation 10.14, it increases with pH, because a basic solution favors the dissolution of an acid.

For a monoacid HA, such as nitric acid, HNO_3:

$$H_{eff} = H\left(1 + \frac{K_1}{[H^+]}\right) \quad (10.15)$$

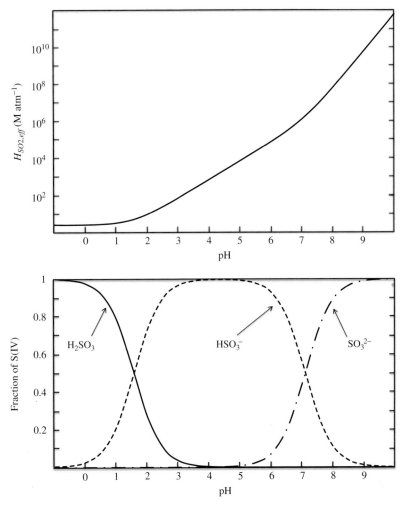

Figure 10.1. Dissolution of SO₂ in droplets. Effective Henry's law constant (top figure) and aqueous-phase chemical composition (bottom figure) of sulfur in oxidation state IV, S(IV), as a function of pH at 5 °C.

For a base, such as ammonia, NH_3 and NH_4OH:

$$H_{eff} = H\left(1 + \frac{K_1}{[OH^-]}\right) = H\left(1 + \frac{K_1[H^+]}{K_{H_2O}}\right) \quad (10.16)$$

This latter equation includes the dissociation of water into protons (H^+) and hydroxide ions (OH^-):

$$H_2O(l) \leftrightarrow H^+ + OH^- \quad (R10.7)$$

At 25 °C, the equilibrium constant for water dissociation is as follows:

$$K'_{H_2O} = \frac{[H^+][OH^-]}{[H_2O(l)]} = 1.81 \times 10^{-16} \text{ M} \quad (10.17)$$

The density of water at 25 °C and 1 atm is 0.997 g cm^{-3}, i.e., 997 g liter^{-1}; the molar mass of water is 18 g mole^{-1}, thus:

$$[H_2O(l)] = \frac{997}{18} = 55.4 \text{ M} \tag{10.18}$$

Therefore:

$$K_{H_2O} = [H^+][OH^-] = 10^{-14} \text{ M}^2 \tag{10.19}$$

10.2.5 pH and Electroneutrality

The pH function is defined as the negative value of the base 10 logarithm of the proton activity:

$$pH = -\log(\gamma_{H^+}[H^+]) \tag{10.20}$$

where γ_{H^+} is the activity coefficient of H^+. In the case of a dilute solution, the solution may be assumed to be ideal. Then, the activity coefficient is unity and one obtains the following relationship, which is most commonly used:

$$pH = -\log([H^+]) \tag{10.21}$$

A neutral solution (for example, distilled water) has a pH of 7 at 25 °C, since there are as many H^+ ions as there are OH^- ions:

$$[H^+] = [OH^-] = 10^{-7} \text{ M} \tag{10.22}$$

An acidic solution will contain more H^+ ions than OH^- ions and its pH will be less than 7. A basic solution will contain more OH^- ions than H^+ ions and its pH will be greater than 7. A very concentrated acidic solution (such as polluted fog droplets) may have a negative pH.

Electroneutrality means that the cation positive charges are balanced by the anion negative charges. For example, for S(IV) in solution:

$$[H^+] = [OH^-] + [HSO_3^-] + 2\,[SO_3^{2-}] \tag{10.23}$$

The factor of 2 used with the sulfite ion concentration, $[SO_3^{2-}]$, accounts for the fact that each sulfite ion has two negative charges.

The pH of a non-polluted rain can be calculated based on electroneutrality and the thermodynamic equilibria of carbon dioxide, CO_2, which has an atmospheric concentration of about 400 ppm. CO_2 is hydrolyzed following its dissolution in the aqueous phase (values of equilibrium constants are given at 25 °C):

$$CO_2(g) \; (+ H_2O(l)) \leftrightarrow H_2CO_3(aq) \quad H = 3.4 \times 10^{-2} \text{ M atm}^{-1} \tag{R10.8}$$

Carbonic acid, H_2CO_3, is a weak diacid (i.e., it does not dissociate much), which leads to bicarbonate ions, HCO_3^-, and carbonate ions, CO_3^{2-}:

$$H_2CO_3(aq) \leftrightarrow HCO_3^- + H^+ \quad K_1 = 4.3 \times 10^{-7} \text{ M} \tag{R10.9}$$

$$HCO_3^- \leftrightarrow CO_3^{2-} + H^+ \quad K_2 = 4.7 \times 10^{-11} \text{ M} \tag{R10.10}$$

The electroneutrality equation is written as follows:

$$[H^+] = [OH^-] + [HCO_3^-] + 2\,[CO_3^{2-}] \quad (10.24)$$

The ion concentrations can be expressed as a function of $[H^+]$ (which is the variable that must be calculated) and of the H_2CO_3 concentration (which can be calculated from the concentration of $CO_2(g)$ and its Henry's law constant). The fraction of CO_2 present in the aqueous phase is assumed to be negligible compared to the amount present in the gas phase (this assumption will be verified at the end of the calculation). According to Henry's law, assuming an atmospheric pressure of 1 atm:

$$[H_2CO_3(aq)] = 3.4 \times 10^{-2}[CO_2(g)] = 3.4 \times 10^{-2} \times 400 \times 10^{-6} = 1.36 \times 10^{-5}\ M$$

Equilibrium R10.9 leads to : $[HCO_3^-] = [H_2CO_3(aq)] \times 4.3 \times 10^{-7} / [H^+]$

Equilibrium R10.10 leads to : $[CO_3^{2-}] = [HCO_3^-] \times 4.7 \times 10^{-11} / [H^+]$

By combining these two equations:

$$[CO_3^{2-}] = [H_2CO_3(aq)] \times 4.3 \times 10^{-7} \times 4.7 \times 10^{-11} / [H^+]^2$$

Given that: $[OH^-] = 10^{-14} / [H^+]$; the electroneutrality relationship may be written as a function of $[H_2CO_3(aq)]$ and $[H^+]$ by expressing $[OH^-]$, $[HCO_3^-]$, and $[CO_3^{2-}]$ as functions of these two concentrations:

$$[H^+] = (10^{-14}/ [H^+]) + \left([H_2CO_3(aq)] \times 4.3 \times 10^{-7}/ [H^+]\right) + \left(4.04 \times 10^{-17}[H_2CO_3(aq)]^2/ [H^+]^2\right)$$

The last term may be neglected (this assumption will be verified at the end of the calculation) and a quadratic equation is obtained for $[H^+]$:

$$[H^+]^2 = (10^{-14}) + \left([H_2CO_3(aq)] \times 4.3 \times 10^{-7}\right)$$

Replacing $[H_2CO_3(aq)]$ by its value, one obtains: $[H^+] = 2.42 \times 10^{-6}$ M. Therefore, the pH of a non-polluted rain is slightly acidic due to the solubility of CO_2.

$$pH = 5.6$$

The second assumption can be verified as follows (the concentration of CO_3^{2-} is negligible compared to that of HCO_3^-):

$$[HCO_3^-] = 2.42 \times 10^{-6}\ M;\ [CO_3^{2-}] = 4.69 \times 10^{-11}\ M$$

The first assumption (CO_2 is mostly present in the gas phase) can also be verified as follows, assuming for example that the liquid water content is 1 g m^{-3}, i.e., about 10^{-3} liter m^{-3}:

$$\begin{aligned}[CO_2(aq)]_t &= 1.36 \times 10^{-5} + 2.42 \times 10^{-6} + 4.69 \times 10^{-11} = 1.60 \times 10^{-5} M \\ &= 1.60 \times 10^{-5} \times 10^{-3} \text{mole per m}^3 \text{of air} = 1.60 \times 10^{-8}/ 40.9\ \text{atm} \\ &= 3.92 \times 10^{-10} \text{atm} = 3.92 \times 10^{-4}\text{ppm} \approx 0.0001\ \% \text{ of gas-phase } CO_2\end{aligned}$$

Therefore, both assumptions have been verified.

10.2.6 Chemical Composition of the Aqueous Phase

A cloud or fog droplet is formed from a hygroscopic particle and, consequently, the soluble fraction of this particle will be present in the aqueous phase. Furthermore, the gas-phase species that are water-soluble will be partially transferred to the aqueous phase according to Henry's law and the corresponding ionic equilibria, if the chemical species is subject to ionic dissociation.

Therefore, one must first define the initial chemical composition of the droplets, which include the particles that acted as condensation nuclei or have collided with a droplet and the gases that are partially dissolved in those droplets. The main chemical species will include water and its ions (H_2O, H^+, and OH^-), sulfuric acid and its bisulfate and sulfate ions (H_2SO_4, HSO_4^-, and SO_4^{2-}) originating from sulfate-containing particles, nitric acid and the nitrate ion (HNO_3 and NO_3^-) originating from nitrate-containing particles and the dissolution of gaseous nitric acid, ammonia and the ammonium ion (NH_4OH and NH_4^+) originating from ammonium-containing particles and the dissolution of gaseous ammonia, carbon dioxide and its bicarbonate and carbonate ions (H_2CO_3, HCO_3^-, and CO_3^{2-}), sulfur dioxide and its bisulfite and sulfite ions (H_2SO_3, HSO_3^-, and SO_3^{2-}), hydrogen peroxide (H_2O_2), ozone (O_3), oxygen (O_2), and some metals, which may act as catalysts for some reactions (iron and manganese, for example). Aqueous-phase chemistry may also take into account water-soluble organic species, such as aldehydes, and radicals (OH, NO_3, HO_2 ...).

The fraction of a chemical species that is present in the aqueous phase depends on its effective Henry's law constant and the atmospheric liquid water content. This mass balance is calculated as follows. Let A be a chemical species present in the gas phase with its concentration [A(g)] expressed in atm and present in the aqueous phase with a total concentration (i.e., including the molecular species and its ionized species, if applicable) [A(aq)] expressed in mole liter^{-1} (M). To calculate the fraction present in the aqueous phase, one must express the concentrations in terms of the same air volume, for example in mole m^{-3}. The gas-phase concentration is converted as follows:

$$C_g = \frac{[A(g)]}{RT} \quad (10.25)$$

where C_g is the gas-phase concentration of A expressed in mole per m^3 of air, T is the temperature in K and R is the ideal gas law constant (8.206×10^{-5} atm m^3 mole^{-1} K^{-1}). The concentration present in the aqueous phase is calculated as follows:

$$C_{aq} = 10^{-3} [A(aq)] L \quad (10.26)$$

where C_{aq} is the aqueous-phase concentration of A expressed in mole per m^3 of air, L is the liquid water content in g per m^3 of air and the factor of 10^{-3} is used to convert grams of liquid water into liters. Therefore, the fraction of A present in the aqueous phase, f_{aq}, is equal to the ratio of C_{aq} and the sum of C_g and C_{aq}. One accounts for the fact that the effective Henry's law constant, H_{eff}, is equal to the ratio of [A(aq)] and [A(g)]:

$$f_{aq} = \frac{C_{aq}}{C_g + C_{aq}} = \frac{10^{-3}\,[A(aq)]\,L}{\frac{[A(g)]}{RT} + 10^{-3}\,[A(aq)]\,L} = \left(1 + \frac{10^3}{L\,H_{\mathit{eff}}\,RT}\right)^{-1} \qquad (10.27)$$

The fraction present in the aqueous phase increases with the Henry's law constant and the liquid water content, but not in a proportional manner. Table 10.1 provides the dissolved fractions of some typical atmospheric chemical species for a liquid water content of 1 g m^{-3} (typical of a cumulus cloud). Some species such as NO, NO$_2$, O$_3$, and CO$_2$ are not very soluble, whereas others such as H$_2$O$_2$, NH$_3$, and HNO$_3$ are mostly present in the aqueous phase.

10.3 Aqueous-phase Chemical Transformations

Some chemical reactions are elementary reactions and, therefore, have chemical kinetics that are expressed similarly to those of the gas phase. Other reactions correspond to more complicated chemical schemes, because they may depend on pH or concentrations of other species acting as catalysts. Oxidation reactions may occur in the aqueous phase. The hydroxyl radical is water-soluble and plays an important role, as it does in the gas phase. Similarly, the nitrate radical can be present in the aqueous phase, although its transformation into the nitrate ion may decrease its concentration significantly. Furthermore, oxidants such as H$_2$O$_2$, O$_3$, and O$_2$ are involved in aqueous chemistry.

In air pollution, clouds play an important role for the formation of two acidic chemical species that are sulfuric acid and nitric acid. These two acids contribute to acid deposition (commonly referred to as acid rain, see Chapter 13). Since the formation of these two acids results from gas-phase reactions as well as from aqueous-phase reactions in clouds, all the reactions relevant to their formation are described here. The aqueous-phase chemistry of organic compounds is only summarized here, because it is described in Chapter 9.

10.3.1 Nitric Acid

Nitric acid (HNO$_3$) can be formed in the gas phase from the following reactions (see Chapter 8). During daytime, the reaction of nitrogen dioxide with the hydroxyl radical is fairly rapid (the NO$_2$ lifetime is on the order of one day):

$$NO_2 + OH + M \rightarrow HNO_3 + M \qquad (R10.11)$$

NO$_2$ may also be oxidized by ozone (O$_3$):

$$NO_2 + O_3 \rightarrow NO_3 + O_2 \qquad (R10.12)$$

During daytime, NO$_3$ is rapidly photolyzed (leading to NO and NO$_2$ according to two photolysis pathways). During nighttime, it can be converted by reaction with NO$_2$ into N$_2$O$_5$ (nitrogen pentoxide):

$$NO_2 + NO_3 \leftrightarrow N_2O_5 \qquad (R10.13)$$

which can dissociate back into its precursors; therefore, an equilibrium among these three species is established. N_2O_5 may also be hydrolyzed by water vapor to form nitric acid:

$$N_2O_5 + H_2O \rightarrow 2\ HNO_3 \tag{R10.14}$$

This reaction is slow in the gas phase and does not contribute significantly to the formation of nitric acid, compared to the oxidation of NO_2 by OH. However, this reaction can also occur with liquid water and if a N_2O_5 molecule gets into contact with a cloud or fog droplet, its hydrolysis rate is fast (e.g., Finlayson-Pitts and Pitts, 2000). Then, a heterogeneous reaction involving a gas-phase molecule and a liquid molecule takes place:

$$N_2O_5(g) + H_2O(l) \rightarrow 2\ HNO_3(aq) \tag{R10.15}$$

On the other hand, NO_3 may also react rapidly when getting into contact with a liquid water molecule:

$$NO_3(g) + H_2O(l) \rightarrow NO_3^- + H^+ + OH(aq) \tag{R10.16}$$

In the presence of clouds, the kinetics of nitric acid formation by NO_3 and N_2O_5 is commensurate with the gas-phase oxidation of NO_2 by OH.

10.3.2 Sulfuric Acid

Sulfuric acid is formed in the gas phase via a set of reactions that also involve the OH radical (Stockwell and Calvert, 1983):

$$SO_2 + OH + M \rightarrow HOSO_2 + M \quad (\text{where M is } O_2 \text{ or } N_2) \tag{R10.17}$$

$$HOSO_2 + O_2 \rightarrow SO_3 + HO_2 \tag{R10.18}$$

$$SO_3 + H_2O \rightarrow H_2SO_4 \tag{R10.19}$$

The last two reactions are very fast and the kinetics of the first reaction governs the overall kinetics of sulfuric acid formation. The rate constant expressed as a bimolecular reaction between SO_2 and OH is 9.6×10^{-13} cm^3 molec^{-1} s^{-1} at 1 atm and 25 °C. This kinetics is, therefore, ten times slower than that of the oxidation of NO_2 by OH. Thus, this chemical pathway does not lead to rapid formation of sulfuric acid. One notes that this set of reactions leads to the formation of a hydroperoxyl radical, which can subsequently react with NO to regenerate the OH radical and produce NO_2.

In the aqueous phase, SO_2 is rapidly oxidized into sulfuric acid, bisulfate or sulfate ions. Hereafter, the term sulfate is used to refer to all three chemical species and the rates are given in terms of [H_2SO_4] representing all forms of sulfate. Three oxidants may convert SO_2 into sulfate in the aqueous phase: hydrogen peroxide (H_2O_2), ozone (O_3), and oxygen (O_2). The latter reaction must be catalyzed by metal ions (such as iron and manganese) to occur rapidly.

The reaction with H_2O_2 is very fast and occurs within a few minutes in a cloud. Thus, it is completed when one of the two reactants (SO_2 or H_2O_2) is entirely consumed. This type of reaction is called a titration reaction. Therefore, the amount of sulfate formed is the initial amount (in mole or ppb) of the reactant present in the smaller amount. For example, if 2 ppb of SO_2 react with 1 ppb of H_2O_2, 1 ppb of sulfate will be formed; 1 ppb of SO_2 will remain and all H_2O_2 will have reacted. This aqueous-phase kinetics can be represented by the following expression (based on Lind et al., 1987):

$$\frac{d[H_2SO_4]}{dt} = 7.2 \times 10^7 \, [H^+] \, [HSO_3^-] \, [H_2O_2] \tag{10.28}$$

where the rate constant is expressed in $M^{-2} \, s^{-1}$ and the concentrations are expressed in M. This expression results from experiments conducted at 18 °C and it applies to pH values between 4 and 5.2; no activation energy data are available for this rate constant.

The reaction with O_3 is slower than that with H_2O_2, but it may nevertheless be faster than that with OH in the gas phase. The O_3 reaction is important in winter when H_2O_2 concentrations are low. The following rate expression ($M \, s^{-1}$) may be used (Hoffmann, 1986):

$$\frac{d[H_2SO_4]}{dt} = (2.4 \times 10^4 \, [H_2SO_3] + 3.7 \times 10^5 \, [HSO_3^-] + 1.5 \times 10^9 \, [SO_3^{2-}]) \, [O_3] \tag{10.29}$$

where the rate constants are given at 25 °C in $M^{-1} \, s^{-1}$ or $M^{-2} \, s^{-1}$, and the concentrations are those of the aqueous-phase species (M). No activation energy is provided for the first rate constant; those of the second and third rate constants are 46 kJ $mole^{-1}$ and 43.9 kJ $mole^{-1}$, respectively.

The O_2 reaction can also be significant, but its kinetics is more uncertain than those of the two previous reactions, because metal concentrations in cloud and fog droplets are highly uncertain. The following rate expression accounts for the catalytic effect of iron and manganese, including their synergistic effect (Martin and Good, 1991):

$$\frac{d[H_2SO_4]}{dt} = \left(2.6 \times 10^3 \, [Fe(III)] + 7.5 \times 10^2 \, [Mn(II)] + 10^{10} \, [Fe(III)][Mn(II)]\right) [S(IV)] \tag{10.30}$$

where the rate constants are given at 25 °C in $M^{-1} \, s^{-1}$ or $M^{-2} \, s^{-1}$, and the concentrations are those of the aqueous-phase species (M). This expression applies to pH values between 3 and 5. No activation energy data are provided for these rate constants.

A comparison of the rates of formation of sulfate, S(VI), from these different oxidation reactions of SO_2 in the aqueous phase is illustrated in Figure 10.2. For the selected oxidant concentrations, the oxidation by H_2O_2 dominates up to a pH of 5, when the kinetics of the oxidation by O_2 catalyzed by iron and manganese becomes more important. The oxidation by O_3 becomes commensurate with that by H_2O_2 for pH values above 6, which are rare in the atmosphere. These results depend of course on the oxidant concentrations and the relative importance of these oxidation mechanisms varies depending on those concentrations.

Example: Kinetics of formation of nitric acid and sulfuric acid in the atmosphere

One is interested in calculating the half-life of NO_2 and SO_2 in the gas phase in the presence of OH radicals at a concentration of $10^6 \, cm^{-3}$. The following kinetic constants are given:

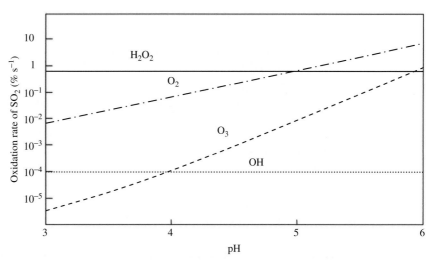

Figure 10.2. Kinetics of the oxidation of SO_2 into sulfate by reaction with H_2O_2, O_3, and O_2 (catalyzed by Fe and Mn) in the aqueous phase as a function of pH and by OH in the gas phase. Conditions for the cloud aqueous phase are as follows: the cloud is at about 3 km altitude, $P = 0.69$ atm, $T = 5$ °C, $L = 0.5$ g m^{-3}. Conditions for the gas phase are as follows: $P = 1$ atm, $T = 25$ °C, clear sky. Concentrations are as follows: $[SO_2(g)] = 2$ ppb, $[H_2O_2(g)] = 2$ ppb, $[O_3(g)] = 40$ ppb, $[Fe(aq)] = 10$ µM, $[Mn(aq)] = 0.5$ µM, $[OH(g)] = 10^6$ cm^{-3}. See text for the pH range to which the rate expressions actually apply.

13600 ppm^{-1} min^{-1} for the oxidation of NO_2 and 1400 ppm^{-1} min^{-1} for the oxidation of SO_2. The initial concentrations are 1 ppb of NO_2 and 1 ppb of SO_2. Calculate how much (in ppb) of HNO_3 and H_2SO_4 will be formed after eight hours at a temperature of 25 °C and a pressure of 1 atm. Next, calculate the half-life of SO_2 for its aqueous-phase oxidation by H_2O_2, present at a concentration of 1 ppb, in a low-level cloud, which has a temperature of 5 °C, an atmospheric pressure of about 1 atm, a pH of 5 and a liquid water content of 0.1 g m^{-3}. The aqueous-phase kinetics of the reaction between S(IV) and $H_2O_2(aq)$ is given by Equation 10.28. The Henry's law constants are given in Table 10.1. The ionic equilibrium constants of SO_2 are as follows: $K_1 = 1.7 \times 10^{-2}$ M, $\Delta H_{R1} = -4.16$ kcal mole^{-1}, $K_2 = 6 \times 10^{-8}$ M, $\Delta H_{R2} = -2.22$ kcal mole^{-1}. The ideal gas law constant is $R = 1.986$ cal mole^{-1} K^{-1}. Compare the formation rate of HNO_3 by oxidation of NO_2 in the gas phase and that of H_2SO_4 by oxidation of SO_2 in the gas and aqueous phases.

The half-life, $t_{1/2}$, corresponds to the time required for half of the initial concentration of NO_2 to react; it is provided by Equation 8.4:

$$t_{1/2}(NO_2(g)) = \frac{\ln(2)}{k_{NO_2}[OH]}$$

One must convert [OH] into ppm; according to Equations 3.9, 7.1, and 7.2: $[OH] = 10^6$ cm$^{-3} \approx 4 \times 10^{-8}$ ppm.

Therefore, for NO_2: $t_{1/2}(NO_2(g)) = 21$ hours 27 min

In less than one day, half of NO_2 initially present will have reacted. After eight hours, the amount of NO_2 remaining after gas-phase oxidation is given by Equation 8.2:

$$[NO_2] = [NO_2]_0 \exp(-k_{NO_2} [OH] t) = 0.77 \text{ ppb}$$

Therefore, the amount of HNO_3 formed is: $[HNO_3] = 1 - 0.77 = 0.23$ ppb

About one quarter of the initial amount of NO_2 has been transformed into nitric acid in eight hours.

The half-life of SO_2 is calculated similarly: $t_{1/2}(SO_2(g)) = 8$ days 16 hours

Similarly, the amount of SO_2 remaining in the gas phase following eight hours of oxidation is:

$$[SO_2] = 0.97 \text{ ppb}; \quad [H_2SO_4] = 1-0.97 = 0.03 \text{ ppb}$$

Only 3 % of the initial amount of SO_2 have been transformed into sulfuric acid after eight hours.

To address the aqueous-phase oxidation of SO_2 (i.e., S(IV)), one must first calculate the Henry's law constants and the ionization constants at 5 °C (278 K) using the law of van't Hoff (Equation 10.7). For SO_2:

$$H_{SO_2}(5 \text{ °C}) = 1.23 \exp\left(\frac{-6.25 \times 10^3}{1.986}\left(\frac{1}{298} - \frac{1}{278}\right)\right) = 2.63 \text{ M atm}^{-1}$$

In order to calculate the effective Henry's law constant, one must calculate the values of the equilibrium constants, K_1 and K_2, corresponding to the formation of HSO_3^- and SO_3^{2-}, respectively, at $T = 278$ K based on their values at $T_{ref} = 298$ K.

$$K_1(5 \text{ °C}) = K_1(T_{ref}) \exp\left(\frac{\Delta H_{R1}}{R}\left(\frac{1}{T_{ref}} - \frac{1}{T}\right)\right)$$

$$= 1.7 \times 10^{-2} \exp\left(\frac{-4.16 \times 10^3}{1.986}\left(\frac{1}{298} - \frac{1}{278}\right)\right) = 2.82 \times 10^{-2} \text{ M}$$

$$K_2(5 \text{ °C}) = K_2(T_{ref}) \exp\left(\frac{\Delta H_{R2}}{R}\left(\frac{1}{T_{ref}} - \frac{1}{T}\right)\right)$$

$$= 6 \times 10^{-8} \exp\left(\frac{-2.22 \times 10^3}{1.986}\left(\frac{1}{298} - \frac{1}{278}\right)\right) = 7.85 \times 10^{-8} \text{ M}$$

Therefore, the effective Henry's law constant of SO_2 is, based on Equation 10.14:

$$H_{SO_2,eff}(5 \text{ °C}) = 7.48 \times 10^3 \text{ M atm}^{-1}$$

The ionic dissociation of SO_2 in the aqueous phase increases its solubility by a factor of about 6,000. For H_2O_2:

$$H_{H_2O_2}(5 \text{ °C}) = 7.45 \times 10^4 \exp\left(\frac{-14.5 \times 10^3}{1.986}\left(\frac{1}{298} - \frac{1}{278}\right)\right) = 4.34 \times 10^5 \text{ M atm}^{-1}$$

The fraction of SO_2, which is dissolved in cloud droplets, is calculated according to Equation 10.27:

$$f_{aq,SO_2} = \left(1 + \frac{10^3}{LH_{eff}\, RT}\right)^{-1} = \left(1 + \frac{10^3}{0.1 \times 7.48 \times 10^3 \times 8.2 \times 10^{-5} \times 278}\right)^{-1} = 1.7\,\%$$

The fraction of H_2O_2, which is dissolved in cloud droplets, is calculated similarly:

$$f_{aq,H_2O_2} = 49\,\%$$

Therefore, the gas-phase concentration of SO_2 is little affected by its dissolution in the cloud, but that of H_2O_2 decreases by about half. The gas-phase concentrations of SO_2 and H_2O_2 are: $[SO_2(g)] = 0.98$ ppb and $[H_2O_2(g)] = 0.51$ ppb.

One seeks the half-life of the oxidation of SO_2 by H_2O_2 in cloud droplets. One is interested in the half-life of total SO_2 (i.e., the sum of gas-phase and aqueous-phase SO_2). First, one calculates the fraction of aqueous-phase SO_2 that is present as HSO_3^-, since it is the species that reacts with H_2O_2. Based on Equation 10.12, the values of K_1 and K_2 calculated previously, and a pH of 5:

$$\frac{[HSO_3^-(aq)]}{[S(IV)(aq)]} = \left(1 + \frac{[H^+]}{K_1} + \frac{K_2}{[H^+]}\right)^{-1} = 0.992$$

Therefore, S(IV) is almost entirely present in the aqueous phase as HSO_3^-, as illustrated in Figure 10.1. HSO_3^- corresponds to 1.7 % of total (gaseous + aqueous) SO_2, since only 1.7 % of SO_2 is present in cloud droplets. Before calculating the oxidation rate of HSO_3^- by H_2O_2, one must calculate the aqueous-phase concentration of H_2O_2, based on its gas-phase concentration (calculated previously) and its Henry's law constant:

$$[H_2O_2(aq)] = H_{H_2O_2}\,[H_2O_2(g)];\ [H_2O_2(aq)] = 2.21 \times 10^{-4}\,M$$

Using the kinetics of the oxidation of HSO_3^- by H_2O_2 given by Equation 10.28, the half-life of HSO_3^- in the aqueous phase is calculated as follows:

$$t_{1/2}(HSO_3^-) = \frac{\ln(2)}{k_{HSO_3^-}\,[H_2O_2(aq)]\,[H^+]} = 4.4\,s$$

Therefore, HSO_3^- (and, therefore, S(IV) since SO_2 is present almost entirely as bisulfite ions in the cloud droplets) is oxidized very rapidly by H_2O_2. However, HSO_3^- represents only 1.7 % of the total amount of SO_2. To calculate the half-life of SO_2, one must take into account the fact that only a small fraction of SO_2 (1.7 %) is present in cloud droplets:

$$t_{1/2}(SO_2(total)) = \frac{4.4\,s}{0.017} = 4\,min\,19\,s$$

Therefore, the oxidation of SO_2 by H_2O_2 in a cloud is very fast due to the kinetics of the aqueous-phase reaction and the high solubility of H_2O_2. This half-life is commensurate with the lifetime of a cloud droplet, which is on the order of a few minutes. Therefore, aqueous-phase oxidation will dominate the transformation of SO_2 into sulfuric acid, since the half-life of the gas-phase oxidation was calculated to be about eight days. In the presence of clouds, the oxidation of SO_2 (via its aqueous-phase reaction) is faster than the gas-phase oxidation of NO_2.

10.3.3 Organic Compounds

Organic chemistry is of interest because the oxidation of some water-soluble organic compounds (for example, aldehydes) may lead to compounds of lower volatility, which will be more conducive to the formation of secondary organic aerosols (SOA) if the cloud of fog evaporates. However, studies conducted so far suggest that the contribution of cloud chemistry to average SOA concentrations is small (<10 %) compared to the gas-phase oxidation (see Chapter 9 for a more detailed discussion of the aqueous-phase reactions of organic compounds); nevertheless, it may be important for some specific locations and periods.

10.4 Emission Control Strategies for Acid Rain

The reduction of the concentrations of sulfate and nitrate, which are the two main constituents of acid rain (or more generally acid deposition), involves the reduction of the emissions of their precursors, which are sulfur dioxide (SO_2) and nitrogen oxides (NO_x).

10.4.1 Non-linearity of the Oxidation of SO_2 into Sulfate

The gas-phase formation of sulfate is proportional to the SO_2 concentration and, therefore, to its emissions. However, this oxidation pathway is generally minor compared to the aqueous-phase oxidation pathways.

The aqueous-phase oxidation of SO_2 by H_2O_2 is fast and may be considered to be a titration reaction. If the SO_2 concentration is greater than that of H_2O_2, a reduction of the SO_2 concentration will not have any effect, because the amount of sulfate formed is limited by the amount of H_2O_2 available. Then, the system is non-linear, because a reduction of the SO_2 concentration will not result in a proportional reduction in sulfate.

The aqueous oxidation of SO_2 by O_3 and O_2 is also slightly non-linear, because these reactions depend on pH: the kinetics of the reaction decreases when the pH decreases. Therefore, these two reactions are self-limiting because the pH decreases as sulfate is being formed (since sulfate is an acid) and the kinetics slows down. Furthermore, if the initial concentration of SO_2 is reduced, the amount of sulfate formed will be less, which implies a greater pH and, consequently, a faster kinetics. Therefore, the reduction in sulfate will be less than the reduction in the SO_2 concentration and the system is non-linear.

Example: Calculate the amount of sulfuric acid formed by titration of SO_2 by H_2O_2

The concentration of sulfur dioxide (SO_2) is 1.5 ppb and the concentration of hydrogen peroxide (H_2O_2) is 1 ppb. The reaction between SO_2 and H_2O_2 occurs rapidly within a cloud. What is the amount of sulfuric acid formed (in ppb)?

The amount of sulfuric acid formed is 1 ppb since it is limited by the amount of H_2O_2.

Now, the emissions of SO_2 are reduced by half and the SO_2 concentration becomes 0.75 ppb. If the H_2O_2 concentration is still 1 ppb, what is the amount of sulfuric acid formed?

The amount of sulfuric acid formed is 0.75 ppb, because now the SO_2 concentration limits the formation of sulfuric acid. Therefore, the amount of sulfuric acid formed has been reduced by 25 % compared to the previous case, whereas the SO_2 emissions were reduced by 50 %.

10.4.2 Non-linearity of the Oxidation of NO_2 into Nitrate

The decrease of NO_x concentrations may have different effects on the concentrations of the oxidants involved in the oxidation of NO_2 into nitric acid (i.e., OH and O_3). In a high-NO_x regime, a decrease of NO_x emissions generally leads to an increase of oxidant concentrations and, therefore, the decrease of the amount of nitric acid being formed will be less than that of the NO_x emissions (in some cases, there may even be a slight increase of the nitric acid concentration if the increase of the oxidant concentration exceeds the decrease of the NO_2 concentration). On the other hand, in a low-NO_x regime, a decrease of NO_x emissions leads to a decrease of oxidant concentrations and the decrease of the nitric acid concentration will be more important than that of the NO_x emissions. Therefore, the system is non-linear and can be antagonistic (in a high-NO_x regime) or synergistic (in a low-NO_x regime) in terms of nitric acid reduction.

10.4.3 Reduction of Acid Rain in the United States

The Clean Air Act imposed SO_2 and NO_x emission controls for coal-fired power plants in order to reduce acid rain in the United States (see Chapter 15). The reduction of sulfate deposition was initially less than proportional compared to the reduction of SO_2 emissions, for the reasons mentioned in Section 10.4.1 on the non-linearity of atmospheric chemistry. Nevertheless, these emission controls became efficient in the long term to reduce atmospheric acid deposition.

Indeed, SO_2 emission controls are not efficient to reduce sulfate levels when the SO_2 levels are greater than those H_2O_2; however, they become efficient once the SO_2 levels become less than those of H_2O_2. Therefore, beyond a certain amount of SO_2 emission controls, the decrease in sulfate concentrations becomes nearly proportional to that of SO_2 emissions (there is a slight non-linearity due to the effect of pH on the oxidation of SO_2 by O_3 and O_2). It is possible that in some regions of China, the current conditions are such that $[SO_2] > [H_2O_2]$.

NO_x emission controls have led mostly to decreases in oxidant concentrations, because the power plants are located in rural areas, i.e., areas that are in a low-NO_x regime (see Chapter 8). Therefore, it is likely that the NO_x emission controls led to decreases in both NO_2 and oxidant levels and, consequently, a decrease in nitrate levels greater than that of the NO_x emissions.

10.5 Numerical Modeling of Aqueous-phase Chemistry

The numerical solution of the concentrations of aqueous-phase chemical species is similar to that presented in Chapter 9 for particulate matter. It is simpler because only the aqueous phase needs to be treated (a solid phase and several liquid phases need to be treated in the case of particulate matter) and the assumption of an ideal solution may be made (thereby, assuming that the activity coefficients are unity). On the other hand, it is appropriate in some cases to account for the kinetics of mass transfer between the gas phase and the droplets. The numerical solution of the aqueous-phase reactions may require an algorithm for stiff ordinary differential equations, depending on which reactions are simulated.

Some numerical models treat only the oxidation of SO_2 to sulfate with some equilibria of water-soluble inorganic species to calculate the droplet pH (e.g., Seigneur et al., 1984; Walcek and Taylor, 1986). Other models treat a greater number of chemical reactions, including for example the oxidation of some species by OH radicals (e.g., Fahey and Pandis, 2003). Some other models include also the oxidation of organic species that can lead to secondary organic aerosol formation (e.g., Hermann et al., 2005; Carlton et al., 2008; Couvidat et al., 2013).

Problems

Problem 10.1 Dissolution of air pollution in fog
The Henry's law constant of nitrogen dioxide (NO_2) at 15 °C is 1.6×10^{-2} M atm^{-1}. There is fog and, therefore, NO_2 is partially dissolved in fog droplets. The gas-phase NO_2 concentration is 3 ppb. What is the NO_2 concentration in fog droplets in moles per liter of water (M)? One assumes that the amount of NO_2 in the aqueous-phase is negligible compared to the amount in the gas phase and that the atmospheric pressure is 1 atm.

Problem 10.2 Reduction of sulfuric acid
a. The atmospheric sulfur dioxide (SO_2) concentration is 2 ppb and the hydrogen peroxide (H_2O_2) concentration is 1 ppb. These two species react rapidly (titration) in clouds to form sulfuric acid (H_2SO_4). How much sulfuric acid is formed?
b. One wants to reduce the sulfuric acid concentration by half. By what percentage should the SO_2 initial concentration be reduced?

References

Ahrens, C.D., 2012. *Essentials of Meteorology – An Invitation to the Atmosphere*, Cengage Learning, Stamford, CT.

Carlton, A. G., B.J. Turpin, K.E. Altieri, S.P. Seitzinger, R. Mathur, S.J. Roselle, and R.J. Weber, 2008. CMAQ model performance enhanced when in-cloud SOA is included: Comparisons of OC predictions with measurements, *Environ. Sci. Technol.*, **42**, 8798–8802.

Couvidat, F., K. Sartelet, and C. Seigneur, 2013. Investigating the impact of aqueous-phase chemistry and wet deposition on organic aerosol formation using a molecular surrogate modeling approach, *Environ. Sci. Technol.*, **47**, 914–922.

Fahey, K.M. and S.N. Pandis, 2003. Size-resolved aqueous-phase atmospheric chemistry in a three-dimensional chemical transport model, *J. Geophys. Res.*, **108**, 4690, D22, doi:10.1029/2003JD003564.

Finlayson-Pitts, B.J. and J.N. Pitts, Jr., 2000. *Chemistry of the Upper and Lower Atmosphere: Theory, Experiments, and Applications*, Academic Press, New York.

Hermann, H., A. Tilgner, P. Barzaghi, Z. Majdik, S. Gligorovski, L. Poulain, and A. Monod, 2005. Toward a more detailed description of tropospheric aqueous phase organic chemistry: CAPRAM 3.0, *Atmos. Environ.*, **39**, 4351–4363.

Hoffmann, M.R., 1986. On the kinetics and mechanism of oxidation of aquated sulfur dioxide by ozone, *Atmos. Environ.*, **20**, 1145–1154.

Lind, J.A., A.L. Lazrus, and G.L. Kok, 1987. Aqueous phase oxidation of sulfur(IV) by hydrogen peroxide, methylhydroperoxide, and peroxyacetic acid, *J. Geophys. Res.*, **92**, 4171–4177.

Martin, L.R. and T.W. Good, 1991. Catalyzed oxidation of sulfur dioxide in solution: The iron-manganese synergism, *Atmos. Environ.*, **25**, 2395–2399.

Seigneur, C., P. Saxena, and P.M. Roth, 1984. Computer simulations of the atmospheric chemistry of sulfate and nitrate formation, *Science*, **225**, 1028–1030.

Seinfeld, J.H. and S.N. Pandis, 2016. *Atmospheric Chemistry and Physics – From Air Pollution to Climate Change*, Wiley, Hoboken, NJ.

Stockwell, W.R. and J.G. Calvert, 1983. The mechanism of the $HO-SO_2$ reaction, *Atmos. Environ.*, **17**, 2231–2235.

Stumm, W. and J.J. Morgan, 1995. *Aquatic Chemistry: Chemical Equilibria and Rates in Natural Waters*, Wiley, Hoboken, NJ.

Walcek, C.J. and G.R. Taylor, 1986. A theoretical method for computing vertical distributions of acidity and sulfate production within cumulus clouds, *J. Atmos. Sci.*, **43**, 339–355.

11 Transfer of Pollutants between the Atmosphere and Surfaces

The atmosphere interacts with the Earth's surface. Thus, air pollutants may be transferred toward surfaces and emitted (or reemitted) from surfaces toward the atmosphere. Atmospheric deposition processes are important because (1) they impact the atmospheric lifetime of air pollutants and (2) they may lead to the contamination of other environmental media. Processes of emission and reemission may contribute significantly to the atmospheric budget of some pollutants and it is, therefore, essential to take those into account. This chapter describes the mechanisms that lead to atmospheric deposition of pollutants, either via dry processes (dry deposition) or via precipitation scavenging (wet deposition). Emissions of particles by the wind (aeolian emissions), waves, and on-road traffic are also described.

11.1 Dry Deposition

Atmospheric pollutants may deposit on buildings, vegetation, soil, and surface waters via "dry" processes, i.e., processes that do not depend on precipitation. The fundamental processes that lead to dry deposition are sedimentation, impacts by inertia or interception, and diffusion (Wesely and Hicks, 2000). The first cases pertain only to particles, whereas diffusion concerns both particles and gases. The concept of dry deposition velocity was first proposed by Gregory (1945). Although sedimentation was the only process mentioned explicitly in his analysis, he noted that atmospheric conditions (turbulence intensity) and the surface roughness affected the dry deposition process.

11.1.1 Sedimentation and Fall Velocity

Sedimentation corresponds to the effect of the Earth's gravity on particles. All particles, regardless of their size, undergo sedimentation. However, only coarse particles (i.e., those particles with a diameter greater than about 2.5 µm) have a sedimentation velocity that is large enough for this process to become commensurate with the process of impact by inertia. Thus, for particles with a diameter of a few tens of microns, sedimentation is the dominant process. For example, a spherical particle with a diameter of 10 µm and a density of 1 g cm^{-3} has a sedimentation velocity of about 0.3 cm s^{-1}. This sedimentation velocity corresponds to the terminal fall velocity of the particle, which results from an equilibrium between the gravitational force and the frictional force. The frictional force is due to interactions between the particle and the surrounding air. The equilibrium between these two forces is represented mathematically by Stokes' law.

When the Reynolds number is very small (Re ≪ 1), the force, Fo_{St}, exerted by the fluid (here the air) on a moving particle or droplet may be written as follows:

$$Fo_{St} = \frac{3\pi \mu_{v,a}\, d_p v_s}{C_c} \qquad (11.1)$$

where $\mu_{v,a}$ is the dynamic viscosity of the air (kg m^{-1} s^{-1}), d_p is the diameter of the particle (or droplet) (m), v_s is the particle sedimentation velocity (m s^{-1}), and c_c is the Cunningham correction factor, which is a function of the particle size. The Cunningham factor is a correction that applies to fine and ultrafine particles (it is equal to 1.01 for a particle with a diameter $d_p = 10$ μm, 1.16 for $d_p = 1$ μm, 2.8 for $d_p = 0.1$ μm, and 22 for $d_p = 0.01$ μm). This formula applies to ultrafine and fine particles and to some coarse particles, because for particles with a diameter less than 10 μm, Re $< 2.5 \times 10^{-3}$.

When the Reynolds number is greater (Re ≫ 1), the following expression applies:

$$Fo_{St} = \frac{\pi c_d\, \rho_a\, d_p^2\, v_s^2}{8\, c_c} \qquad (11.2)$$

where c_d is the drag factor (or drag coefficient), ρ_a is the air density, and the other terms are the same as in the previous formula. The drag coefficient is a function of Re and, therefore, depends on the particle size since: Re $= \rho_a\, d_p\, v_s / \mu_{v,a}$. This formula applies to very large particles and droplets ($d_p > 50$ μm), for example raindrops and some dust particles.

The acceleration of a particle (or droplet) moving in the atmosphere is governed by gravity, the buoyant force due to displacement of the fluid (Archimedes' principle), and the air resistance:

$$m_p \frac{dv_s}{dt} = m_p\, g - m_{air}\, g - Fo_{St} \qquad (11.3)$$

The mass of the particle or droplet (m_p) is about 1,000 times greater than that of the displaced air (m_{air}) and the corresponding buoyant force ($m_{air}\, g$) may, therefore, be neglected. The particle or droplet reaches its terminal fall velocity ($dv_s/dt = 0$) when the friction due to the air resistance compensates gravity:

$$m_p\, g = Fo_{St} \qquad (11.4)$$

Thus, the terminal sedimentation velocity (v_s^f) may be calculated for the two regimes defined in Equations 11.1 and 11.2 for Stokes' law. If Re ≪ 1 (particles and fog droplets):

$$v_s^f = \frac{m_p\, g\, c_c}{3\, \pi\, \mu_{v,a}\, d_p} \qquad (11.5)$$

The particle mass may be expressed as a function of its diameter and density:

$$m_p = \rho_p \frac{\pi\, d_p^3}{6} \qquad (11.6)$$

Then, the terminal fall velocity of the particle as a function of its diameter and density is as follows:

$$v_s^f = \frac{\rho_p d_p^2 g c_c}{18 \mu_{v,a}} \quad (11.7)$$

For an atmospheric particle or fog droplet, the final fall velocity is proportional to the square of its diameter (however, the Cunningham correction modifies this simple relationship); therefore, it decreases as its diameter decreases.

If Re $\gg$ 1 (large dust particles and raindrops):

$$v_s^f = \left(\frac{8 m_p g c_c}{\pi c_d \rho_a d_p^2} \right)^{\frac{1}{2}} \quad (11.8)$$

$$c_d = 18.5 \, \text{Re}^{-0.6}; \quad \text{Re} = \frac{v_s^f d_p}{v_{v,a}}$$

where $v_{v,a}$ is the kinematic viscosity of the air. Therefore, the terminal fall velocity must be calculated by iteration since it depends on the drag coefficient, which depends via the Reynolds number on the fall velocity. Expressing the mass of a particle or drop as a function of its density and diameter:

$$v_s^f = \left(\frac{4 \rho_p g c_c d_p}{3 c_d \rho_a} \right)^{\frac{1}{2}} \quad (11.9)$$

Thus, the terminal fall velocity is proportional to the square root of the particle diameter (with a correction due to the drag coefficient). For raindrops, the simplified Kessler formula may be used (Kessler, 1969):

$$v_s^f = 130 \sqrt{d_r} \quad (11.10)$$

where v_s^f is the raindrop fall velocity (m s^{-1}) and d_r is the raindrop diameter (m). The difference between these two equations is <10 % for a raindrop of 1 mm: 4.5 m s^{-1} with Equation 11.9 and 4.1 m s^{-1} with Equation 11.10. These formulas tend to overestimate the terminal fall velocity of drops greater than 2.5 mm. More detailed parameterizations are available that better represent the fall velocity of raindrops over a wide range of sizes (see the review by Duhanyan and Roustan, 2011). Note that raindrops rarely have diameters greater than 6 mm, because they tend to break into several smaller drops when they get bigger.

11.1.2 Interception and Inertia

Deposition by interception and inertia concerns atmospheric particles that may interact with surfaces via these processes and then deposit. When the air flow must go around an obstacle, particles present in the air may get into contact with the obstacle either because of their size or mass. The size of a particle is much greater than that of an air molecule and the particle may, therefore, interact with the obstacle because of its size. This interception process is roughly proportional to the cross-section of the particle, i.e., ($\pi\, d_p^2$). On the other hand, the mass of a particle leads to inertia when the air flow changes direction to go around an obstacle. This inertia is roughly proportional to the particle mass, i.e., proportional to its volume, ($\pi\, d_p^3 / 6$). Therefore, these two processes are negligible for ultrafine particles and are only relevant for fine particles (0.1 μm < d_p < 2.5 μm) and coarse particles (d_p > 2.5 μm).

These processes of impact by inertia and interception are also essential for technologies of particulate matter emission control (see Chapter 2). For example, cyclones use inertia and baghouses use interception and inertia to capture particles. Also, atmospheric particles with a diameter greater than a few microns (μm) are captured efficiently by these processes in the nose and upper airways and, therefore, are not transported deeply into the lungs, unlike fine and ultrafine particles (see Chapter 12).

11.1.3 Diffusion

The diffusion of a gas molecule or particle in the atmosphere toward a surface may be considered to include several steps (Wesely, 1989). For all pollutants (gases and particles), two successive steps bring the pollutant in contact with the surface: (1) turbulent mass transfer in the atmosphere toward the surface and (2) diffusive mass transfer within a very thin layer (on the order of a millimeter) in direct contact with the surface. This latter layer is not affected by atmospheric turbulence and is, therefore, considered to be in a quasi-laminar regime. Mass transfer within that layer occurs by molecular diffusion for gases and via brownian diffusion for particles. In addition, the processes of impact by inertia and interception for particles take place within that layer. Typically, one considers that particles deposit on a surface once they get into contact with that surface (a bouncing coefficient may be used in some cases to account for the fact that some particles may bounce and, therefore, will not remain on the surface). For gases, a third step is included. It determines the deposition rate by absorption into the surface (for example, dissolution into a dew layer), adsorption onto the surface (for example, adsorption onto activated carbon) or chemical reaction at the surface (for example, an acid gas may react with an alkaline surface). A combination of absorption or adsorption followed by chemical reaction may also occur. For particles, inertia and interception processes are important in the case of deposition on surfaces with complex geometries, such as vegetation, for particles with a diameter between about 1 and 10 μm.

This series of processes represents the overall dry deposition phenomenon by diffusion and it may be seen as a series of resistances to mass transfer, by analogy with an electrical circuit.

Aerodynamic Resistance

The first step is associated with an aerodynamic resistance and pertains to the turbulent mass transfer of the pollutant (gas molecule or particle) in the atmosphere toward a layer near the surface. Therefore, it is governed by the vertical transport processes, i.e., vertical atmospheric dispersion. This transfer resistance is large when the atmosphere is stable. It is small when the atmosphere is unstable; then, turbulence readily brings pollutants into contact with the surface via vertical mixing.

The vertical mass flux, F_d, due to this turbulent process may be calculated using a first-order closure of atmospheric turbulence. A K type turbulence representation is generally used (the negative sign is used here because the flux corresponds to a sink for the atmospheric compartment):

$$F_d = -K_z \frac{dC}{dz} \tag{11.11}$$

The vertical dispersion coefficient may be expressed as follows (see Chapter 6):

$$K_z = \kappa u_* z \tag{11.12}$$

where κ is the von Kármán constant ($\kappa = 0.4$) and u_* is the friction velocity. This expression is valid for a neutral atmosphere. If the atmosphere is stable or unstable, the vertical profile of the concentrations in the surface layer differs and this equation must be modified using the parameter $\Phi_H(z)$. The corresponding parameter $\Phi_M(z)$ for momentum is sometimes used; however, mass transfer is better associated with heat transfer (both are scalars) than with momentum transfer (a vector). Therefore, the temperature profile seems more appropriate for mass transfer than the vertical profile of the horizontal wind speed (see Chapter 4):

$$K_z = \frac{\kappa u_* z}{\Phi_H(z)} \tag{11.13}$$

$$F_d \frac{\Phi_H(z)\, dz}{\kappa u_* z} = -dC \tag{11.14}$$

Integrating between the reference height (located within the surface layer), z_r, which corresponds to the height of the reference concentration, C, and the bottom of this turbulent layer (taken to be by definition the roughness length, z_0):

$$F_d \int_{z_0}^{z_r} \frac{\Phi_H(z)}{\kappa u_* z}\, dz = -\int_{C_0}^{C} dC \tag{11.15}$$

Thus:
$$F_d = -\frac{\Delta C}{\int_{z_0}^{z_r} \frac{\Phi_H(z)\, dz}{\kappa u_* z}} \tag{11.16}$$

For a neutral atmosphere ($\Phi_H(z) = 1$):

$$F_d = -\frac{\kappa u_* \Delta C}{\ln\left(\frac{z_r}{z_0}\right)} \tag{11.17}$$

The vertical flux may be expressed as a function of the aerodynamic resistance, r_a:

$$F_d = -\frac{\Delta C}{r_a} \tag{11.18}$$

Thus, r_a may be written as a function of the variables that govern the turbulent mass transfer:

$$r_a = \int_{z_0}^{z_r} \frac{\Phi_H(z)\,dz}{\kappa u_* z} \tag{11.19}$$

In the case of a neutral atmosphere, $\Phi_H(z) = 1$, therefore:

$$r_a = \frac{\ln\left(\frac{z_r}{z_0}\right)}{\kappa u_*} \tag{11.20}$$

In the case of stable or unstable atmospheres, the solution is more complicated, because one must integrate using the profile of $\Phi_H(z)$. The profile given in Equation (4.68) is used for a stable atmosphere and the integration leads to the following solution for r_a:

$$r_a = \frac{0.95 \ln\left(\frac{z_r}{z_0}\right) + 7.8\,\frac{(z_r - z_0)}{L_{MO}}}{\kappa u_*} \tag{11.21}$$

where L_{MO} is the Monin-Obukhov length (see Chapter 4). The profile given in Equation (4.69) is used for an unstable atmosphere and the integration leads to the following solution:

$$r_a = \frac{0.95}{\kappa u_*}\left[\ln\left[\frac{(1 - 11.6\frac{z_r}{L_{MO}})^{\frac{1}{2}} - 1}{(1 - 11.6\frac{z_r}{L_{MO}})^{\frac{1}{2}} + 1}\right] - \ln\left[\frac{(1 - 11.6\frac{z_0}{L_{MO}})^{\frac{1}{2}} - 1}{(1 - 11.6\frac{z_0}{L_{MO}})^{\frac{1}{2}} + 1}\right]\right] \tag{11.22}$$

These formulations imply several assumptions and they apply in theory only within the surface layer, i.e., the layer where the vertical fluxes for momentum, heat, and mass are constant with height. This surface layer ranges from the surface up to a height that varies between a few tens of meters to about 100 m (~10 % of the planetary boundary layer, see Chapter 4).

Resistance Due to Diffusion and Impact Processes

The second step corresponds to a resistance by diffusion in a very thin layer in contact with the surface. This layer (with a thickness on the order of 1 mm) is characterized by a quasi-laminar flow. Therefore, resistance is mostly due to diffusion and is a function of the molecular diffusion coefficient for gases and brownian diffusion coefficient for particles. Molecular diffusion coefficients depend on the physico-chemical properties of the molecule; however, they vary within a rather limited range of values. For example, the diffusion coefficient of carbon monoxide (CO) in the air is 0.12 cm^2 s^{-1}, that of ozone (O_3) is 0.14 cm^2 s^{-1}, and that of nitrogen dioxide (NO_2) is also 0.14 cm^2 s^{-1}.

The resistance to diffusion in the quasi-laminar layer, r_b, is defined as follows:

$$r_b = \frac{5\,Sc^{\frac{2}{3}}}{u_*} \tag{11.23}$$

where u_* is the friction velocity ($u_*^2 = -\overline{u'w'}$) and Sc is the Schmidt number, which characterizes the relative importance of advection and diffusion (it corresponds to the Prandtl number, which is used in heat transfer):

$$\text{Sc} = \frac{v_{v,a}}{D_m} \qquad (11.24)$$

where $v_{v,a}$ is the kinematic viscosity of the air (m^2 s^{-1}; the kinematic viscosity is related to the dynamic viscosity via the fluid density: $v_{v,a} = \mu_{v,a}/\rho_a$) and D_m is the molecular diffusion coefficient of the gaseous pollutant in the air (m^2 s^{-1}).

For a particle, mass transfer within the quasi-laminar layer depends on brownian diffusion. However, inertia and interception processes also take place within that layer and they are, therefore, treated jointly with diffusion. The following empirical formula may be used (Zhang et al., 2001):

$$r_b = \frac{1}{3u_*(E_B + E_{IM} + E_{IN})f_p} \qquad (11.25)$$

where E_B, E_{IM}, and E_{IN} are collection efficiencies of brownian diffusion, impact by inertia, and interception, respectively, and f_p is a correction factor representing the fraction of particles that stick to the surface after contact:

$$\begin{aligned} E_B &= \frac{\text{Sc}^{-2/3}}{15} \\ E_{IM} &= \left(\frac{\text{St}}{\alpha + \text{St}}\right)^\beta \\ E_{IN} &= \frac{1}{2}\left(\frac{d_p}{d_f}\right)^2 \end{aligned} \qquad (11.26)$$

where Sc is defined as in Equation 11.24 for gaseous molecules, but using the brownian diffusion coefficient, D_p, instead of the molecular diffusion coefficient, St is the Stokes number, d_p is the particle diameter, d_f is a characteristic dimension of the surface (vegetation leaf, filter ...), and α and β are empirical parameters that depend on the surface type. Zhang et al. (2001) proposed some values for these parameters depending on surface type: α ranges between 0.6 and 100, $\beta = 2$, and d_f ranges between 2 and 10 mm. Here, the brownian diffusion term, E_B, has been modified to be consistent with (1) the formulation of the molecular diffusion for gaseous pollutants and (2) the experimental data of Möller and Schumann (1970). The Stokes number is a characteristic of both the particle and the flow; the smaller the particle, the smaller St. It is defined differently for vegetation-covered surfaces and smooth surfaces:

$$\begin{aligned} \text{St}_{\text{vegetation}} &= \frac{v_s\, u_*}{g\, d_f} \\ \text{St}_{\text{smooth}} &= \frac{v_s\, u_*^2}{g\, v_{v,a}} \end{aligned} \qquad (11.27)$$

where v_s is the particle sedimentation velocity, u_* is the friction velocity, d_f is the characteristic length of the surface, and $v_{v,a}$ is the kinematic viscosity of the air.

The brownian diffusion coefficient of a particle decreases with increasing particle size, whereas the mass transfer velocities for inertia, interception, and sedimentation increase with increasing particle size. As a result, ultrafine particles (i.e., those with a diameter less than 0.1 µm) and coarse particles (i.e., those with a diameter greater than 2.5 µm) have lower resistances than fine particles (i.e., those with a diameter less than 2.5 µm but greater than 0.1 µm), because ultrafine particles are rapidly deposited via brownian diffusion and coarse particles are deposited effectively via sedimentation, inertia, and interception. Figure 11.1 shows the contributions of these different deposition processes as a function of particle diameter. Conditions used in this figure correspond to deposition over grassland under neutral atmospheric conditions with a friction velocity of 0.5 m s^{-1}, a roughness length of 0.01 m, a particle density of 1.5 g cm^{-3}, and a characteristic length of 2 mm for grass. Under these conditions, dry deposition via interception is negligible compared to the other processes. The aerodynamic resistance limit is indicated. Note that sedimentation is not affected by the aerodynamic resistance, whereas all other processes are. Therefore, the process of deposition via inertia, which is commensurate with sedimentation up to a particle diameter of about 10 µm, becomes capped for larger particles.

Surface Resistance

For gaseous pollutants, a third resistance is taken into account. It corresponds to the transfer of the molecule from the air to the surface and, accordingly, it is called the surface resistance, r_c. It may be a complicated formula, for example, in the case of deposition of gaseous pollutants on vegetation. For particles, this surface resistance is incorporated implicitly into the diffusion resistance via the correction factor, f_p.

Total Resistance

The total resistance to atmospheric deposition is simply the sum of these two or three resistances. It is expressed in s m^{-1}. The deposition velocity is the inverse of the resistance and it is expressed in m s^{-1}. The vertical mass flux is constant in the surface layer and for gaseous pollutants the following relationships may be written (for particles, the last equation is not taken into account):

$$F_d = -\frac{\Delta_a C}{r_a} = -\frac{\Delta_b C}{r_b} = -\frac{\Delta_c C}{r_c} \tag{11.28}$$

where the subscript of Δ represents the atmospheric layer of interest for the corresponding resistance. The total resistance is obtained from the overall concentration difference:

$$F_d = -\frac{\Delta_t C}{r_t} \tag{11.29}$$

$$\Delta_t C = \Delta_a C + \Delta_b C + \Delta_c C \tag{11.30}$$

Thus, for the overall concentration difference:

$$\Delta_t C = -F(r_a + r_b + r_c) \tag{11.31}$$

Therefore:

$$r_t = r_a + r_b + r_c \tag{11.32}$$

It may be noted that, in the resistance approach, the atmospheric layer between the roughness height, z_0, and the quasi-laminar layer is not treated, which implies that mass transfer through that layer is not limiting compared to mass transfer in the other layers, i.e., the quasi-laminar layer in the case of unstable atmospheric conditions or the planetary boundary layer in the case of stable conditions.

11.1.4 Dry Deposition Velocities and Fluxes

For particles, sedimentation must be combined with the diffusion, interception, and inertia processes to obtain the full dry deposition process. The correct expression is obtained by combining the deposition fluxes to derive the total deposition velocity, v_d (Venkatram and Pleim, 1999):

$$v_d = \frac{v_s}{\left(1 - \exp(-r_t \, v_s)\right)} \tag{11.33}$$

where v_s is the sedimentation velocity and r_t is the resistance to deposition by diffusion, interception, and inertia (i.e., the inverse of the dry deposition velocity due to these three processes). Thus, as v_s tends toward 0 (ultrafine or fine particle), v_d tends toward $(1/r_t)$; as $v_s \gg 1 / r_t$ (coarse particle), then, v_d tends toward v_s. Figure 11.1 shows the

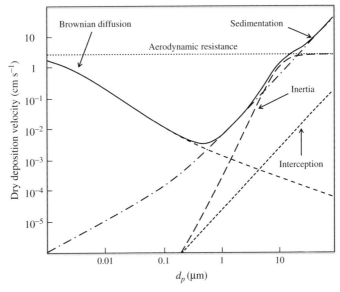

Figure 11.1. Dry deposition velocities of particles as a function of diameter ($u_* = 0.5$ m s^{-1}, $z_0 = 0.01$ m, $\rho_p = 1.5$ g cm^{-3}, $d_f = 2$ mm).

importance of sedimentation for coarse particles, because it is not limited by the aerodynamic resistance.

The dry deposition velocities differ among pollutants (gaseous molecules and particles of various sizes), atmospheric conditions (which affect turbulence), and surface type (soil, vegetation, buildings, water, ...). A chemical species that is highly water soluble, such as nitric acid, may have a dry deposition velocity that is several cm s^{-1} during daytime (i.e., when atmospheric turbulence is significant). On the other hand, fine particles have deposition velocities that are on the order of 0.1 cm s^{-1} during daytime. At night, stable atmospheric conditions may reduce deposition velocities by an order of magnitude or more. Generally, dry deposition velocities are greater in forested areas than over grassland, because the available surface area of tree leaves is much greater than the projected ground surface area.

The deposition flux is the mass of pollutant (g) that is deposited per unit surface area (m^2) per unit time (s). Therefore, the dry deposition flux is equal to the dry deposition velocity multiplied by the pollutant concentration, taken to be negative as it is a sink term for the atmosphere:

$$F_d = -v_d\, C \tag{11.34}$$

The change with respect to time of the atmospheric concentration of a pollutant that is affected only by dry deposition may be calculated via a mass balance on the atmospheric layer affected by dry deposition. As a first approximation, the planetary boundary layer (PBL), which during daytime is typically a well-mixed layer, may be used to carry out this mass balance. Therefore, given a volume of air of surface area A and height z_i (i.e., the PBL height), the pollutant mass, M, present in that volume is:

$$M = C A z_i \tag{11.35}$$

The quantity of pollutant that is deposited per unit time is: $A\, F_d = -A\, v_d\, C$. Therefore, the change with respect to time of the pollutant concentration is governed by the following ordinary differential equation:

$$\frac{dM}{dt} = A F_d \tag{11.36}$$

Therefore:

$$\frac{d(C A z_i)}{dt} = -A\, v_d\, C; \text{ thus : } \frac{dC}{dt} = -\frac{v_d}{z_i} C \tag{11.37}$$

The solution is as follows:

$$C = C_0 \exp\left(-\frac{v_d}{z_i} t\right) \tag{11.38}$$

where C_0 is the initial concentration at $t = 0$. The term (v_d / z_i) is equivalent to a kinetic constant for a first-order process.

Example: Calculation of the fraction of nitric acid that is removed from the atmosphere by dry deposition.

The calculation is performed for a daytime period of 12 hours, using a dry deposition velocity of 4 cm s^{-1} and a PBL height of 2,000 m.

After 12 hours: $\frac{C}{C_0} = \exp\left(-\frac{0.04}{2000} \times 12 \times 3600\right) = 0.42$

Therefore, 42 % of nitric acid are still present in the PBL and 58 % have been deposited on surfaces.

11.2 Wet Deposition

11.2.1 Processes

Wet deposition includes several processes. Generally, a distinction is made between wet deposition associated with precipitation (rain, snow, hail) and occult wet deposition associated with the impact of cloud droplets on elevated terrain (e.g., a mountain) and the sedimentation of fog droplets. Wet deposition by precipitation scavenging dominates and rain scavenging tends to be the most important process, because raindrops can absorb larger amounts of pollutants than snow flakes or ice crystals.

During a precipitation event, two main processes contribute to wet deposition. One process takes place within the cloud, whereas the other process takes place below the cloud base. Within a cloud, during its formation, particles act as condensation nuclei for the formation of cloud droplets (which may later become raindrops) and the pollutants present initially in those particles become incorporated into those droplets. Furthermore, gaseous and particulate pollutants are captured by cloud droplets and raindrops within the cloud. Gaseous pollutants are captured by dissolution in the aqueous phase. Particles are captured when the particle collides with a cloud droplet or raindrop. These processes taking place within the cloud are considered to constitute an overall process called rainout.

Then, precipitation scavenges a fraction of the pollutants present in the atmosphere between the Earth's surface and the cloud base. This scavenging occurs via dissolution of water-soluble gaseous pollutants and capture of particles via collisions between a particle and a raindrop. This below-cloud scavenging is called washout. Thus, water solubility of a gaseous pollutant increases its scavenging efficiency. For particles, brownian diffusion, interception, and impact by inertia contribute to the precipitation scavenging process to various extents depending on particle size. Brownian diffusion dominates for ultrafine particles, interception tends to be the major process for fine particles between 0.1 and 1 μm, and impact by inertia dominates for the larger particles (typically, those with a diameter greater than 1 μm). In the case of interception, a larger particle has a greater probability to enter into contact with a raindrop than a small particle. In the case of inertia, a light particle is more likely to follow the air flow around a raindrop than a heavy particle, which may then collide with the raindrop. The collection efficiency of particles by raindrops is minimum for fine particles (those with a diameter between 0.1 and 2.5 µm) compared to those of ultrafine particles (diameter less than 0.1 µm) and coarse particles (diameter greater than 2.5 µm). The efficiency of this scavenging may be represented quantitatively by a scavenging coefficient, which can be calculated theoretically or estimated empirically. This scavenging coefficient depends on the water solubility of gaseous pollutants and on the size and density of particles. It depends also on precipitation intensity and on the raindrop size distribution.

11.2.2 Parameterizations

The mass transfer of gaseous and particulate pollutants from the gas phase to cloud droplets or raindrops comprises several processes including the activation of hygroscopic particles, collision of particles with droplets, and diffusion of molecules and gas/liquid equilibrium for gaseous pollutants. The parameterization of these processes is described here. Scavenging of pollutants by snow is less efficient than scavenging by rain and, consequently, the former has been little studied. Therefore, only rain scavenging is presented here.

Parameterization of Rainout

This parameterization consists of identifying the hygroscopic particles and parameterizing the fraction that is activated as cloud condensation nuclei (e.g., Lance et al., 2013). The water-soluble fraction of those particles is transferred to the aqueous phase, whereas the insoluble fraction remains as a solid particle within the droplet. In the case of gaseous pollutants, Henry's law equilibrium is generally assumed. The dynamic treatment of a mass transfer step from the gas phase to the droplet is neglected, because it is assumed that the cloud droplet lifetime is sufficiently long (a few minutes) to reach equilibrium.

Parameterization of Washout for Gaseous Pollutants

The dynamic mass transfer of a molecule from the gas phase to the raindrop surface may limit pollutant scavenging because gas/raindrop equilibrium may not be reached. The terminal fall velocity of a raindrop of 5 mm is about 9 m s^{-1}. If the cloud base is at 2,000 m above ground level, the raindrop lifetime is less than four minutes. This time period is commensurate with that mentioned previously for cloud droplets. However, mass transfer of a gas molecule to a cloud droplet with a diameter of a few tens of microns is faster than that of a molecule to a raindrop with a diameter of a few millimeters, as shown in the following equations.

The mass transfer flux, F_m (g m^{-2} s^{-1}), of the pollutant from the gas phase to the raindrop is expressed as a function of (1) the concentration gradient between the gas phase and the raindrop surface and (2) a mass transfer coefficient:

$$F_m = k_m(C_g - C_{g,e}) \qquad (11.39)$$

where k_m is the mass transfer coefficient (m s^{-1}), C_g is the bulk gas-phase concentration, and $C_{g,e}$ is the gas-phase concentration at the surface of the drop (i.e., in thermodynamic equilibrium with the aqueous-phase concentration according to Henry's law). These concentrations are expressed here in g m^{-3}. The mass transfer coefficient may be written as a function of dimensionless numbers:

$$k_m = \text{Sh} \frac{D_m}{d_r} \quad (11.40)$$

where Sh is the Sherwood number (which characterizes the ratio of total mass transfer and transfer by diffusion), D_m is the molecular diffusion coefficient of the gaseous pollutant in the air (m^2 s^{-1}), and d_r is the raindrop diameter (m). The Sherwood number is related to the Reynolds number, which characterizes the flow as the ratio of advection (or convection) and viscosity, and the Schmidt number, which characterizes the ratio of viscosity and diffusion:

$$\text{Sh} = 2 + 0.6 \, \text{Re}^{1/2} \, \text{Sc}^{1/3} \quad (11.41)$$

where Re and Sc were defined in Section 11.1. Therefore:

$$k_m = \frac{D_m}{d_r} \left(2 + 0.6 \, (v_{s,r} \, d_r)^{1/2} \left(\frac{\rho_a}{\mu_{v,a}} \right)^{1/6} D_m^{-1/3} \right) \quad (11.42)$$

where $v_{s,r}$ is the fall velocity of the raindrop (m s^{-1}). The mass transfer coefficient, k_m, is inversely proportional to the raindrop diameter when the raindrop fall velocity is low or to the square root of the raindrop diameter when the raindrop fall velocity is high. In any case, mass transfer is slower for a larger raindrop.

A mass balance on a raindrop leads to the following equation:

$$\frac{dM_{aq}}{dt} = F_m \, s_r \quad (11.43)$$

where M_{aq} is the pollutant mass in the raindrop (i.e., in the aqueous phase, in g), which changes as a function of the mass transferred from the gas phase to the raindrop, and s_r is the surface of the raindrop (m^2). Therefore, this mass transfer rate is proportional to the mass transfer flux, F_m, and the raindrop surface area. The pollutant mass in the raindrop may be expressed as a function of the pollutant concentration in the raindrop and the raindrop volume:

$$M_{aq} = C_{aq} \, v_r \quad (11.44)$$

where C_{aq} is the concentration of the pollutant in the raindrop (g m^{-3}) and v_r is the raindrop volume (m^3). By definition:

$$v_r = \frac{\pi \, d_r^3}{6}; \quad s_r = \pi \, d_r^2 \quad (11.45)$$

In addition, the aqueous-phase concentration is related to the gas-phase concentration at the raindrop surface by Henry's law (using the standard Henry's law constant or the effective Henry's law constant depending on the pollutant, expressed in M atm^{-1}):

$$C_{aq} = H (C_{g,e} RT) \times 10^3 \qquad (11.46)$$

where R is the ideal gas law constant (atm m^3 mole^{-1} K^{-1}), T is the temperature (K) and the factor 10^3 is used to convert the volume from liters to m^3. If the atmosphere below the cloud base is significantly polluted compared to the air where the cloud is located (for example, if the cloud base is above the PBL), then $C_g \gg C_{g,e}$, and the mass transfer flux may be simplified as follows: $F_m = k_m C_g$. Thus:

$$\frac{dC_{aq}}{dt} = \frac{6 k_m C_g}{d_r} \qquad (11.47)$$

Assuming that the atmosphere below the cloud base is well mixed (C_g = constant):

$$C_{aq} = \frac{6 k_m C_g}{d_r} t \qquad (11.48)$$

where t corresponds here to the lifetime of the raindrop. It is assumed here that $C_{aq}(t=0) = 0$. The raindrop lifetime depends on the raindrop fall velocity, $v_{s,r}$: $t = h_c / v_{s,r}$, where h_c is the height of the cloud base above ground level:

$$C_{aq} = \frac{6 k_m C_g h_c}{d_r v_{s,r}} \qquad (11.49)$$

It is assumed here that the raindrop does not become saturated with the pollutant. The wet deposition flux, F_w, is the quantity of pollutant deposited per unit surface area per unit time. It corresponds to the product of the pollutant concentration in the raindrops, C_{aq}, and the precipitation intensity (m of precipitation per second), I_p (a negative sign is used here to treat the wet deposition flux as an atmospheric sink):

$$F_w = -C_{aq} I_p \qquad (11.50)$$

Therefore:
$$F_w = -\frac{6 k_m C_g h_c I_p}{d_r v_{s,r}} \qquad (11.51)$$

The scavenging coefficient, $\Lambda(s^{-1})$, is defined as the kinetic constant corresponding to the loss of air pollutant via wet deposition in the air column below the cloud base:

$$\frac{dC_g}{dt} = -\Lambda t \qquad (11.52)$$

Therefore:
$$C_g(t) = C_{g0} \exp(-\Lambda t) \qquad (11.53)$$

where C_{g0} is the pollutant concentration at the beginning of the precipitation event. A mass balance on an air volume of surface area A and height h_c (from the surface to the cloud base, i.e., the air volume scavenged by rain) leads to the mass of pollutant, M, available for scavenging by precipitation:

$$M = C_g A h_c = C_{g0} \exp(-\Lambda t) A h_c \qquad (11.54)$$

The wet deposition flux for a time period dt is the quantity of pollutant scavenged per unit surface area and per unit time. Therefore, deriving Equation 11.54 with respect to t:

$$F_w = \frac{1}{A}\frac{dM}{dt} = -C_g(t)\,\Lambda\,h_c \tag{11.55}$$

Reconciling the two formulations for F_w, Equations 11.51 and 11.55, the scavenging coefficient may be written as a function of the precipitation intensity, mass transfer coefficient, raindrop diameter, and raindrop fall velocity:

$$\Lambda = \frac{6\,k_m\,I_p}{d_r\,v_{s,r}} \tag{11.56}$$

The scavenging coefficient appears to be proportional to the precipitation intensity. However, a more intense precipitation tends to involve larger raindrops, which have greater fall velocities. Therefore, the scavenging coefficient is not proportional to precipitation intensity. For example, the scavenging coefficient of nitric acid is estimated to be 5×10^{-5} s^{-1} for a precipitation intensity of 1 mm h^{-1}, whereas it is estimated to be only 2×10^{-4} s^{-1} (i.e., four times greater) for a precipitation intensity of 10 mm h^{-1} (i.e., ten times greater).

Parameterization of Particle Scavenging

Particle scavenging results from the collision of raindrops with particles and the incorporation of the particle within the raindrop. It is generally assumed that the incorporation process is 100 % efficient.

The collision rate between a raindrop and a particle is proportional to the cross-section area of both the particle and raindrop and to the difference between their fall velocities, $v_{s,r}$ and $v_{s,p}$:

$$\Lambda = \frac{\pi}{4}(d_r + d_p)^2\,(v_{s,r} - v_{s,p})\,E(d_r, d_p)\,N_{rp} \tag{11.57}$$

where the subscripts r and p refer to raindrop and aerosol particle, respectively, $E(d_r, d_p)$ is the collision efficiency resulting from the diffusion, interception, and inertia processes, and N_{rp} is the number concentration of raindrops for that precipitation event. In these collision processes, the raindrop plays the role of the surface in the formulations presented in Section 11.1 for dry deposition. This collision rate is equivalent to the scavenging rate if each collision results in the incorporation of the particle within the raindrop. The following formulation (Slinn, 1983) may be used to represent the collision processes between a raindrop and an atmospheric particle. These processes include brownian diffusion, interception, and impact by inertia:

$$\begin{aligned}E(d_r, d_p) = &\frac{4}{\text{Re Sc}}(1 + 0.4\,\text{Re}^{1/2}\,\text{Sc}^{1/3} + 0.16\,\text{Re}^{1/2}\,\text{Sc}^{1/2})\\ &+ 4\frac{d_p}{d_r}\left(\frac{\mu_{v,w}}{\mu_{v,a}} + (1 + 2\,\text{Re}^{1/2})\frac{d_p}{d_r}\right) + (1 - 0.9\,\text{St}^{1/2})\left(\frac{\rho_p}{\rho_w}\right)^{1/2}\end{aligned} \tag{11.58}$$

where Re is the Reynolds number of the air flow around the raindrop (defined in Equation 11.8), μ_v is the dynamic viscosity, ρ is the density, the subscript a refers to air, and the subscript w refers to water. The first term on the right-hand side represents the collision of particles with raindrops due to brownian motion of particles. The second term represents the interception process. The third term represents collision by inertia. Here, the simplification of Jung et al. (2003) is used for the inertia term. Figure 11.2 shows the collision efficiency for monodispersed precipitation (i.e., with a single raindrop size) of various intensities. It appears that the collision efficiency is minimum for fine particles, i.e., those particles of diameter between 0.1 and 2.5 µm. This range is called the Greenfield gap, after Stanley Greenfield (1957) of Rand Corporation, who first identified experimentally this minimum in the scavenging efficiency of atmospheric particles by rain.

For a strong precipitation intensity, the following assumptions may be introduced: $d_r \gg d_p$; $v_{s,r} \gg v_{s,p}$. Then:

$$\Lambda = \frac{\pi}{4} d_r^2 \, v_{s,r} \, E(d_r, d_p) \, N_{rp} \tag{11.59}$$

The precipitation intensity may be expressed by conducting a mass balance on the rain water (i.e., the volume of all raindrops per volume of air multiplied by the raindrop fall velocity). For a monodispersed rain:

$$I_p = \frac{\pi}{6} d_r^3 \, v_{s,r} \, N_{rp} \tag{11.60}$$

Therefore, the scavenging coefficient may be written as a function of the precipitation intensity, the scavenging efficiency, and the raindrop diameter:

$$\Lambda = \frac{3}{2} \frac{I_p E(d_r, d_p)}{d_r} \tag{11.61}$$

Several formulations are available to represent the relationship between the rain intensity and the raindrop size distribution (e.g., Duhanian and Roustan, 2011). For the sake of simplicity, a monodispersed precipitation may be used with a simple relationship between the rain intensity and the mean raindrop diameter: $d_r = 0.976 \, I_p^{0.21}$, where I_p is in mm h^{-1} and d_r is in mm (Pruppacher and Klett, 1998). Figure 11.2 shows the raindrop/particle collision efficiency and the scavenging coefficient as a function of the particle diameter for different raindrop diameters and rain intensities. The scavenging rate is a function of the rain intensity. However, it is inversely proportional to the raindrop diameter (which increases with the rain intensity) and proportional to the collision efficiency (which is a complicated function of the raindrop and particle diameters). For example, the scavenging coefficient of a particle with a diameter of 2.5 µm is 3.7×10^{-7} s^{-1} for a rain intensity of 1 mm h^{-1} and it is 2.1×10^{-6} s^{-1} (i.e., seven times greater) for a rain intensity of 25 mm h^{-1} (i.e., 25 times greater).

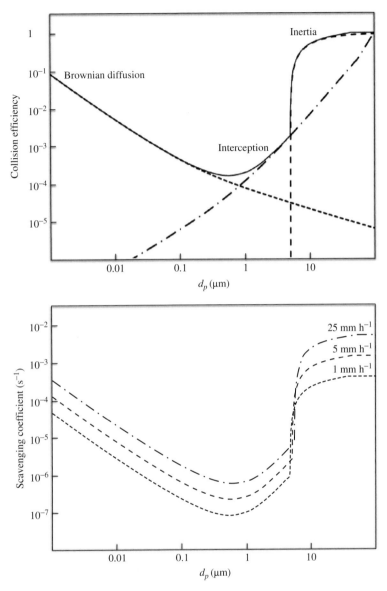

Figure 11.2. Scavenging of particles by raindrops. Top figure: collision efficiency of particles by raindrops with a diameter of 1.37 mm and a rain intensity of 5 mm h^{-1}, as a function of particle diameter; bottom figure: scavenging coefficient for several rain intensities as a function of particle diameter.

Example: Calculation of the fraction of nitric acid that is removed from the atmosphere via wet deposition

It is assumed that the rain event lasts one hour and that the scavenging coefficient of nitric acid (HNO_3) is 1.5×10^{-4} s^{-1}. It is assumed that there is no nitric acid within the cloud and that nitric acid is uniformly mixed within a PBL of 2,000 m.

The change with respect to time of the nitric acid concentration in the PBL due to rain scavenging is given by Equation 11.53:

$$[HNO_3(g)](t) = [HNO_3(g)]_0 \exp(-\Lambda t)$$

Therefore, after one hour (3,600 s):

$$\frac{[HNO_3(g)](t = 1 \text{ h})}{[HNO_3(g)]_0} = 0.58$$

Therefore, 58 % of the initial nitric acid remain in the atmosphere and 42 % have been removed by rain. The amount of nitric acid scavenged by rain in one hour is commensurate with that deposited via dry processes in 12 hours. Note that the PBL height does not come into play in the calculation because scavenging affects the entire column below the cloud base.

11.3 Reemissions of Pollutants Deposited on Surfaces and Natural Emissions of Particles

11.3.1 General Considerations

Some pollutants that have been deposited on surfaces may be reemitted into the atmosphere. It is important to take these reemission processes into account in order to (1) develop complete emission inventories of air pollution and (2) develop a better understanding for some pollutants of the net balance of their mass transfer between the atmosphere and some ecosystems.

This reemission process may occur in different ways. In its simplest form, the pollutant is reemitted under the same form as it was deposited. Semi-volatile pollutants such as persistent organic pollutants (POP) have a volatility that varies with temperature. They deposit more readily when the temperature is low as they are mostly present in the particulate phase. When the ambient temperature increases, their gas/particle partitioning tends to shift toward the gas phase and they are reemitted as gases. A succession of deposition and reemission steps of a semi-volatile pollutant is called the "grasshopper effect," because the pollutant may be transported over long distances (several thousands of kilometers) via a series of deposition/reemission "hops."

The reemission of a pollutant may also occur following a chemical transformation. This is the case, for example, for the emission of some nitrogen oxides (NO and N_2O) from soils following denitrification of nitrate compounds. This is also the case for mercury, which may deposit as gaseous oxidized mercury (for example, as mercury chloride, which is highly water soluble) and, following reduction in soil or surface waters, it may be reemitted as elementary mercury, which is little water soluble and very volatile.

Particles that have been deposited may be reemitted either because of wind-related processes (aeolian resuspension) or because of anthropogenic activities (agricultural activities, on-road traffic ...). Also, particles may be emitted naturally into the atmosphere via aeolian erosion (e.g., from highly erodible soils, such as deserts) and via the evaporation of droplets from waves leading to the emission of sea salt particles into the atmosphere.

Currently, the processes leading to the reemission of gases and particles and to the natural emissions of particles from surfaces are poorly characterized and there are significant associated uncertainties. Three major processes of reemissions and natural emissions of particles are described in the following sections: (1) the resuspension of particles by on-road traffic, (2) the natural emission of particles from soils via aeolian erosion, and (3) the natural emission of sea salt particles.

11.3.2 Resuspension of Particles by On-road Traffic

Empirical estimates and numerical modeling suggest that the resuspension of particles by on-road traffic may contribute significantly to ambient PM_{10} concentrations in the vicinity of major roadways. Therefore, it is essential to account for this particle resuspension term in PM emission inventories, especially for urban areas. Although the size distribution of resuspended particles suggests that they are mostly large particles, the coarse fraction with diameters between 2.5 and 10 μm may contribute significantly to ambient PM_{10} concentrations. This resuspension process depends on several factors including the types of vehicles (a heavy-duty vehicle will lead to greater amounts of particle resuspension than a passenger vehicle), the vehicle speed, the traffic flow, and the erodibility of the particles present on the roadway. This erodibility is closely related to humidity, because a dry roadway is more conducive to particle resuspension than a wet roadway. In addition, the amount of particles potentially available for resuspension is an important factor. Several empirical, semi-empirical, and process-based models have been developed to simulate quantitatively the resuspension rate of particles by on-road traffic. The Nortrip model developed by Denby et al. (2013) appears to be the most complete model, because it takes into account all the relevant factors. In particular, Nortrip uses a mass balance to estimate the quantity of particles present on the roadway and potentially available for resuspension. The change with respect to time of the mass of particles per unit length present on the roadway, $M_{l,r}$ (g km^{-1}), is calculated as follows:

$$\frac{dM_{l,r}}{dt} = F_{l,r} - S_{l,r} \tag{11.62}$$

where t is time (h), $F_{l,r}$ represents the addition of particles onto the roadway, and $S_{l,r}$ represents the loss (sink) of particles from the roadway (g km^{-1} h^{-1}). The addition of particles includes processes that are directly related to on-road traffic, such as brake wear, tire wear, and road wear. These wear processes lead to the emission of particles into the atmosphere, but also to the formation of coarse particles that may deposit immediately on the roadway rather than being emitted to the atmosphere. The addition of particles may also include activities such as sanding and salting of roadways during snow episodes. Finally, atmospheric deposition of particles (see Section 11.1) could contribute to the addition of particles to roadways, although the small dry deposition velocities of atmospheric particles suggest that the direct contribution of traffic-related processes should dominate. The loss terms include mostly the resuspension of particles by traffic and the removal of particles during rain events via water runoff.

Taking into account only the addition of particles due to traffic:

$$F_{l,r} = \sum_{v=1}^{2} N_v \sum_{w=1}^{3} W_{v,w} \quad (11.63)$$

where the subscript v represents the vehicle type (here, heavy-duty vehicles and light-duty vehicles, including passenger cars) and the subscript w represents the wear processes for brakes, tires, and road surface, N_v is the number of vehicles per hour (traffic flow), and $W_{v,w}$ is the wear rate in g of particles per vehicle per km. The wear rates for tires and road surface are expressed as a function of vehicle speed. The brake wear rate depends on the type of driving (road, urban, congested traffic ...). Denby et al. (2013) provide data to calculate those wear rates.

The loss term corresponding to particle resuspension may be written as follows:

$$S_{l,r} = M_{l,r} \sum_{v=1}^{2} N_v f_v(v_v) \quad (11.64)$$

where f_v is the particle mass fraction present on the roadway that is resuspended by a vehicle. This fraction depends on the vehicle type (subscript v) and its speed, v_v. Denby et al. suggest using $f_{v,ref} = 5 \times 10^{-6}$ per vehicle for light-duty vehicles (including passenger cars) and $f_{v,ref} = 5 \times 10^{-5}$ per vehicle for heavy-duty vehicles at a reference speed of 70 km h^{-1}. A loss term may also be calculated for particle removal during rain events; however, this calculation requires detailed information on the rain event (duration and rain intensity). As a first approximation, the particle mass present on the roadway can be reset to zero after each significant rain event.

The equation of change of the particle mass present on the roadway may be written as follows for dry periods between rain events:

$$\frac{dM_{l,r}}{dt} = F_{l,r} - M_{l,r} \sum_{v=1}^{2} N_v f_{v,ref} \frac{v_v}{v_{v,ref}} \quad (11.65)$$

where the subscript ref corresponds to values at the reference speed and particle resuspension is assumed to be proportional to vehicle speed. At steady state, assuming $M_{l,r}(t=0) = 0$:

$$M_{l,r}(t) = \frac{F_{l,r}}{\sum_{v=1}^{2} N_v f_{v,ref} \frac{v_v}{v_{v,ref}}} \left(1 - \exp\left(-\sum_{v=1}^{2} N_v f_{v,ref} \frac{v_v}{v_{v,ref}} t\right)\right) \quad (11.66)$$

The particle resuspension emission rate by on-road traffic is obtained by combining Equations 11.63 and 11.66. Figure 11.3 shows the resuspension rates for an urban freeway and a suburban boulevard in France. They are compared to the direct particulate emissions due to exhaust and wear processes (brakes, tires, and road surface) (Thouron et al., 2018). Direct emissions were calculated using the Copert 4 emission factors (see Chapter 2). These

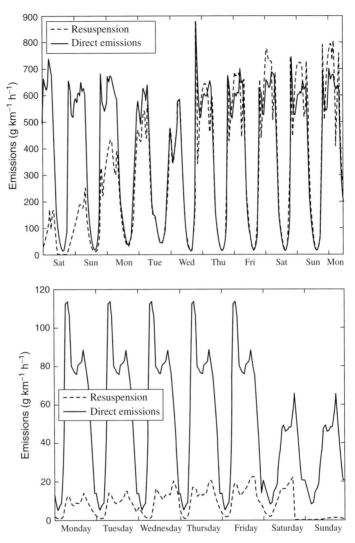

Figure 11.3. PM$_{10}$ direct emission rates and traffic-related resuspension emission rates. Top figure: An urban freeway near Grenoble, France; bottom figure: a suburban boulevard in the Paris region. Source: Adapted from Thouron et al. (2018).

two roadways have very different characteristics. The urban freeway in the Grenoble region in southeastern France has an average traffic flow of 4,000 vehicles per hour with 4 % of heavy-duty vehicles and an average traffic speed of 69 km h^{-1}. The suburban boulevard in the Paris region has an average traffic flow of 800 vehicles per hour with 3 % of heavy-duty vehicles and an average traffic speed of 32 km h^{-1}. In addition, precipitation was negligible during the Grenoble case study, whereas there were several rain events during the suburban Paris case study. The results show that the contribution of resuspension by traffic to PM$_{10}$ emissions is much greater for the urban freeway than for the suburban boulevard, because of greater traffic flow, greater traffic speed, and less precipitation (i.e., a dryer roadway).

Under these conditions, resuspension is commensurate with direct emissions from traffic for this urban freeway. On the other hand, lower traffic flow and speed, as well as frequent rain events, in the suburban Paris case study lead to a contribution of resuspension by traffic, which is only about 10 % of direct PM_{10} emissions.

The formulas presented previously in this section may be refined to take into account tire types (for example, studded snow tires) and the effect of ambient humidity on the resuspension rate. In addition, the size distribution of resuspended particles must be estimated as a function of the processes involved and, in the case of road wear, traffic speed. Denby et al. (2013) provide parameterizations to estimate particle size distributions.

11.3.3 Particle Resuspension by Wind Erosion

Resuspension of particles from an erodible surface by wind erosion (aeolian resuspension) may contribute significantly to atmospheric particle concentrations and these particles may be transported over long distances (Knippertz and Stuut, 2014). For example, wind erosion episodes in the Sahara desert or the Gobi desert may lead to large ambient PM concentrations in Europe or in Asia, respectively. The aeolian resuspension process consists first of a hydrodynamic lifting of particles available at the surface. When the air flow encounters an obstacle, an acceleration of the flow occurs near the obstacle, which induces a pressure drop. This pressure drop leads to a lift of some of the particles potentially available at the surface. These particles may subsequently deposit back to the surface, especially if they are large particles: this is known as the saltation process. This saltation corresponds to a horizontal particle flux and leads to impacts of particles on the surface. These impacts may fragment the surface and generate particles, which become available for resuspension. Then, a vertical flux of particle resuspension occurs. Models that simulate this vertical flux of particle aeolian resuspension must represent the generation of particles by impact of larger particles via saltation.

A review of the theory and experimental data obtained in wind tunnels and in field programs led to the following expression for the vertical flux of particles resuspended by wind erosion, F_e (Kok et al., 2014):

$$F_e = c_e f_{er} f_{clay} \frac{\rho_a (u_{*s}^2 - u_{*t}^2)}{u_{*st}} \left(\frac{u_{*s}}{u_{*t}}\right)^{2.7 \frac{u_{*st} - u_{*st,ref}}{u_{*st,ref}}} \tag{11.67}$$

for $\quad u_{*s} > u_{*t}$

with $\quad c_e = 4.4 \times 10^{-5} \exp\left(-2 \frac{u_{*st} - u_{*st,ref}}{u_{*st,ref}}\right)$

where c_e is the dimensionless coefficient for the resuspension of particles via wind erosion, f_{er} is the erodible fraction of the soil surface, f_{clay} is the clay fraction of the surface (clay content), which corresponds to particles of diameter less than 2 µm (i.e., once resuspended by wind erosion, fine particles), ρ_a is the air density, u_{*s} is the surface friction velocity, u_{*t} is the threshold surface friction velocity above which wind erosion occurs, u_{*st} is the threshold surface friction velocity at standard conditions (here for an air density at 1 atm and 15 °C, i.e., $\rho_a = 1.225$ kg m^{-3}), and $u_{*st,ref}$ is the reference value of the standard threshold surface friction velocity (i.e., for a surface with optimal erodibility). The threshold friction velocities depend on the nature of the erodible surface and may be measured experimentally. They generally vary between 0.2 and 1.5 m s^{-1}. Thus, the vertical resuspension flux depends on the surface friction velocity, which varies according to the air flow and the surface roughness, and on the nature of the erodible surface, which defines the threshold friction velocity leading to erodibility.

The surface friction velocity corresponds to the component of the friction velocity (see Chapter 4) that pertains to the erodible soil fraction. The surface friction velocity may be calculated from data on surface type (roughness length, size distribution of roughness elements). Okin (2008) proposed an exponential formulation to estimate u_{*s} from u_* and the height of non-erodible elements, h_{ne} (for example, vegetation). Measurements suggest that, just downwind of the non-erodible elements, the friction velocity of the surface is 32 % of the friction velocity and that farther downwind the surface friction velocity tends asymptotically toward the friction velocity (being about 90 % of its value at a downwind distance of 10 h_{ne}). Therefore:

$$u_{*s} = u_* \left(1 - 0.68 \exp\left(-\frac{x}{4.8 h_{ne}}\right)\right) \quad (11.68)$$

where x is the distance downwind of the non-erodible element.

An upper limit of the resuspension vertical flux may be estimated for fine particles ($u_{*st} = u_{*st,ref}$):

$$F_e = 4.4 \times 10^{-5} f_{er} f_{clay} \frac{\rho_a (u_*^2 - u_{*t}^2)}{u_{*st}} \quad (11.69)$$

The comparison of the model of particle resuspension by wind erosion with measurements obtained from field experiments leads to better agreement than earlier models for a range of measured resuspension fluxes spanning more than two orders of magnitude (from 10 to 2,000 µg m^{-2} s^{-1}). Most of the calculated fluxes are within a factor of 5 of the measurements and half of the fluxes are within a factor of 2. The uncertainty associated with this model may reach an order of magnitude; however, one should note that the uncertainty associated with the measurements is typically a factor of two. This formulation does not provide any information on the size distribution of the resuspended particles. Particle size distributions for resuspension by wind erosion are available in the scientific literature.

11.3.4 Emissions of Sea Salt Particles

Sea and ocean waves lead to the formation of sea salt droplets in suspension in the air. The evaporation of those droplets may lead to the formation of sea salt aerosol particles. Wave formation depends strongly on wind speed. Different processes are involved in the formation of sea salt droplets. The bursting of bubbles leads to the formation of fine particles ($d_p < 1$ µm) originating from the bubble surface. The airflow that follows the vacuum generated by the bursting of a bubble leads to the production of larger particles (on the order of 1 to 10 µm). Finally, the separation of the foam from the wave crest leads to coarse particles (>10 μm); this latter process only occurs for very high wind speeds. Therefore, a wide spectrum of sea salt particles results from these various processes. Sea salt particle formation may be important and may affect PM concentrations in coastal areas. Accordingly, several algorithms have been developed to simulate sea salt particle emissions. Monahan et al. (1986) have developed one of the first algorithms that have been widely used to simulate sea salt particle formation. The emission flux may be written as the distribution of the particle vertical flux emitted from the sea surface, F_s, as a function of particle diameter:

$$\frac{dF_s}{dd_p} = \text{function}(u_{10},\ d_p,\ T_{ss},\ s_s) \tag{11.70}$$

This function depends on the wind speed at 10 m, u_{10}, the sea salt aerosol particle diameter, d_p, the sea surface temperature, T_{ss} (also known as SST for sea surface temperature), and in some cases the sea salinity, s_s. Traditionally, the sea salt particle size is represented by the particle radius at 80 % relative humidity. Since this radius is about twice that of the dry particle, it can be replaced by the dry particle diameter. Grythe et al. (2014) compared about twenty formulations for quantifying sea salt particle emissions. They concluded that the global spatio-temporal variability of sea salt particle emissions was best represented when the emission rate depends not only on wind speed, but also on sea surface temperature.

The influence of the wind speed on sea salt particle emissions is intuitive, since a high wind speed generates greater waves. Typically, functions such as u^{α_*} are used, with α_* ranging between 2 and 3.5 depending on the formulation. However, a conceptual representation of wave formation should use the shear stress of the turbulent airflow, τ_*, rather than the average wind speed at 10 m:

$$\tau_* = u_*^2\, \rho_a \tag{11.71}$$

where u_* is the friction velocity and ρ_a is the air density. The friction velocity is related to the wind speed via the following equation under neutral conditions (see Chapter 4):

$$u_* = \kappa\, \frac{u(z)}{\ln\left(\frac{z}{z_0}\right)} \tag{11.72}$$

where κ is the von Kármán constant ($\kappa = 0.4$) and z_0 is the roughness length. Therefore, the formation of sea salt particles would be proportional to $(\tau_*)^{\alpha_*/2}$.

Laboratory experiments have shown that the emission of sea salt particles depends on temperature. Therefore, taking into account the sea surface temperature is important in order to correctly represent sea salt emissions, for example in the tropics. The physical explanation of this temperature dependence is not obvious. Jaeglé et al. (2011) proposed several explanations. First, the terminal vertical velocity of a bubble is proportional to the kinematic viscosity of sea water and this viscosity depends on temperature. Thus, more bubbles reaching the sea surface lead to more sea salt particle emissions. Second, temperature could affect the mass transfer between bubbles and the surrounding water, which would influence the size distribution of those bubbles reaching the sea surface and, consequently, particle formation. The dependence of coarse mode sea salt particle concentrations on sea surface temperature was approximated by Jaeglé et al. (2011) using a third-order polynomial, thereby leading to the following formula for the sea salt particle emission flux (expressed as a particle-size distributed flux):

$$\frac{dF_s}{dd_p} = (0.3 + 0.1\ T_{ss,C} - 0.0076\ T_{ss,C}^2 + 0.00021\ T_{ss,C}^3)$$
$$\times 1.373\ u_{10}^{3.41}\ d_p^{-A}\ (1 + 0.057 d_p^{3.45}) \times 10^{1.607\ \exp(-B^2)} \quad (11.73)$$

where

$$A = 4.7(1 + 30\ d_p)^{-0.017 d_p^{-1.44}}$$
$$B = 1 - 2.3\log_{10}(d_p)$$

$T_{ss,C}$ is the sea surface temperature in °C ($T_{ss,c} = T_{ss} - 273$) and the temperature-dependent term is equal to 1 at $T_{ss,C} \approx 21°C$.

11.4 Numerical Modeling of Atmospheric Deposition and Emissions

Numerical modeling of atmospheric deposition and emissions does not present any particular difficulties and standard algorithms may be used to solve the time integration of deposition and emission fluxes.

Modeling of dry deposition of gaseous and particulate pollutants in non-urban areas has been widely studied. The parameterizations proposed by Zhang et al. (2001, 2003) are used, for example, in a large number of air pollution models. In urban areas, the modeling approach of Cherin et al. (2015) may be used in order to take into account the presence of buildings and to calculate deposition fluxes according to surface type (roofs, walls, streets ...).

Modeling wet deposition requires simulating both rainout and washout processes. Duhanyan and Roustan (2011) provide a comprehensive review of existing algorithms and associated uncertainties for the scavenging by precipitation (washout) of gaseous pollutants and atmospheric particles. In particular, the importance of the raindrop size distribution affects strongly the scavenging coefficient.

Emissions of particles by wind erosion, waves, and on-road traffic may be modeled with the numerical algorithms described in Section 11.3. Numerical modeling of other emission processes is described in Chapter 2.

Problems

Problem 11.1 Calculation of dry deposition
The average dry deposition velocity of fine particles is assumed to be 0.02 cm s^{-1}. The PBL height is given to be 500 m. What fraction of fine particles will be deposited over a

week (7 days) and what fraction will remain within the PBL? It is assumed that fine particle concentrations are affected here only by dry deposition.

Problem 11.2   Calculation of wet deposition

The rain scavenging coefficient of fine particles is assumed to be 8×10^{-7} s^{-1} for a precipitation intensity of 25 mm h^{-1}. Calculate the fraction of fine particles removed by wet deposition during a three-hour rain event and the fraction remaining in the atmosphere at the end of the rain event.

References

Cherin, N., Y. Roustan, L. Musson-Genon, and C. Seigneur, 2015. Modelling atmospheric dry deposition in urban areas using an urban canopy approach, *Geosci. Model Dev.*, **8**, 893–910.

Denby, B.R., I. Sundvor, C. Johansson, L. Pirjola, M. Ketzel, M. Norman, K. Kuplainen, M. Gustafsson, G. Blomqvist, and C. Omstedt, 2013. A coupled road dust and surface moisture model to predict non-exhaust road traffic induced particle emissions (NORTRIP). Part 1: Road dust loading and suspension modelling, *Atmos. Environ.*, **77**, 283–300.

Duhanyan, N. and Y. Roustan, 2011. Below-cloud scavenging by rain of atmospheric gases and particulates, *Atmos. Environ.*, **45**, 7201–7217.

Greenfield, S.M., 1957. Rain scavenging of radioactive particulate matter from the atmosphere, *J. Meteor.*, **14**, 115–125.

Gregory, P.H., 1945. The dispersion of air-borne spores, *Trans. Br. Mycol. Soc.*, **28**, 26–72.

Grythe, H., J. Ström, R. Krejei, P. Quinn, and A. Stohl, 2014. A review of sea-spray aerosol source functions using a large global set of sea salt aerosol concentration measurements, *Atmos. Chem. Phys.*, **14**, 1277–1297.

Jaeglé, L., P.K. Quinn, T.S. Bates, B. Alexander, and J.-T. Lin, 2011. Global distribution of sea salt aerosols: new constraints from in-situ and remote sensing observations, *Atmos. Chem. Phys.*, **11**, 3137–3157.

Jung, C.H., Y.P. Kim, and K.W. Lee, 2003. A moment model for simulating raindrop scavenging of particles, *J. Aerosol Sci.*, **34**, 1217–1233.

Kessler, E., 1969. On the distribution and continuity of water substance in atmospheric circulations, *Meteorological Monographs*, **16**, American Meteorological Society, Boston.

Knippertz, P. and J.-B.W. Stuut, 2014. *Mineral Dust – A Key Player in the Earth System*, 508 pp., Springer, Dordrecht, The Netherlands.

Kok, J.F., N.M. Mahowald, G. Fratini, J.A. Gillies, M. Isizuka, J.F. Leys, M. Mikami, M.-S. Park, S.-U. Park, R.S. van Pelt, and T.M. Zobeck, 2014. An improved dust emission model – Part I: Model description and comparison against measurements, *Atmos. Chem. Phys.*, **14**, 13023–13041.

Lance, S., T. Raatikainen, T.B. Onasch, D.R. Worsnop, X.-Y. Yu, M.L. Alexander, M.R. Stolzenburg, P.H. McMurry, J.N. Smith, and A. Nenes, 2013. Aerosol mixing state, hygroscopic growth and cloud activation efficiency during MIRAGE 2006, *Atmos. Chem. Phys.*, **13**, 5049–5062.

Möller, U. and G. Schumann, 1970. Mechanisms of transport from the atmosphere to the Earth's surface, *J. Geophys. Res.*, **75**, 3013–3019.

Monahan, E.C., D.E. Spiel, and K.L. Davidson, 1986. A model of marine aerosol generation via whitecaps and wave disruption in oceanic whitecaps, in *Oceanic Whitecaps and Their Role in Air-Sea Exchange Processes*, Monahan, E.C. and G.M., Niocaill, eds., 167–174, Reidel Publishing, Dordrecht, The Netherlands.

Okin, G.S., 2008. A new model of wind erosion in the presence of vegetation, *J. Geophys. Res.*, **113**, F02S10.

Pruppacher, H.R. and J.D. Klett, 1998. *Microphysics of Clouds and Precipitation*, Kluwer Academic Publishers, Dordrecht, The Netherlands.

Slinn, W.G.N., 1983. *Precipitation Scavenging in Atmospheric Sciences and Power Production*, U.S. Department of Energy, Washington, DC.

Thouron, L., C. Seigneur, Y. Kim, F. Mahé, M. André, D. Lejri, D. Villegas, B. Bruge, H. Chanut, and Y. Pellan, 2018. Intercomparison of three modeling approaches for traffic-related road dust resuspension using two experimental data sets, *J. Transp. Res. Part D*, **58**, 108–121.

Venkatram, A. and J. Pleim, 1999. The electrical analogy does not apply to modeling dry deposition of particles, *Atmos. Environ.*, **33**, 3075–3076.

Wesely, M.L., 1989. Parameterization of surface resistances to gaseous dry deposition in regional-scale numerical models, *Atmos. Environ.*, **23**, 1293–1304.

Wesely, M.L. and B.B. Hicks, 2000. A review of the current status of knowledge on dry deposition, *Atmos. Environ.*, **34**, 2261–2282.

Zhang, L., S. Gong, J. Padro, and L. Barrie, 2001. A size-segregated particle dry deposition scheme for an atmospheric aerosol module, *Atmos. Environ.*, **35**, 540–560.

Zhang, L., J.R. Brook, and R. Vet, 2003. A revised parametrization for gaseous dry deposition in air-quality models, *Atmos. Chem. Phys.*, **3**, 2067–2082.

12 Health Effects

The health effects of air pollution are difficult to characterize because of the large number of air pollutants present in the atmosphere and the relatively small contribution of their health effects compared to all other causes. In addition, air pollution does not affect all people in the same way. Some persons are more sensitive than others: for example, those suffering from asthma, chronic obstructive pulmonary disease (COPD) or cardiovascular problems, the elderly, and children. Also, some individuals are more vulnerable than others: those include, for example, workers and residents who tend to be in locations where air pollution exposure is greater than average. This chapter describes first how adverse health effects of air pollution can be identified and quantified using toxicological and epidemiological studies. Next, methods commonly used to conduct health risk assessments related to air pollution are presented. The use of such information to set up air quality regulations is presented in Chapter 15.

12.1 Identification and Characterization of Health Effects

12.1.1 Toxicology and Epidemiology

Two large categories of health studies are used to understand and quantify the health effects of air pollution. Those are toxicological studies and epidemiological studies. The former are conducted in laboratories under controlled conditions and are useful to study specific pollutants. The latter are field studies that estimate the adverse health effects of air pollution in the ambient environment and aim to isolate the components of the individual air pollutants using statistical methods. These two types of studies present advantages and shortcomings and it is preferable to obtain results from both types of studies in order to properly characterize the health effects of an air pollutant. Brief descriptions of toxicological and epidemiological studies are presented in Sections 12.1.2 and 12.1.3, respectively. Next, the approaches used to combine the results obtained from such distinct types of health effect studies are summarized in Section 12.1.4.

12.1.2 Toxicological Studies

The term "toxicological studies" is used here to represent an ensemble of studies characterized by controlled experimental conditions. However, these studies may be extremely different: for example, in vitro versus in vivo studies. Furthermore, in vivo studies may be conducted with human volunteers or animals, such as rats, mice, monkeys or dogs (e.g., Wichers Stanek et al., 2011).

In vitro studies allow one to formulate hypotheses concerning the toxicity of a chemical substance and also to investigate the mechanisms leading to adverse health effects at the cellular or even molecular level (e.g., Devlin et al., 2005). For example, the Ames test allows one to characterize the mutagenicity of chemical substances and, therefore, to identify the substances that could be carcinogenic, since most carcinogenic substances are mutagenic (Ames et al., 1975). These in vitro studies are conducted with cell cultures in the laboratory. In the case of atmospheric particles, for example, cell cultures of the epithelium of the lung are used to understand how particle size may affect the transfer of those particles in such tissues. The oxidative stress of cells and tissues, which could lead, for example, to the production of inflammatory mediators, has also been studied in vitro with cell cultures.

In vivo studies conducted with human volunteers typically involve exposing several individuals to various concentration levels of an air pollutant in order to determine at which concentration level some adverse health effect is observed. These studies are performed in a clinical laboratory and the exposure conditions are, therefore, well controlled. However, such studies are limited to short exposure durations (referred to as acute exposure). In addition, sensitive individuals are generally not exposed, because of the significant health risks that could be associated with their exposure to moderate and high air pollutant concentrations.

Toxicological studies conducted on animals offer several advantages: more hypotheses may be tested, the reproducibility of the results can be tested, longer exposure durations can be used (referred to as chronic exposure), and sensitive animals can be exposed. Furthermore, exposure can be continued until death of the animal so that autopsies may then be performed to better understand the biological phenomena involved. Therefore, more information may be obtained with animal studies than with human studies, but the study protocols must be defined so that animal suffering is minimized. However, the results obtained from animal studies must be extrapolated to humans. Models are used to perform these extrapolations, but there are many associated uncertainties, particularly when the physiological functions of the animal differ significantly from those of humans. For example, the dosimetry of fine particles in the respiratory system may be different in some animals (e.g., rodents, monkeys, dogs) and humans. Therefore, the extrapolation of such animal studies to humans would be adversely affected by this different dosimetry if it were not explicitly taken into account. Indeed, particles deposit within the respiratory system and only a fraction penetrates deeply to reach the lungs and deposit there. These particle deposition processes depend on particle size, on the characteristics of the airflow within the respiratory system, and on the surfaces available for deposition (see Chapter 11). Figure 12.1 illustrates the efficiency of particle deposition in the human respiratory system as a function of particle size. Ultrafine particles (diameter <0.1 μm) are those that deposit the most in the lungs. Coarse particles (diameter >2.5 μm) deposit in the upper part of the respiratory system (nose, mouth) and, therefore, do not penetrate deeply into the respiratory system. Fine particles (diameter between 0.1 and 2.5 μm) deposit little in the upper part of

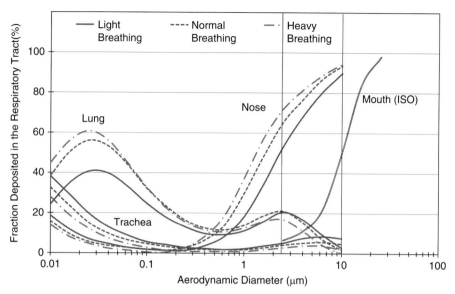

Figure 12.1. Fraction of particles deposited in different regions of the human respiratory system as a function of particle size for different breathing intensity levels (light, normal, and heavy). Source: Cao et al. (2013). Reproduced with authorization from Aerosol and Air Quality Research.

the respiratory system and, therefore, may reach the lungs; however, they do not deposit significantly within the lungs compared to ultrafine particles.

The advantage of toxicological studies is that a specific pollutant may be studied under controlled and reproducible conditions. In the case of human studies, the concentration levels above which an adverse health effect is observed can be quantified for short-term exposure of healthy individuals. In the case of animal studies, chronic exposure studies may be conducted and higher concentration levels may be used. After autopsy, detailed information on the causes of death may be obtained, which could not be obtained otherwise.

Shortcomings of toxicological studies include the fact that the concentrations used are generally greater than those observed in the ambient atmosphere and that the effect of an individual air pollutant may differ from its effect in the presence of other air pollutants. In addition, although some studies may be conducted with sensitive animals, that is typically not the case with sensitive human individuals.

12.1.3 Epidemiological Studies

In an epidemiological study, the objective is to quantify the statistical relationship between the concentration of a pollutant and an adverse health effect (Rothman, 2012). The large number of pollutants present in the ambient air makes the characterization of the effects of a single pollutant difficult because of interference by other pollutants that may have similar effects. These are called confounding factors. Other confounding factors are present due to personal behavior (e.g., smoking) or environmental conditions (e.g., heat wave, cold weather). In addition, the adverse health effects due to air pollution are generally small in terms of excess relative risk (a few %), which makes their quantification in the presence of confounding factors even more difficult. Therefore, the statistical analysis plays a major role in epidemiological studies, because it must be sufficiently powerful and robust to isolate and quantify the relationship between the pollutant concentration and the corresponding adverse health effect.

Epidemiological studies provide quantitative results that are statistical associations between a pollutant concentration and an adverse health effect. However, they do not provide any information on a cause-effect relationship. Therefore, cause-effect relationships must be obtained from toxicological studies. In addition, there is a large number of uncertainties associated with the pollutant concentration measurements, the exposure of the population or individuals, confounding factors, etc., which must be taken into account when presenting the results of an epidemiological study. For example, individuals may be exposed to pollutants outdoor and indoor (at home, at work, etc.). Pollutant concentrations may vary considerably among those various microenvironments, for example, because of indoor pollution sources, various indoor penetration rates for outdoor pollutants, loss processes for indoor pollutants by deposition on walls, furniture, etc. (e.g., Abt et al., 2000). Outdoor ambient pollutant concentrations are generally used in epidemiological studies, because they are the most readily available via air quality monitoring networks. Therefore, the characterization of the actual exposure of the subjects is approximate if exposure to indoor air is significantly different from exposure to outdoor ambient air. All these uncertainties must be quantified and the results of an epidemiological study are presented with error bounds (for example, the 95 % confidence interval; i.e., the interval that has a 95 % probability of including the true value).

There are several types of epidemiological studies. Two categories of epidemiological studies that are the most widely used in air pollution are briefly described next. They are (1) longitudinal studies, where the adverse health effects are analyzed as a function of air pollution levels that vary with time and (2) cross-sectional studies, where health risks are quantified in terms of their spatial variation.

Longitudinal Studies

In a longitudinal study, the occurrence of adverse health effects is observed as a function of time and correlation with exposure to air pollution is analyzed. Generally, ambient air pollutant concentrations are used as a surrogate for exposure.

Longitudinal studies may be ecological studies. Then, the adverse health effects are studied for a whole population (hospital admissions, deaths ...) as a function of air pollutant concentrations. A latency period for the health effect (one day or a few days) may be used to account for the fact that the health effect may not be maximum shortly after exposure to the air pollution, but instead after some time. In an ecological study, there is no information on the specific exposure of individuals to the air pollution (some may have been exposed to greater air pollutant concentrations than others, depending on their activity, location of residence, etc.). Therefore, the statistical power of the ecological study must be sufficiently large to smooth out the exposure variability.

Longitudinal studies may follow a cohort of individuals during a specific period. In such cases, the exposure of the individuals can be characterized with some level of accuracy. A cohort study is more expensive than an ecological study since it requires collecting additional information on the cohort members, but it provides more accurate data in terms of the individual exposure and specific health effects.

An approach used to minimize the cost associated with a longitudinal study, while taking into account the exposure and health effects of specific individuals, is to conduct a case-control study. A case-control study is particularly useful in cases where the health risks related to air pollution are low compared to those of other causes. Identifying and quantifying comparatively low health risks in a cohort requires following a large number of individuals in order to have sufficient statistical power; therefore, large resources are needed and the associated costs may be significant. In a case-control study, the individuals that show adverse health effects are identified first: they are the "case" group. Next, individuals with similar behavior, but who do not show those adverse health effects are identified. These individuals must be at least as many as those of the case group; they are generally selected to be in greater number to obtain a larger statistical sample. These individuals are the "control" group. The exposures of the case and control groups to air pollution are then estimated retrospectively.

There are two major differences between a cohort study and a case-control study. In the cohort study, the exposure of the individuals constituting the cohort are estimated first and the adverse health effects are identified next and correlated with the estimated exposure. In the case-control study, the individuals with adverse health effects are identified first and the exposures of individuals with and without the health effects are estimated next. The other difference is that a case-control study uses a smaller number of individuals than a cohort study. Thus, a case-control study is less costly than a cohort study. However, its statistical power is less and the results are less robust. In addition, exposure is estimated, but is generally not documented with precision. Therefore, the incidence rates (see definition in the section on Definitions of Various Terms) of the health effect cannot be determined, because there is no information on the population corresponding to the selected individuals with the adverse health effects. Therefore, absolute risks cannot be calculated and only relative risks are estimated. The result of such a calculation in a case-control study is called an odds ratio (OR): the ratio of the odds of an adverse health effect for exposed individuals and the odds of the same adverse health effect for non-exposed individuals (see the calculation of the odds ratio in the section on Definitions of Various Terms).

Cross-sectional Studies

In a cross-sectional study, the difference in adverse health effects is analyzed between two populations that have different exposures to air pollution. A cross-sectional study may be conducted, for example, as an ecological study at the level of urban areas. Then, the exposure of the population in each urban area may be estimated from air pollutant concentrations measured by the air quality monitoring network and the adverse health effects may be estimated from hospital admissions (morbidity studies) and death certificates (mortality studies). Cross-sectional studies may also be conducted with smaller areas and populations, such as populations living at different distances of a major roadway or industrial site. Cross-sectional studies pertain to chronic exposure to air pollution.

Definitions of Various Terms Used in Epidemiology: Incidence, Risk, and Odds-ratio

The main terms used in epidemiology to quantify adverse health effects are defined further in this section. In the examples provided, one refers to exposed and non-exposed groups for the sake of clarity. Actually, air pollutants are always present in some amount in the atmosphere, but at different concentration levels. Therefore, epidemiological studies that pertain to air pollution compare groups exposed to air pollution levels that are significantly different, rather than groups exposed to zero and non-zero air pollutant levels. The results of air pollution epidemiological studies are presented either in terms of relative risk or odds-ratio per increment of air pollutant concentration; for example, for an increment of 10 µg m^{-3} or 10 ppb. An example of results from epidemiological studies conducted to characterize the adverse health effects due to fine particles, $PM_{2.5}$, in terms of concentration differences of 10 µg m^{-3} is shown in Figure 12.2.

Incidence: The incidence rate, I_H (or simply incidence), is defined as the number of individuals who develop an adverse health effect (N_H) divided by the duration of the exposure to the pollutant for all individuals in the cohort (T_N):

$$I_H = \frac{N_H}{T_N} \tag{12.1}$$

If all individuals (total number: N_T) have been exposed over the same duration D_E:

$$I_H = \frac{N_H}{N_T D_E} \tag{12.2}$$

Risk: The health risk, R_H, is defined as the incidence rate multiplied by the exposure duration:

$$R_H = I_H D_E \tag{12.3}$$

The health risk is therefore a unitless fraction. If the exposure duration is identical for all exposed individuals:

$$R_H = \frac{N_H}{N_T} \tag{12.4}$$

The excess relative risk, ER, of an air pollutant is defined as the excess risk due to the pollutant (i.e., the difference in the risk for two distinct levels of exposure to the pollutant, R_{H1} and R_{H0}) divided by the risk for the lowest level of exposure to the pollutant (R_{H0}):

$$ER = \frac{(R_{H1} - R_{H0})}{R_{H0}} = \frac{R_{H1}}{R_{H0}} - 1 \tag{12.5}$$

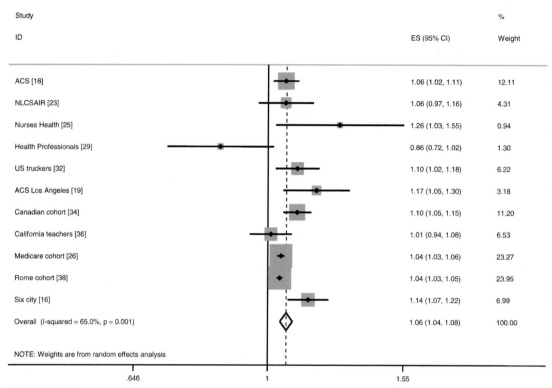

Figure 12.2. Relative risk for the statistical association between premature death (all causes) and an increase in PM$_{2.5}$ concentration of 10 μg m^{-3} from the meta-analysis of Hoek et al. (2013). The relative risk is noted ES for effect estimate. The 95 % confidence intervals are indicated in parentheses and by the solid black lines; the weight given to each study is listed in % in the far-right column and illustrated by the gray squares. The result of the meta-analysis is indicated by the diamond. See Hoek et al. (2013) for details on the different studies and the method used for the meta-analysis.

The relative risk, RR, may be written in terms of a risk ratio or an incidence ratio (assuming identical exposure durations):

$$RR = \frac{R_{H1}}{R_{H0}} = \frac{I_{H1}}{I_{H0}} \tag{12.6}$$

Therefore:

$$ER = RR - 1 \tag{12.7}$$

As mentioned previously, the excess relative risk is typically calculated for an increment in pollutant concentration, rather than with respect to a zero concentration of the pollutant.

The etiologic fraction of risk (EFR) is defined as the difference between the risks for two distinct levels of exposure to the pollutant (the current level of exposure of the population to the pollutant and a lower level of exposure, typically considered safe) divided by the risk of the population exposed to the current level of the pollutant (i.e., it is the health risk fraction that is due to the pollutant):

$$EFR = \frac{(R_{H1} - R_{H0})}{R_{H1}} = 1 - \frac{1}{RR} = \frac{RR - 1}{RR} = \frac{ER}{ER + 1} \tag{12.8}$$

If $RR \approx 1$, then: $EFR \approx ER$.

Odds ratio: In the case of case-control studies, the case group (individuals showing an adverse health effect) includes a subgroup exposed to the pollutant, which is represented here by a number of individuals a_N, and a subgroup not exposed to the pollutant (or exposed to a significantly lower concentration level of that pollutant), represented by a number of individuals b_N. The incidence rates are respectively:

$$I_{H1} = \frac{a_N}{T_{N1}}$$
$$I_{H0} = \frac{b_N}{T_{N0}} \tag{12.9}$$

There is no information available to define T_{N1} and T_{N0}. Therefore, they must be estimated. If one selects in the control group an exposed subgroup with a number of individuals c_N and a non-exposed subgroup (or a subgroup exposed to a lower concentration level) with a number of individuals d_N, such that:

$$\frac{c_N}{d_N} = \frac{T_{N1}}{T_{N0}} \tag{12.10}$$

Then:

$$OR = \frac{I_{H1}}{I_{H0}} = \frac{a_N \, d_N}{b_N \, c_N} \tag{12.11}$$

The odds ratio is the ratio of the odds that the exposed individuals show an adverse health effect (i.e., the ratio of the probability to show an adverse health effect and the probability not to show any adverse health effect), (a_N / c_N), and the odds that the non-exposed individuals show this same adverse health effect, (b_N / d_N). Therefore:

$$OR = RR \tag{12.12}$$

However, this relationship is verified only if the control group represents correctly the statistics of the whole population to which the control group belongs. If the control group represents only approximately the population exposure, then:

$$OR \approx RR \tag{12.13}$$

As an example, Table 12.1 displays the results of an epidemiological case-control study where benzene is the pollutant of interest and the adverse health effect is child acute myeloblastic leukemia.

The odds ratio is: $OR = \dfrac{a_N \, d_N}{b_N \, c_N} = \dfrac{59 \times 2909}{33 \times 3238} = 1.6$

The 95 % confidence interval given by Houot et al. for this odds ratio ranges from 1.0 to 2.4. The study by Houot et al. (2015) is actually more complete and includes an analysis of the distances between the exposed individuals and the roadway and of the road segment

Table 12.1. Contingence table showing some aspect of a case-control study for exposure to benzene in close proximity to on-road traffic and incidence of child acute myeloblastic leukemia. Source: Houot et al. (2015).

Exposure to benzene (concentration level in parentheses)	Acute myeloblastic leukemia	
	Yes (cases)	No (controls)
Yes (>1.3 µg m^{-3})	$a_N = 59$	$c_N = 3\,238$
No (<1.3 µg m^{-3})	$b_N = 33$	$d_N = 2\,909$

lengths. The whole set of results suggest some association between benzene exposure and the incidence of child acute myeloblastic leukemia.

12.1.4 Analysis of Adverse Health Effects

Determination of the Cause-Effect Relationship

A list of the various aspects used to determine a cause-effect relationship was initially proposed by Hill (1965) for epidemiological studies. The U.S. Environmental Protection Agency (EPA) has extended this list to include toxicological studies (i.e., clinical human exposure studies, toxicological animal studies, and in vitro studies). It is as follows:

- Consistency: An inference of causality is strengthened when a pattern of elevated risk is observed across several independent epidemiological studies. However, statistical significance is not the only factor to be considered, as discrepancies among epidemiological studies may result from different exposures, confounding factors, and statistical power of the studies.
- Strength of the observed association between exposure and health effect: Large and precise adverse health effects increase confidence that the observed statistical association is not random or due to experimental biases.
- Specificity of the statistical association: A given pollutant leads to a specific adverse health effect (it is, however, seldom the case).
- Temporal relationship between exposure and the health effect: The effect occurs after the exposure.
- Biological gradient: The effect increases with the exposure level.
- Biological plausibility: The demonstration of a biological process linking the exposure to the effect reinforces the possibility of a relationship between exposure and effect.
- Coherence: Qualitatively similar results between clinical or laboratory (toxicological) studies and epidemiological studies reinforce the interpretation of a causal relationship.
- Experimental evidence: It is provided by the epidemiological studies.
- Analogy: Analogies with other chemical compounds having similar chemical structures may suggest some type of health effect.

It is generally accepted that the determination of an adverse health effect due to a pollutant requires that several of these factors be verified (but not necessarily all of them). In the case of air pollutants, this goal is generally attained by combining toxicological and epidemiological studies. The following aspects predominate:

- Plausibility of the health effect due to the air pollutant: this plausibility requires that a biological mechanism be identified that can relate the inhalation or ingestion of the pollutant to an observed adverse health effect. This plausibility is typically demonstrated using toxicological studies.
- Coherence between toxicological studies and epidemiological studies: this coherence is established, for example, via similar effects being obtained for concentrations of the same order of magnitude. The effects obtained in an epidemiological study will generally be obtained for lower exposure levels than those of toxicological studies, because in the former sensitive individuals will be exposed, whereas such sensitive individuals would have been eliminated from a clinical study. In addition, an epidemiological study may show an effect for chronic exposure, whereas only acute exposure can be studied in clinical studies. Coherence is, therefore, qualitative or in some cases semi-quantitative.
- Consistency among epidemiological studies: Consistency is obtained when several epidemiological studies lead to similar results. These studies may give results that are quantitatively different, but it is inappropriate if some studies show a significant effect while several others show no effect.

The results of an analysis of adverse health effects (combining toxicology and epidemiology) may be grouped as follows in terms of causation between the health effect and the exposure to the pollutant (EPA, 2008b):

- Sufficient to conclude that there is a causal relationship
- Sufficient to conclude that there is a likely causal relationship
- Suggestive but not sufficient to conclude that there is a likely causal relationship
- Inadequate to conclude that there is or not a causal relationship
- Suggestive of the lack of a causal relationship

A regulation is generally proposed in the first two cases (the precaution principle is implicitly applied in the second case).

Table 12.2 summarizes the analyses of relationships between pollutant exposure and health effects conducted by the U.S. EPA for air pollutants regulated in the United States and Europe. It appears that there is sufficient evidence for causal relationships concerning adverse health effects (morbidity) for chronic exposure to lead (Pb) and benzene, and acute exposure to sulfur dioxide (SO_2), nitrogen dioxide (NO_2), and ozone (O_3), and both acute and chronic exposures to fine particles ($PM_{2.5}$). On the other hand, evidence is less certain (likely or only suggestive of a likely causal relationship) for acute and chronic exposure to carbon monoxide (CO), and chronic exposure to NO_2 and O_3. The evidence is inadequate for chronic exposure to SO_2. For inhalable coarse particles (i.e., those particles of diameter between 2.5 and 10 µm) and ultrafine particles (those particles with a diameter less than 100 nm), the evidence is only suggestive of an adverse health effect for acute exposure and is inadequate for chronic exposure.

Table 12.2. Adverse health effects of air pollutants regulated in the United States and in Europe. Sources: U.S. Environmental Protection Agency; EPA, 2008a, 2008b, 2009, 2010, 2013a, 2013b, 2016.

Pollutant	Duration	Cause-effect relationship[a]	
		Morbidity[b]	Mortality[b]
Lead (Pb)	Long term	Causal (children: cognitive deficit; adults: hypertension, cardiovascular diseases, reproduction …)	Causal (cardiovascular diseases)
Carbon monoxide (CO)	Short term	Likely (cardiovascular diseases)	Suggestive
	Long term	Suggestive (child birth and development)	None
Sulfur dioxide (SO_2)	Short term	Causal (respiratory system)	Suggestive
	Long term	Inadequate	Inadequate
Nitrogen dioxide (NO_2)	Short term	Causal (respiratory system)	Suggestive
	Long term	Suggestive	Inadequate
Ozone (O_3)	Short term	Causal (respiratory system), likely (cardiovascular diseases)	Likely
	Long term	Likely (respiratory system)	Suggestive
Fine particles ($PM_{2.5}$)	Short term	Causal (cardiovascular diseases), likely (respiratory system)	Causal
	Long term	Causal (cardiovascular diseases), likely (respiratory system), suggestive (reproduction, cancers)	Causal
Coarse particles ($PM_{10-2.5}$)	Short term	Suggestive (cardiovascular diseases, respiratory system)	Suggestive
	Long term	Inadequate	Inadequate
Ultrafine particles ($PM_{0.1}$)	Short term	Suggestive	Inadequate
	Long term	Inadequate	Inadequate
Benzene	Long term	Causal (lower number of lymphocytes)	Causal (leukemia)

(a) Definitions of the pollutant exposure-health effect relationships:
  Causal: Sufficient to conclude that there is a causal relationship with relevant pollutant exposures
  Likely: Sufficient to conclude that a causal relationship is likely to exist with relevant pollutant exposures
  Suggestive: Suggestive but not sufficient to conclude that there is a likely causal relationship
  Inadequate: Inadequate to conclude that there is or not a causal relationship
  None: Suggestive of the lack of a causal relationship
(b) Effects are mentioned in parentheses if the information is available.

Causal relationships are conclusive for deaths (mortality) due to chronic exposure to lead (cardio-vascular diseases) and benzene (leukemia) and due to both acute and chronic exposures to fine particles. There is sufficient evidence of a likely causal relationship (mortality) due to acute exposure to ozone. However, the experimental data are only suggestive of a causal relationship (mortality) for acute exposure to CO, SO_2, NO_2, and inhalable coarse particles, and for chronic exposure to ozone. For other exposure durations and other pollutants, the data are inadequate to conclude to a relationship between exposure and mortality.

A regulation is generally proposed when the causal relationship is conclusive, whether it is for morbidity or mortality. A regulation may also be proposed when the causal relationship is likely or if the information is suggestive of a causal relationship; however, it will not always be the case and the regulatory context then plays a role (history of regulations, existing similar regulations ...). The setup of air pollutant regulations is described in Chapter 15.

Classification of Carcinogenic Chemical Substances

Health agencies such as the International Agency for Research on Cancer (IARC, which is part of the World Health Organization, WHO) and the U.S. Environmental Protection Agency develop lists of carcinogenic and non-carcinogenic substances. Chemical substances may be classified as follows:

- Category 1: carcinogenic to humans
- Category 2A: probably carcinogenic to humans
- Category 2B: possibly carcinogenic to humans
- Category 3: not classifiable as to carcinogenicity to humans
- Category 4: probably not carcinogenic to humans

For example, benzene, which is regulated as an air pollutant in Europe, is listed in Category 1 (cause of leukemia for chronic exposure).

The list of chemical substances evaluated by the IARC is available at: http://monographs.iarc.fr

The list of chemical substances evaluated by the U.S. EPA is available at: http://www.epa.gov/iris

12.2 Health Risk Assessment

Health risk assessment (HRA) provides quantitative estimates of the health risks associated with exposure to pollutants of an individual or a population. These pollutants may be present in the air or may have been transferred from the atmosphere to other environmental media (water, soil, vegetation, animals ...) following atmospheric deposition to an ecosystem. Multi-media modeling is required in the latter case (see Chapters 11 and 13). Therefore, several exposure pathways must be taken into account. Exposure to pollutants present in the atmosphere occurs via inhalation. Exposure to pollutants present in other media may occur via ingestion and/or dermal absorption.

An HRA may be seen as a four-step process as described by the U.S. National Research Council (NRC, 1994) and recommended by the U.S. Environmental Protection Agency (EPA). These four steps are the following:

- Hazard identification: It pertains to the identification of the chemical substances that are likely to lead to adverse health effects either through inhalation or through other exposure pathways after their transfer to environmental media other than the air. Typically, it consists of identifying the chemical substances emitted from a source or present in the environment and assessing whether they are toxic.
- Dose-response assessment: It involves the selection of quantitative relationships between the dose of a chemical substance absorbed by an individual and the corresponding adverse health effects (i.e., the response). They must be defined for each chemical substance identified in the previous step and may be defined for short- and long-term exposure as well as for carcinogenic and non-carcinogenic effects.
- Exposure assessment: It pertains to the quantitative estimate (calculation or measurement) of the exposure of the individuals or population studied to the chemical substances identified in the first step and the calculation of the corresponding doses absorbed by these individuals or the population.
- Risk characterization: It combines the results of the exposure and dose-response assessments for the chemical substances selected in the hazard identification to provide quantitative estimates of the adverse health effects associated with exposure to those chemical substances. This step should also include some discussion or quantitative assessment of the associated uncertainties.

Identifying whether a chemical substance is toxic involves evaluating whether it may lead to adverse health effects for the exposure periods considered. Both carcinogenic and non-carcinogenic risks must be considered. In addition, risks are typically evaluated for acute exposure (i.e., short-term exposure, generally ranging from one hour to one day) and chronic exposure (i.e., long-term exposure, generally ranging from a few months to a lifetime). There are also subacute and subchronic exposures, which correspond to intermediate duration periods; however, those types of exposures are not considered in standard air pollution studies. Several databases exist that provide detailed information that may be used for this hazard identification step. One may mention, for example, the Integrated Risk Information System (IRIS) of the U.S. EPA (www.epa.gov/iris) and the chemical database of the California Office of Environmental Health Hazard Assessment (OEHHA; https://oehha.ca.gov/chemicals). These databases also provide quantitative information useful for the dose-response assessment.

In the case of non-carcinogenic health effects, one typically assumes that a threshold level exists below which there is no adverse health effect. This threshold level is expressed as a reference concentration, RfC ($\mu g \ m^{-3}$), for air pollutants present in the atmosphere (i.e., being inhaled) and as a reference dose, RfD ($mg \ kg^{-1} \ day^{-1}$), for pollutants that have been transferred to other media (i.e., those being ingested or absorbed through the skin). The use of a threshold may in some cases be an approximation, because such a threshold may not exist or may not have been identified for some air pollutants. It is the case, for example, for fine particles: some sensitive individuals may develop some adverse health effects at very low concentrations.

In the case of carcinogenic effects, it is assumed that there is no threshold below which there would be no risk of cancer. Furthermore, it is assumed that the relationship between the effect (i.e., the probability of developing cancer) and the dose (or the exposure concentration in the case of inhalation) is linear. The slope of this relationship between the dose (or exposure concentration) and the cancer risk is called the unit risk factor, UR (μg^{-1} m^3), in the case of inhalation (i.e., for a concentration-response relationship) and a cancer slope factor, CSF (mg^{-1} kg day), in the case of ingestion and dermal absorption (i.e., for a dose-response relationship). UR corresponds to the additional probability of getting cancer per additional unit concentration of the carcinogenic compound. Similarly, CSF corresponds to the additional probability of getting cancer per additional unit dose of the carcinogenic compound. Units of UR and CSF are the inverse of the units of concentration and dose, respectively.

The reason for using a concentration rather than a dose for inhalation is that there is a single medium to be considered for exposure (namely, the air). Therefore, a default dose/concentration relationship is assumed (typically, an inhalation rate of 20 m^3 of air per day) to get the reference concentration and the calculation of the inhalation dose can then be skipped during the HRA. This simplifying approach cannot be used for ingestion because there are several exposure pathways to be considered (drinking water and milk, eating meat and vegetables, etc.), which are associated with different chemical concentrations and ingestion rates. Therefore, the complete dose calculation must be performed for ingestion. Dermal absorption is not a significant exposure pathway in the case of air pollutants.

The exposure assessment involves calculating (or measuring) the concentrations of the chemical substances in the air and, in the case of ingestion and dermal absorption, the dose via multimedia exposure. This step is conducted using the methods described in Chapter 6 to calculate the transport and dispersion of pollutants in the atmosphere. For multimedia exposure, the transfer of air pollutants to other media via dry and wet deposition must be calculated, as described in Chapter 11.

Next, the results of the dose-response assessment and exposure assessment are combined to perform the risk characterization.

The non-carcinogenic health risk is quantified by the hazard quotient, HQ, which is the ratio of the exposure concentration, C (or dose, D_H), and the reference concentration, RfC (or reference dose, RfD):

$$HQ = \frac{C}{RfC}$$
$$HQ = \frac{D_H}{RfD} \quad (12.14)$$

The carcinogenic health risk is calculated as an excess individual risk, which is the product of the exposure concentration (or dose) and unit risk factor (or cancer slope factor):

$$R_H = UR \times C$$
$$R_H = CSF \times D_H \quad (12.15)$$

The sum of the health risks due to exposure via different exposure pathways may be calculated. For non-carcinogenic health risks, the hazard index, HI, is calculated as the sum of the hazard quotients for inhalation, ingestion, and dermal absorption. Similarly, a cancer risk is calculated for all exposure pathways:

$$HI = \sum_{i=1}^{3} HQ_i$$
$$R_H = \sum_{i=1}^{3} R_{H,i} \quad (12.16)$$

where HQ_i refers to the hazard quotient for exposure pathway i and $R_{H,i}$ refers to the excess cancer risk for exposure pathway i.

Next, the total risk for exposure to multiple pollutants may be calculated as the sum of the risks of the individual pollutants: HI_t for non-carcinogenic health risks and $R_{H,t}$ for carcinogenic risks:

$$HI_t = \sum_{j=1}^{N_c} HI_j$$
$$R_{H,t} = \sum_{j=1}^{N_c} R_{H,j} \quad (12.17)$$

where the subscript j refers to individual pollutants and N_c is the total number of pollutants (chemicals) included in the HRA. If $HI_t < 1$, there is no non-carcinogenic risk. If $HI_t > 1$, the calculation may be performed again by separating the adverse health effects according to different target organs and values of HI_t that are organ-specific are then obtained. If all those organ-specific HI_t values are <1, then, one may consider that there is no non-carcinogenic risk.

The acceptable carcinogenic risk varies depending on the application. For cancer risk related to radioactivity, a value up to 10^{-4} is generally considered acceptable (i.e., a probability of 1 in 10,000 to get cancer during lifetime). For cancer risk due to chemical exposure, an upper value ranging from 10^{-6} to 10^{-5} is generally considered acceptable. For example, the U.S. EPA targets an objective of 10^{-6} for the excess cancer risk due to air pollution; i.e., air pollution should not increase the probability of getting cancer by more than 10^{-6} over lifetime (i.e., a probability of one in one million). For comparison, the risk for the whole U.S. population to develop cancer is on average 0.4 (American Cancer Society, 2019). However, the fraction of cancer-related deaths is smaller: 0.22 (Xu et al., 2018), because many cancer cases are cured.

The sum of the individual excess cancer risks over the whole population is called the population burden. It is generally supposed to be less than 1 for air pollution: i.e., less than one person in the population should develop cancer due to exposure to air pollution.

A health risk assessment for non-carcinogenic air pollutants may also be conducted by assuming that there is a continuous relationship between the cause and the effect instead of using a threshold reference concentration. For example, given some concentration levels for ozone, fine particles or nitrogen dioxide in an urban area, one may estimate the health risk attributable to that pollution (the etiologic fraction, see Section 12.1.3) in terms of morbidity or mortality. Quantitative dose-response relationships obtained from epidemiological studies (relative risk or odds ratio) are used. Then, one may include cases where no threshold has been observed. For example, in the case where the relative risk, RR, is expressed according to a Poisson regression (commonly used in environmental epidemiological studies, also called log-linear relationship), the incidence rate, I_{H1}, for a given concentration, C_1, is written as follows:

$$\ln\left(\frac{I_{H1}}{I_{H0}}\right) = \beta_S \, C_1 \tag{12.18}$$

where I_{H0} is the incidence rate for a zero concentration and β_S is the slope of the concentration-risk relationship. For an epidemiological study where two incidence rates are, I_{H1} and I_{H2}, which correspond to two different concentrations of the air pollutant, C_1 and C_2, the relative risk may be written as the ratio of the two incidence rates ($RR_{\Delta C} = I_{H1} / I_{H2}$):

$$\ln(RR_{\Delta C}) = \beta_S \, \Delta C \tag{12.19}$$

where ΔC is the pollutant concentration difference, ($C_1 - C_2$), corresponding to the relative risk of the two exposed populations. For example, a concentration difference of 10 µg m^{-3} may be used to express the values of RR (and OR) obtained from different epidemiological studies on a same scale. The etiologic fraction may be calculated to estimate the health risk due to a given air pollution level. For example, the number of deaths that would be avoided if the annual PM$_{2.5}$ concentration, C, were reduced to the value recommended by the World Health Organization (WHO), i.e., $C_{WHO} = 10$ µg m^{-3}, is calculated as follows:

$$EFR = \frac{I_H - I_{H,WHO}}{I_H} = \frac{\exp(\beta_S \, C) - \exp(\beta_S \, C_{WHO})}{\exp(\beta_S \, C)} = 1 - \exp(\beta_S \, (C_{WHO} - C)) \tag{12.20}$$

Thus, it may be written as a function of the relative risk expressed for a concentration difference ΔC (Equation 12.19):

$$EFR = 1 - \exp\left(\frac{\ln(RR_{\Delta C})(C_{WHO} - C)}{\Delta C}\right) \tag{12.21}$$

As an example, a relative risk of 1.06 is used here for $\Delta C = 10$ µg m^{-3} (see Figure 12.2; Hoek et al., 2013), which was used in a HRA performed by the French national public health agency, "Santé publique France" (Pascal et al., 2016). The annual PM$_{2.5}$ concentration for the urban background is assumed to be 15 µg m^{-3}, which is typical of an urban area such as Paris (Airparif, 2015). The fraction of deaths attributable to PM$_{2.5}$ concentrations above the value recommended by WHO is calculated as follows:

$$EFR = 1 - \exp\left(\frac{\ln(1.06)(10 - 15)}{10}\right) \approx 2.9\,\% \tag{12.22}$$

Given an annual death rate in the Paris region of 70,500, fine particle air pollution would correspond to about 2,000 annual premature deaths in the Paris region (a premature death is defined as a death occurring prior to the average life expectancy). Actually, PM$_{2.5}$ concentrations vary spatially and a more detailed analysis would be needed. The study conducted by "Santé publique France" (Pascal et al., 2016) for the whole metropolitan France provides more detailed information; it estimates that 17,000 deaths would be avoided annually in France if PM$_{2.5}$ concentrations were reduced to the concentration recommended by WHO. Note that the relative risk used in this study ($RR = 1.06$) has an associated uncertainty and that the 95 % confidence interval leads to a relative risk ranging between 1.04 and 1.08 (Hoek et al., 2013).

The calculation of a health risk involves a myriad of uncertainties resulting from the dose-response assessment and exposure assessment. Therefore, it is recommended, at the minimum, to discuss those uncertainties or, to the extent possible, to quantify those uncertainties. The U.S. National Research Council provided some recommendations to address uncertainties in an HRA (NRC, 1994, 2009). The most comprehensive approach consists of assigning probability distributions to the various sources of uncertainty and calculating the propagation of those uncertainties through the HRA to obtain a probability distribution of the resulting health risk. In addition, one may want to distinguish between epistemic uncertainties and aleatory uncertainties. Epistemic uncertainties are those that could be reduced with additional information. Aleatory uncertainties are those that cannot be reduced. One example of aleatory uncertainty is the meteorological variability, which affects the pollutant concentrations and, therefore, the exposure assessment. Generally, epistemic uncertainties are simply referred to as uncertainty and aleatory uncertainties are called variability. A comprehensive analysis of the uncertainties in an HRA may treat both uncertainty and variability jointly providing a single probability distribution of the resulting health risk. Alternatively, uncertainty and variability may be treated separately in a two-dimensional analysis, thereby providing more detailed information on the main sources of the overall uncertainty affecting the calculated health risk (e.g., Lohman et al., 2000).

Problems

Problem 12.1 Health effect regulations
An epidemiological study shows a correlation between high concentrations of a chemical compound and death rates of older people during a heat wave. On the other hand, toxicological studies do not show any health effect even at concentrations greater than those observed in the atmosphere. Should this chemical compound be regulated as an air pollutant?

Problem 12.2 Carcinogenic health risk assessment
Calculate the cancer risk due to a benzene concentration of 5 $\mu g\ m^{-3}$, which is the European air quality standard. The benzene unit risk factor, UR, is $2.2 \times 10^{-6}\ (\mu g\ m^{-3})^{-1}$.

Problem 12.3 Carcinogenic and non-carcinogenic health risk assessment
Calculate the non-carcinogenic and carcinogenic health risks due to an annual concentration of hexavalent chromium (Cr(VI)) of 0.003 $\mu g\ m^{-3}$ in the air and a chronic dose by ingestion of Cr(VI) of $2 \times 10^{-6}\ mg\ kg^{-1}\ day^{-1}$. For non-carcinogenic health risk, the chronic reference concentration for inhalation, RfC, is $8 \times 10^{-6}\ mg\ m^{-3}$ and the chronic reference dose for ingestion, RfD, is $3 \times 10^{-3}\ mg\ kg^{-1}\ day^{-1}$. The unit risk factor, UR, for carcinogenic risk via inhalation is $1.2 \times 10^{-2}\ (\mu g\ m^{-3})^{-1}$.

References

Abt, E., H.H. Suh, P. Catalano, and P. Koutrakis, 2000. Relative contribution of outdoor and indoor particle sources to indoor concentrations, *Environ. Sci. Technol.*, **34**, 3579–3587.

Airparif, 2015. *Surveillance & Information sur la Qualité de l'Air en Île-de-France – Bilan année 2015*, Airparif, Paris.

American Cancer Society, 2019. Cancer Facts & Figures 2019. www.cancer.org/content/dam/cancer-org/research/cancer-facts-and-statistics/annual-cancer-facts-and-figures/2019/cancer-facts-and-figures-2019.pdf

Ames, B.N., J. McCann, and E. Yamasaki, 1975. Methods for detecting carcinogens and mutagens with the salmonella/mammalian-microsome mutagenicity test, *Mutation Res.*, **31**, 347–364.

Cao, J., J.C. Chow, F.S.L. Lee, and J.G. Watson, 2013. Evolution of $PM_{2.5}$ measurements and standards in the U.S. and future perspectives for China, *Aerosol Air Quality Res.*, **13**, 1197–1213.

Devlin, R.B., M.L. Frampton, and A.J. Ghio, 2005. In vitro studies: What is their role in toxicology, *Exp. Toxicol. Pathology*, **57**, 183–188.

EPA, 2008a. *Integrated Science Assessment for Sulfur Oxides – Health Criteria,* EPA 600/R-08/047 F, U.S. Environmental Protection Agency, Research Triangle Park, NC.

EPA, 2008b. *Integrated Science Assessment for Oxides of Nitrogen – Health Criteria*, EPA 600/R-08/071, U.S. Environmental Protection Agency, Research Triangle Park, NC.

EPA, 2009. *Integrated Science Assessment for Particulate Matter*, EPA 600/R-08/139 F, U.S. Environmental Protection Agency, Research Triangle Park, NC.

EPA, 2010. *Integrated Science Assessment for Carbon Monoxide*, EPA 600/R-09/019 F, U.S. Environmental Protection Agency, Research Triangle Park, NC.

EPA, 2013a. *Integrated Science Assessment for Ozone and Related Photochemical Oxidants,* EPA 600/R-10/076 F, U.S. Environmental Protection Agency, Research Triangle Park, NC.

EPA, 2013b. *Integrated Science Assessment for Lead*, EPA 600/R-10/075 F, U.S. Environmental Protection Agency, Research Triangle Park, NC.

EPA, 2016. *Integrated Risk Information System*, www.epa.gov/iris.

Hill, A.B., 1965. The environment and disease: Association or causation?, *Proc. R. Soc. Med.*, **58**, 295–300.

Hoek, G., R.M. Krishnan, R. Beelen, A. Peters, B. Ostro, B. Brunekreef, and J.D. Kaufman, 2013. Long-term air pollution exposure and cardio-respiratory mortality: A review, *Environ. Health*, **12**, 43–57.

Houot, J., F. Marquant, S. Goujon, L. Faure, C. Honoré, M.-H. Roth, D. Hémon, and J. Clavel, 2015. Residential proximity to heavy-traffic roads, benzene exposure, and childhood leukemia – The GEOCAP study, 2002–2007, *Am. J. Epidemiol.*, **182**, 685–693.

Lohman, K., P. Pai, C. Seigneur, D. Mitchell, K. Heim, K. Wandland, and L. Levin, 2000. A probabilistic analysis of regional mercury impacts on wildlife, *Hum. Ecol. Risk Assess.*, **6**, 103–130.

NRC, 1994. *Science and Judgment in Risk Assessment*, National Research Council, National Academy Press, Washington, DC.

NRC, 2009. *Science and Decisions, Advancing Risk Assessment*, National Research Council, National Academy Press, Washington, DC.

Pascal M., P. de Crouy Chanel, M. Corso, S. Medina, V. Wagner, S. Goria et al., 2016. *Impacts de l'exposition chronique aux particules fines sur la mortalité en France continentale et analyse des gains en santé de plusieurs scénarios de réduction de la pollution atmosphérique.* Santé publique France, Saint-Maurice, France, available at http://invs.santepubliquefrance.fr/Dossiers-thematiques/Environnement-et-sante/Air-et-sante/Publications.

Rothman, K.J., 2012. *Epidemiology: An Introduction*, Oxford University Press, Oxford.

Wichers Stanek, L., J.S. Brown, J. Stanek, J. Gift, and D.L. Costa, 2011. Air pollution toxicology – A brief review of the role of the science in shaping the current understanding of air pollution health risks, *Toxicol. Sci.*, **120** (S1), S8–S27.

Xu, J.Q., S.L. Murphy, K.D. Kochanek, B. Bastian, and E. Arias, 2018. Deaths: Final data for 2016. National Vital Statistics Reports; **67**, N° 5. National Center for Health Statistics, Hyattsville, MD.

13 Environmental Impacts

Air pollutants may be transferred via dry and wet deposition to other media such as soil, surface waters, vegetation, and buildings. These pollutants may then contaminate these surfaces and have adverse impacts on ecosystems, vegetation, and the built environment. In addition, some chemical species that do not have any adverse health effects via inhalation may become toxic via bioaccumulation in the food chain and subsequent ingestion. This chapter describes briefly the impacts of air pollutant deposition on ecosystems, agricultural crops, and buildings, as well as the indirect adverse effects on human health via the food chain.

13.1 Ozone

Ozone is an air pollutant that has adverse health effects on humans (see Chapter 12). It also has adverse impacts on vegetation (EPA, 2007). These effects may be significant on crop yields and, therefore, may translate into economic impacts due to reduced agricultural crop production. Ozone may also show impacts on forests, as is the case, for example, in California.

The first symptoms of ozone deposition on vegetation appear on the outer layer of leaves. They are more important on leaves that are exposed to sunlight. The effects vary depending on the type of vegetation (e.g., deciduous trees, conifers, agricultural crops). The main symptoms are a discoloration of the leaves exposed to sunlight (photobleaching), small spots on the leaf surface (stippling and molting), and/or a brownish coloration on the upper parts of the leaves (bronzing). The mechanism leading to those symptoms may be summarized as follows. Ozone is absorbed by the plant stomata and reacts with organic molecules, such as isoprene and ethylene, which are present in the extracellular plant fluid. Oxidizing organic compounds are formed, which react with the proteins of the cellular membrane of the plant. This deterioration of the membrane leads to the visible effects corresponding to the exposure of the vegetation to high atmospheric ozone concentrations. Some other secondary effects also occur, such as a reduction in CO_2 fixation (either via a perturbation of the enzymatic function or via stomata damage). The perturbation of the photosynthesis of the plant leads to accelerated aging of the leaves and a decrease in the root/shoot ratio and grain/biomass ratio. Therefore, there is a decrease in the grain yield and number of healthy leaves.

Exposure of vegetation to ozone must be quantified with a function that represents exposure above harmful levels. To that end, the AOT40 function (Accumulated ozone exposure over a threshold of 40 ppb) has been widely applied. The calculation of AOT40 is generally limited to daylight hours (solar radiation >50 W m^{-2}). It corresponds to the cumulative exposure of vegetation to ozone concentrations greater than 40 ppb during daylight hours over a three-month period (from May to July) for agricultural crops and over a six-month period (April to September) for trees:

$$\text{AOT40} = \sum_{i=1}^{N_{dh}} \text{Max}([O_3]_i - 40;\ 0) \qquad (13.1)$$

where N_{dh} represents the total number of daylight hours during the period of interest and $[O_3]$ is expressed in ppb. For example, the European Union has a target value of 3,000 ppb h for crops based on an AOT40 calculated using a daylight-hour period of 8 am to 8 pm Central European Time. Other exposure threshold values may be used. For example, a threshold value of 60 ppb is used in the SUM60 function. It is calculated in the same way as AOT40, but with an ozone concentration threshold of 60 ppb. The U.S. EPA currently uses a sigmoidal cumulative function, W126, which does not use an adverse exposure concentration threshold, but instead gives more weight to the higher concentrations. It is calculated for daylight hours between 8 am and 8 pm, i.e., for a twelve-hour period, as follows:

$$W126_j = \sum_{i=8am}^{7pm} [O_3]_i \left(\frac{1}{1 + \left(4403 \exp(-126\ [O_3]_i)\right)} \right) \qquad (13.2)$$

Next, the function is calculated over a three-month period:

$$W126 = \sum_{j=1}^{N_w} W126_j \qquad (13.3)$$

where N_w is the total number of days during the three-month period (i.e., $N_w = 91$ or 92). (Adjustments are made for periods with missing O_3 data.) W126 is calculated as a moving three-month average between April and October and the maximum value of the three-month averaged W126 is selected. Next, the average of these maximum values is calculated over a three-year period. The U.S. EPA currently uses W126 to estimate vegetation exposure to ozone and a range of 13,000 to 17,000 ppb h is considered acceptable to protect vegetation from ozone damage. Although there is no direct relationship between the U.S. health-based air quality standard (a concentration of 0.070 ppm averaged over 8 hours, referred to as the primary ambient air quality standard) and W126, the U.S. EPA considers that the primary standard should be protective of vegetation and has not proposed a secondary standard that would be specific to vegetation exposure.

The adverse effects of ozone on vegetation depend on the types of crops and trees, as some species are more resistant than others. Nevertheless, the impact of ozone on vegetation translates into significant economic impacts for agriculture. For example, ozone levels for the year 2000 have been estimated to lead to decreases in global crop yields ranging from 2 to 15 % depending on crop type (soybean, wheat or maize) and the methodology used (Avnery et al., 2011).

13.2 Acid Rain

The acidity of water is represented on a logarithmic scale by the pH. It is the negative value of the base 10 logarithm of the activity of the proton H^+ (for a dilute solution, the activity is equivalent to the concentration; see Chapter 10):

$$\text{pH} = -\log([H^+]) \tag{13.4}$$

Pure water has a neutral pH of 7 at 25 °C, since H^+ and OH^- are present in identical concentrations. However, liquid water in the atmosphere does not have a neutral pH, even in a pristine atmosphere. Carbon dioxide (CO_2), which is naturally present in the atmosphere, is a weak acid, which is soluble in water. For an atmospheric CO_2 concentration of 400 ppm (parts per million), the pH of a cloud droplet or raindrop is calculated to be 5.6 (see Chapter 10). Therefore, acid rain has a pH that is less than 5.6.

Chemical species that lead to acid rain are strong acids, i.e., they dissociate quasi totally in water into H^+ and anions. In the atmosphere, the main strong acids are sulfuric acid (H_2SO_4), nitric acid (HNO_3), and hydrochloric acid (HCl). The discovery of acid rain goes back to the beginning of the industrial era. Although the original references given in the English literature list mostly the work of R.A. Smith (1852, 1872), the first scientific article on that topic was published by a French pharmacist, M. Ducros (1845), who proposed that the acidity of the rain was due to nitric acid, resulting, for example, from the formation of nitrogen oxides during thunderstorms. A reanalysis of the data obtained by Smith on precipitation in the region of Manchester suggested that acidity was mostly due to HCl. It is possible that coal combustion in that region led to high atmospheric concentrations of HCl (chlorine is one of the compounds present in coal and it is emitted mostly as HCl during coal combustion).

One should note that acid rain is generally used as a term that covers more generally both wet and dry deposition. Wet deposition includes scavenging by rain, but also by snow and hail, as well as occult wet deposition (from mountain clouds and fogs). Dry deposition affects both particles (which may contain sulfate and nitrate) and gases (such as nitric acid); see Chapter 11. Therefore, the correct scientific term is "acid deposition," rather than "acid rain."

In the 1970s, regions such as Scandinavia in Europe and Canada in North America showed significant modifications in their lakes with significant decreases in their fish population. In addition, some forests such as the Black Forest in Germany showed important signs of decline. An analysis of the causes of these environmental changes showed that the pH of lakes had decreased significantly (i.e., lake acidification). In the forests, the rain acidity led to a change in the geochemistry with a modification of the chemical equilibria of the various inorganic chemical species (e.g., Reuss and Johnson, 1986). In particular, species such as magnesium (Mg) and calcium (Ca), which are nutrients for vegetation, became soluble ions. Thus, they were washed out by the rain and were no longer available for absorption by the roots of the trees. In addition, some elements that are toxic to vegetation, such as aluminum (Al), became available for absorption by roots. As a result, these different factors led to the death of a large number of trees in some regions.

The main acids responsible for the acidification of lakes and soils were identified to be sulfuric acid and nitric acid.

These two acids are not emitted in the atmosphere in significant amounts, but are formed in the atmosphere by the chemical oxidation of primary pollutants emitted in the atmosphere (see Chapters 8 and 10):

$$SO_2 + \text{oxidant} \rightarrow H_2SO_4 \tag{R13.1}$$

$$NO_2 + \text{oxidant} \rightarrow HNO_3 \tag{R13.2}$$

Therefore, sulfuric acid and nitric acid are secondary pollutants.

Sulfur is present in coal, oil, and various minerals. Therefore, it is emitted during the combustion of coal (power plants, residential heating ...), oil-derived fuels (diesel, gasoline ...), and some industrial activities (smelters ...), mostly in the form of sulfur dioxide (SO_2). Nitrogen oxides (nitric oxide, NO, and nitrogen dioxide, NO_2) are emitted during all combustion processes because nitrogen (N_2) and oxygen (O_2) present in the air react at the high temperatures of the combustion process to form these nitrogen oxides (mostly NO; NO_x is used to represent the sum of NO and NO_2). Since it takes some time for those primary pollutants to be oxidized into acids, acid rain pollution tends to cover long distances (several hundreds of kilometers). In addition, there are some natural sources of SO_2 and NO_x. Volcanic eruptions lead to significant emissions of SO_2. Oceans emit dimethyl sulfide (DMS), which gets oxidized slowly into SO_2. Soils emit nitrogen oxides (NO_x, as well as nitrous oxide, N_2O, a greenhouse gas). Lightning produces NO_x in the high atmospheric layers (as suggested by Ducros, 1845). Table 2.1 summarizes the main categories of SO_2 and NO_x sources (see Chapter 2).

Another effect of acid rain is the degradation of buildings and statues (Brimblecombe, 2003). Some stones are calcareous (i.e., calcite, some marbles, freestone ...). This calcareous compound is calcium carbonate, which reacts, for example, with sulfuric acid to lead to calcium sulfate (gypsum). This reaction leads to a change in the stone cohesiveness; the stone is no longer homogeneous in its chemical composition and some erosion of the stone occurs.

Acid deposition simulations are performed with three-dimensional (3D) atmospheric chemical-transport models. Such models were initially developed during the 1980s. The chemistry of the oxidation of SO_2 into sulfuric acid and of NO_x into nitric acid is rather well known (see Chapters 8 and 10). Figure 13.1 shows a comparison between wet deposition fluxes (precipitation) of simulated and measured sulfate at different stations of the National Acid Deposition Program (NADP) monitoring network in the United States. The spatial correlation between the simulation and measurements is satisfactory (determination coefficient of 0.77) and the model bias is low (8 %).

SO_2 and NO_x emissions have been regulated in the 1980s in order to reduce acid deposition. For example, in the United States, SO_2 and NO_x emission regulations have been promulgated using a cap-and-trade approach (see Chapter 15). At the international level, the Göteborg protocol was introduced in 1999. It is officially called the "Gothenburg Protocol to Abate Acidification, Eutrophication and Ground-level Ozone"; therefore, it is also known as the "Multi-effect Protocol," because it addresses several forms of air

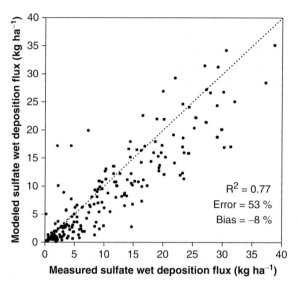

Figure 13.1. Comparison of simulated and measured wet deposition fluxes of sulfate in the United States for the year 1996. Source: Reproduced with authorization from Vijayaraghavan et al. (2007), © American Meteorological Society.

pollution (see Chapter 15). As of August 2017, 25 countries (including the U.S.) and the European Union had ratified this protocol, which defines ceiling targets for national emissions that must be attained by a given date (currently 2020). SO_2 regulations have targeted mostly coal-fired power plants, smelters, and the sulfur content of gasoline and diesel fuels. NO_x regulations have been mostly driven by the contribution of NO_x species to ozone formation. Regulations have targeted emissions from on-road traffic, refineries, and fossil-fuel fired power plants. Long-term monitoring of sulfate and nitrate, as well as rain pH, has shown that these regulations were efficient and led to a decrease in acid deposition downwind of those sources. Most of the lakes and soils have seen their pH increase. However, some ecosystems recover faster than others, depending on their physico-chemical characteristics, and some lakes and forests are still under recovery. Nevertheless, it appears that overall the public policies introduced to reduce SO_2 and NO_x emissions have been effective in North America and in Europe, so that ecosystems that were adversely impacted in the 1970s are recovering and, in some cases, have recovered to their pre-industrial status. However, the acid deposition problem may now be present in some regions of Asia due to a fast industrial growth over the past several years.

13.3 Eutrophication

Vegetation needs nitrogen to grow. Plants obtain nitrogen naturally from the atmosphere and the soil. Agriculture uses nitrogenous fertilizers (for example, ammonium nitrate) to increase the yields of some crops. An increase in nitrogenous compound concentrations in water bodies leads, therefore, to an increase in aquatic vegetation. This increase in vegetation at the surface of the water body leads to a decrease in the amount of sunlight reaching the lower layers of the water body. This decrease in the solar radiation flux has a negative impact on plant photosynthesis (which converts carbon dioxide to oxygen) and the lower production of oxygen by plants leads to anoxic conditions (i.e., lack of oxygen). As a result, the concentrations of living organisms (e.g., phytoplankton, zooplankton) decrease, because of a lack of oxygen in the lower layers of the river, lake or sea. This phenomenon is called eutrophication. Its consequences are an increase in plants at the surface of the water body and a decrease in the aquatic life in the deep layers of that water body (e.g., Carpenter et al., 1998).

The main cause of eutrophication for most water bodies is the input of phosphorus and nitrogen originating from land-based human activities (e.g., use of fertilizers). Nevertheless, atmospheric deposition of nitrogen species contributes also to eutrophication (e.g., Bergström and Jansson, 2006). The main contributors to nitrogenous atmospheric deposition are nitric acid (which is also a contributor to acid deposition) and ammonia. Nitrogen oxides, NO_x, are not very soluble in water and have low contributions to the dry and wet deposition fluxes of nitrogen overall. The contribution of organic nitrates is estimated to be commensurate to that of NO_x and is, therefore, significantly less than those of nitric acid and ammonia (e.g., Vijayaraghavan et al., 2010). Anthropogenic sources of precursors (NO_x) of nitric acid have been mentioned in Section 13.2. The main sources of atmospheric ammonia are agriculture-related activities: (1) the use of fertilizers for crops and (2) farm animals (cattle, pigs ...). Table 2.1 summarizes the main source categories for nitrogen oxides and ammonia (see Chapter 2). The emissions of those pollutants are mostly regulated because of their air pollutant status in the case of NO_2 and their role as precursors of ozone (NO_x), acid deposition (NO_x), and secondary fine particulate matter (NO_x and NH_3). Therefore, they are typically not targeted because of their eutrophication impacts. Nevertheless, this aspect is mentioned in some regulations, such as the Göteborg Protocol, which defines emission limits by country in Europe in order to reduce acid deposition, eutrophication, and ozone formation (see Chapter 15).

13.4 Persistent Organic Pollutants

Persistent organic pollutants (POP) include a large number of organic species that (1) have harmful health effects on humans (and, for many of those species, also on fauna) and (2) are not degraded rapidly in the environment. Environmental problems related to POP have been presented to the general public by Rachel Carson in her book titled *Silent Spring*, published in 1962. In that book, she linked the use of pesticides such as DDT to the decrease of the population of some bird species (i.e., the silent spring that would result from a lack of singing birds). Scientific studies have been conducted to confirm and document quantitatively what Rachel Carson had identified as a major environmental problem. Among these

Table 13.1. Persistent organic pollutants targeted by the Stockholm Convention.

Persistent Organic Pollutants	Source and/or use
Aldrin ($C_{12}H_8Cl_6$)	Insecticide
Chlordane ($C_{10}H_6Cl_8$)	Insecticide
Dichlorodiphenyltrichloroethane (DDT, $C_{14}H_9Cl_5$)	Insecticide
Dieldrin ($C_{12}H_8Cl_6O$)	Insecticide
Endrin ($C_{12}H_8Cl_6O$)	Insecticide
Mirex ($C_{10}Cl_{12}$)	Insecticide
Heptachlor ($C_{10}H_5Cl_7$)	Insecticide
Hexachlorobenzene (C_6Cl_6)	Fungicide
Polychlorinated biphenyls (PCB, several compounds with x chlorine atoms, x = 1 to 10, chemical formula: $C_{12}H_{10-x}Cl_x$)	Dielectric compounds used in transformers and other industrial processes; combustion products
Toxaphene (mixture of compounds $C_{10}H_{11}Cl_5$ to $C_{10}H_6Cl_{12}$, on average $C_{10}H_8Cl_8$)	Insecticide
Polychlorinated dibenzodioxins (PCDD, several compounds among which the most toxic is 2,3,7,8-tetrachlorodibenzo-p-dioxin, $C_{12}H_4O_2Cl_4$)	Combustion products
Polychlorinated dibenzofurans (PCDF, several compounds among which the most toxic is 2,3,7,8-tetrachlorodibenzofuran, $C_{12}H_4OCl_4$)	Combustion products

scientific articles that establish a link between some POP and their adverse impacts on ecosystems, one may mention those of Ratcliffe (1967) and Prest et al. (1970).

The Stockholm Convention, which was signed in 2001 by 151 countries and was adopted in 2004, identified twelve POP or POP categories (the "dirty dozen"), for which atmospheric emissions should be controlled. These pollutants include several insecticides, a fungicide, polychlorinated biphenyls (PCB, compounds with interesting dielectric properties, which are used in electric transformers), and dioxins and furans (compounds produced mostly during waste combustion). These POP are listed in Table 13.1. They all contain chlorine. However, all POP do not contain chlorine. For example, polycyclic aromatic hydrocarbons (PAH) emitted by combustion (for example, by diesel and gasoline vehicles) are POP and some PAH are carcinogenic.

Most POP are semi-volatile or non-volatile compounds. Only a few POP are volatile at ambient temperature and pressure (for example, naphthalene, which is the simplest PAH). POP with high molecular weights are non-volatile and, therefore, will be emitted as particles. A large number of POP have a saturation vapor pressure in a range that favors their partitioning between the gas phase and the particulate phase as a function of ambient temperature and pressure. This semi-volatile property of many POP leads to the "grasshopper effect," where a POP may deposit to the Earth's surface as particles under cold conditions (for example, in winter) and be reemitted later as gas molecules under warm conditions (for example, the following spring), thereby becoming available for atmospheric transport (see the description of reemissions in Chapter 11). Therefore, pollution by POP may be a long-distance problem because of this grasshopper effect. Thus, one may find significant POP concentrations in regions that are not directly associated with anthropogenic activities. Most POP are little reactive chemically and, therefore, they have a long environmental residence time (Jones and de Voogt, 1999; Franklin et al., 2000; Keyte et al., 2013).

POP have adverse impacts on the environment, because they tend to concentrate in the food chain. They are present in the atmosphere at very low background concentrations (on the order of 10^{-12} atm). However, once they have been transferred to another environmental medium, they may reach high concentrations via bioconcentration, bioaccumulation, and biomagnification in animal species. Bioconcentration characterizes the ratio between the pollutant concentration in the animal species (or a plant) and that in its environment at equilibrium (for example, concentrations in a fish and water). Bioaccumulation characterizes the increase of that concentration through the food chain, for example via a predator-prey relationship. Biomagnification includes all these processes. Biomagnification factors may be on the order of a million for animal species at the top of the food chain, which may lead to concentrations on the order of 1 ppm (1 part per million, i.e., 1 mg of pollutant per kg of fish for example).

Emissions of some POP have been controlled significantly. For example, the use of some pesticides has been forbidden in some countries in North America and Europe. POP, which are produced during some combustion processes (in incinerators for example) may be regulated at the source. It is the case, for example, for dioxin and furan emissions from incinerators, which are regulated in several countries in North America and Europe.

13.5 Heavy Metals

The term "heavy metals" does not have a scientific definition and there is no unique classification of elements grouped within that term. Densities greater than 4 to 5 g cm^{-3} are sometimes mentioned for heavy metals. Also, some metalloids, such as arsenic, are sometimes considered to be heavy metals. Several heavy metals have adverse health effects. For example, lead poisoning may be the cause of learning disabilities and behavioral problems in children and other effects in adults (kidney problems, anemia ...). Cadmium may have various adverse effects (kidney problems, bone weakness, perturbation of the digestive system). Other heavy metals, such as chromium and mercury, have various adverse health effects depending on their chemical form, as described in Sections 13.5.1 and 13.5.2, respectively.

13.5.1 Chromium

Chromium is present in the atmosphere in two main oxidation states: trivalent, Cr(III), and hexavalent, Cr(VI). Only hexavalent chromium is carcinogenic. Therefore, it is important to understand the chemical transformations that occur between these two main chemical forms of atmospheric chromium. The major atmospheric reactions of chromium are shown

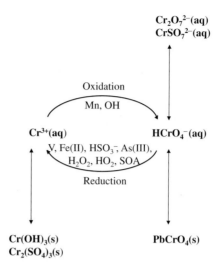

Figure 13.2. Chemical transformations of atmospheric chromium. These transformations occur within atmospheric particles and fog/cloud droplets. Source: After Seigneur and Constantinou (1995), see text for details.

in Figure 13.2. The oxidation of Cr(III) to Cr(VI) tends to be slow and the reduction of Cr(VI) to Cr(III) dominates in the atmosphere (Seigneur and Constantinou, 1995; Lin, 2002). In addition to the chemical species identified in the chemical kinetic mechanism of Seigneur and Constantinou (1995), hydrogen peroxide (H_2O_2), hydroperoxyl radicals (HO_2/O_2^-), and secondary organic aerosols (SOA) may also contribute to the reduction of Cr(VI) in droplets and aerosol particles (Hug et al., 1996; Pettine et al., 2002; Huang et al., 2013), and hydroxyl radicals (OH) may contribute to the oxidation of Cr(III) (Lin, 2002).

13.5.2 Mercury

Mercury presents some adverse health effects in its elemental from (hydrargyrism, which includes symptoms such as shaking and memory loss), but at concentrations much greater than those observed in the ambient environment. On the other hand, once mercury has been transformed into an organic form, it can bioaccumulate in the food chain and have adverse health effects on animals that rely on fish for their diet (e.g., some birds, mink ...), as well as on other animals on top of the food chain that eat piscivorous animals (e.g., puma). In the case of humans, the main risk is for newborns via exposure of the mother during pregnancy. Then, the effect is damage of the central nervous system of the child, which may become permanent. Mercury is used here as an example for the description of the environmental cycle of a heavy metal, because its multi-scale atmospheric transport, its chemical transformations in various environmental media, and its bioaccumulation in the food chain make it an interesting species to study (e.g., Selin, 2009).

Mercury is emitted from anthropogenic and natural sources. Anthropogenic activities are estimated to contribute about two-thirds of the current atmospheric mercury levels.

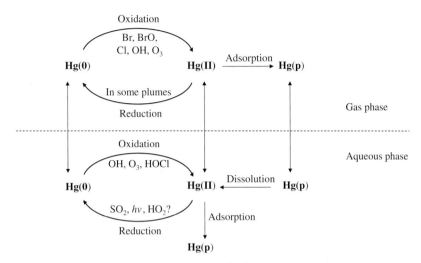

Figure 13.3. Chemical transformations of atmospheric mercury. See text for details.

Anthropogenic emissions contribute directly only one-third to the global emission inventory. However, a fraction of the mercury deposited to the Earth's surface is reemitted to the atmosphere. About one-third of the global mercury emissions correspond to anthropogenic mercury reemitted after atmospheric deposition.

Mercury is present mostly under three forms in the atmosphere: elemental mercury (Hg°), gaseous divalent mercury (Hg^{II}) (also called oxidized mercury or reactive mercury), and particulate mercury. Particulate mercury is divalent (i.e., the oxidized form of mercury); it may be a solid mercury species, such as mercury oxide (HgO) or mercury sulfide (HgS), or it may be a gaseous oxidized mercury species adsorbed on atmospheric particles, such as mercury chloride, $HgCl_2$, or mercury hydroxide, $Hg(OH)_2$. In the atmosphere, several chemical reactions oxidize elemental mercury to divalent mercury, while others reduce divalent mercury to elemental mercury (see Figure 13.3). Some reaction pathways are well established (for example, the oxidation of Hg° by bromine), whereas others such as the heterogeneous oxidation of Hg° by O_3 or the heterogeneous reduction of Hg^{II} in some power plant plumes are still uncertain (Ariya et al., 2015; Horowitz et al., 2017). On average, elemental mercury dominates in the atmosphere (>90 %), whereas gaseous or particulate oxidized mercury account for only a few percent. However, in some cases (e.g., near a source, in the high atmosphere), oxidized mercury may dominate. Elemental mercury is little water soluble and its atmospheric removal occurs mostly via oxidation to divalent mercury. Divalent mercury species such as $HgCl_2$ and $Hg(OH)_2$ are highly water soluble and, therefore, they are readily scavenged by precipitation.

In other environmental media such as surface waters and soils, mercury undergoes chemical transformations. Elemental mercury may be oxidized into divalent mercury and divalent mercury may be reduced to elemental mercury. Divalent mercury may be transformed by bacteria into organic mercury (monomethyl mercury). Organic mercury is the form that accumulates in the aquatic food chain and may subsequently have adverse health effects on fauna and human health. Therefore, mercury exposure occurs mostly via consumption of fish with high mercury concentrations. Fish with high concentrations of mercury are typically those at the top of the food chain. These high mercury concentrations may result from direct releases of mercury into the surface water body (e.g., the contamination by a chemical factory of Minamata Bay in Japan) or from atmospheric deposition. In addition, the physico-chemical and biological properties of the ecosystem play an important role in the transformation of inorganic mercury into organic mercury and the subsequent bioaccumulation of mercury in the aquatic food chain. For example, the formation of organic mercury is faster if sulfate concentrations in water are high. Therefore, a decrease in atmospheric sulfate deposition (needed to reduce acid rain) has a beneficial effect on the contamination of ecosystems by mercury, since it reduces the amount of the mercury species that bioaccumulates in the food chain.

Atmospheric deposition of mercury to ecosystems is simulated similarly to acid deposition with chemical-transport models. However, there are more uncertainties associated with atmospheric mercury simulations, because (1) emissions are not as well known as those of sulfur and nitrogen oxides (in particular natural emissions and reemissions from soils and surface waters) and (2) mercury atmospheric chemistry is complex and some reactions are still poorly understood. An example of simulations of the cycle of mercury in the atmosphere and lakes is presented in Figure 13.4. The atmospheric simulation consisted of a global simulation using emissions over the continents and the oceans, followed by a simulation at a finer spatial resolution over North America. Atmospheric deposition fluxes obtained from that simulation were then used as input data for the simulation of mercury cycling in a Michigan lake in the U.S. The atmospheric mercury deposition simulation is satisfactory since half of the spatial variability of the wet deposition of mercury in the U.S. is explained by the model ($r^2 = 0.50$) and the overall mean error between simulated and measured wet deposition fluxes is 25 %. The simulation of transformations and bioaccumulation of mercury in surface waters is also satisfactory. The simulation of mercury concentrations in various fish species of different ages leads to a determination coefficient of 0.56 and a low mean bias of 3 %. Therefore, it is appropriate to study the effect of emission control scenarios on ecosystems with this type of numerical models. Results of such scenario studies showed that atmospheric mercury is mostly global and atmospheric deposition of mercury results in great part from global emissions. For example, less than half of mercury atmospheric deposition in the U.S. results on average from North American anthropogenic emissions. Nevertheless, local or regional emissions may be significant contributors to atmospheric deposition of mercury near sources of oxidized mercury (which is highly water soluble and, therefore, readily scavenged by rain).

The Minamata Convention (named after the Japanese village where industrial releases had contaminated a bay of Kyûshû Island) was set to incite the signing countries to reduce their mercury emissions. Currently, 128 countries (including the U.S., France, China, and India) have signed this convention. This convention targets mostly the regulation of manufactured goods containing mercury and, therefore, it is not very constraining for atmospheric emissions from major source categories such as incinerators, coal-fired

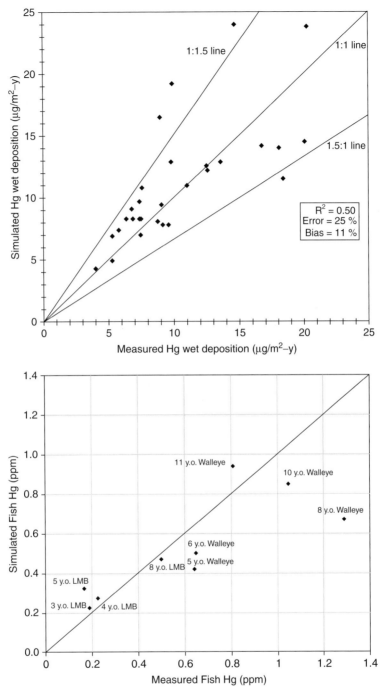

Figure 13.4. Comparisons of simulations and measurements of atmospheric wet deposition fluxes of mercury over the U.S. (top figure) and mercury concentrations in fish in Lake Mitchell in Michigan (bottom figure). Top figure: wet deposition fluxes are measured and simulated at stations of the Mercury Deposition Network (MDN) of the U.S. National Atmospheric Deposition Program (NADP). Bottom figure: The fish species are largemouth bass (LMB) and walleye ranging in age from 3 to 11 year old (y.o.). Sources: The top figure is reproduced with authorization from Seigneur et al. (2004), copyright 2004 American Chemical Society; the bottom figure is reproduced with authorization from Lohman et al. (2000), © Elsevier.

power plants, and cement kilns. Some national regulations have been introduced concerning those atmospheric emissions (see Chapter 15). Nevertheless, an estimation of a reduction of atmospheric mercury deposition due to those regulations suggests that the Minamata Convention leads to twice as much reduction over the U.S. than national regulations, thereby reflecting the global nature of atmospheric mercury pollution (Giang and Selin, 2016). However, national regulations are more efficient in some U.S. regions where local or regional emissions contribute significantly to atmospheric deposition.

Problems

Problem 13.1 Atmospheric mercury

Annual global emissions of mercury are assumed to be 2,000 Mg originating from natural sources ($S_{Hg,N}$), 2,000 Mg from anthropogenic sources ($S_{Hg,A}$), and 2,000 Mg from reemission of anthropogenic mercury deposited from the atmosphere on land surfaces and oceans ($S_{Hg,R}$). The average concentration of mercury in the atmosphere near the Earth's surface is stable at about 1.6 ng m^{-3}. One would like to reduce this atmospheric concentration to 1.2 ng m^{-3}. What should the anthropogenic emission reduction be to attain this objective?

Problem 13.2 Persistent organic pollutants

An industrial source in the northern hemisphere emits PCB (polychlorinated biphenyls). Some of those PCB contain three chlorine atoms, some six, and some eight. Which one of those PCB will turn up preferentially at the high latitudes? Why?

References

Ariya, P.A., M. Amyot, A. Dastoor, D. Deeds, A. Feinberg, G. Kos, A. Poulain, A. Ryjkov, K. Semeniuk, M. Subir, and K. Toyota, 2015. Mercury physicochemical and biogeochemical transformation in the atmosphere and at atmospheric interfaces: A review and future directions, *Chem. Rev.*, **115**, 3760–3802.

Avnery, S., D. Mauzerall, J. Liu, and L.W. Horowitz, 2011. Global crop yield reductions due to surface ozone exposure: 1. Year 2000 crop production losses and economic damage, *Atmos. Environ.*, **45**, 2284–2296.

Bergström, A.-K. and M. Jansson, 2006. Atmospheric nitrogen deposition has caused nitrogen enrichment and eutrophication of lakes in the northern hemisphere, *Global Change Biol.*, **12**, 635–643.

Brimblecombe, P., ed., 2003. *The Effects of Air Pollution on the Built Environment, Air Pollut. Rev.*, **2**, Imperial College Press, London.

Carpenter, S.R., N.F. Caraco, D.L. Correll, R.W. Horwarth, A.N. Sharpley, and V.H. Smith, 1998. Nonpoint pollution of surface waters with phosphorus and nitrogen, *Ecological Applications*, **8**, 559–568.

Ducros, M., 1845. Observation d'une pluie acide, *J. Pharmacie Chimie Série 3*, **7**, 273–277.

EPA, 2007. *Technical Report on Ozone Exposure, Risk, and Impact Assessments for Vegetation*, EPA 452/R-07–002, U.S. Environmental Protection Agency, Research Triangle Park, NC.

References

Franklin, J., R. Atkinson, P.H. Howard, J.J. Orlando, C. Seigneur, T.J. Wallington, and C. Zetzsch, 2000. Chapter 2: Quantitative determination of persistence in air, pp. 7–62, *Persistence and Long-Range Transport of Organic Chemicals in the Environment*, G. Klecka et al., eds., Society for Environmental Toxicology and Chemistry, Pensacola, FL.

Giang, A. and N.E. Selin, 2016. Benefits of mercury controls for the United States, *Proc. National Acad. Sci.*, **113**, 286–291.

Horowitz, H.M., D.J. Jacob, Y. Zhang, T.S. Dibble, F. Slemr, H.M. Amos, J.A. Schmidt, E.S. Corbitt, E.M. Marais, and E.M. Sunderland, 2017. A new mechanism for atmospheric mercury redox chemistry: implications for the global mercury budget, *Atmos. Chem. Phys.*, **17**, 6353–6371.

Huang, L., Z. Fan, C.H. Yu, P.K. Hopke, P.J. Lioy, B.T. Buckley, L. Lin, and Y. Ma, 2013. Interconversion of chromium species during air sampling: Effects of O_3, NO_2, SO_2, particle matrices, temperature, and humidity, *Environ. Sci. Technol.*, **47**, 4408–4415.

Hug, S.J., H.-U. Laubscher, and B.R. James, 1996. Iron(III) catalyzed photochemical reduction of chromium(VI) by oxalate and citrate in aqueous solutions, *Environ. Sci. Technol.*, **31**, 160–170.

Jones, K.C. and P. de Voogt, 1999. Persistent organic pollutants (POPs): state of the science, *Environ. Pollut.*, **100**, 209–221, 1999.

Keyte, I.J., R.M. Harrison, and G. Lammel, 2013. Chemical reactivity and long-range transport potential of polycyclic aromatic hydrocarbons – A review, *Chem. Soc. Rev.*, **42**, 9333–9391.

Lin, C.-J., 2002. The chemical transformations of chromium in natural waters – A model study, *Wat. Air Soil Pollut.*, **139**, 137–158.

Lohman, K., P. Pai, C. Seigneur, and L. Levin, 2000. Sensitivity analysis of mercury human exposure, *Sci. Total Environ.*, **259**, 3–11.

Pettine, M., L. Campanella, and F. Millero, 2002. Reduction of hexavalent chromium by H_2O_2 in acidic solutions, *Environ. Sci. Technol.*, **36**, 901–907.

Prest, I., D.J. Jefferies, and N.W. Moore, 1970. Polychlorinated biphenyls in wild birds in Britain and their avian toxicity, *Environ. Pollut.*, **1**, 3–26.

Ratcliffe, D.A., 1967. Decrease in eggshell weight in certain birds of prey, *Nature*, **215**, 208–210.

Reuss, J.O. and D.W. Johnson, 1986. *Acid Deposition and the Acidification of Soils and Waters, Ecological Studies*, **59**, Springer Science + Business Media, LLC, New York.

Seigneur, C. and E. Constantinou, 1995. Chemical kinetic mechanism for atmospheric chromium, *Environ. Sci. Technol.*, **29**, 222–231.

Seigneur, C., K. Vijayaraghavan, K. Lohman, P. Karamchandani, and C. Scott, 2004. Global source attribution for mercury deposition in the United States, *Environ. Sci. Technol.*, **38**, 555–569.

Selin, N.E., 2009. Global biogeochemical cycling of mercury: A review. *Annual Rev. Environ. Resources*, **34**, 43–63.

Smith, R.A., 1852. On the air and rain of Manchester, *Memoirs of the Literary and Philosophical Society of Manchester, Series 2*, **10**, 207–217.

Smith, R.A., 1872. *Air and Rain: The Beginnings of a Chemical Climatology*, Longmans Green, London.

Vijayaraghavan, K., C. Seigneur, P. Karamchandani, and S.-Y. Chen, 2007. Development and application of a multipollutant model for atmospheric mercury deposition, *J. Appl. Meteor. Climat.*, **46**, 1341–1353.

Vijayaraghavan, K., C. Seigneur, R. Balmori, S.-Y. Chen, P. Karamchandani, J.T. Walters, J.J. Jansen, J.E. Brandmeyer, and E. Knipping, 2010. A case study of the relative effects of power plant NO_x and SO_2 emission reductions on atmospheric nitrogen deposition, *J. Air Waste Manage. Assoc.*, **60**, 287–293.

14 Climate Change and Air Pollution

Climate change is a major issue regarding the atmospheric environment. It differs from air pollution in spatial and temporal scales. Climate change is due mostly to greenhouse gases that have long atmospheric lifetimes and, therefore, are distributed relatively uniformly in the atmosphere. To the contrary, air pollution shows great spatio-temporal variability and offers a relatively short response time in terms of the relationship between emissions and atmospheric concentrations. Nevertheless, there are many links between climate change and air pollution. First, some air pollutants contribute to climate change. Second, climate change may affect air pollution. Finally, some sources emit both air pollutants and greenhouse gases, whereas other sources may emit mostly air pollutants or mostly greenhouse gases. Therefore, it is essential to identify all the emissions associated with a given source in order to avoid creating one problem when trying to solve another. This chapter first summarizes in general terms the main aspects of climate change and then describes the interactions between climate change and air pollution.

14.1 General Considerations on Climate Change

As explained in Chapter 5, the Earth's climate is due in part to the natural presence of greenhouse gases (GHG), which absorb part of the infrared radiation emitted by the Earth and, consequently, lead to an average temperature of about 15 °C at the Earth's surface. It would be about −19 °C without those GHG.

Climate change results from an increase of the concentrations of GHG either via an increase in concentrations of natural GHG, such as carbon dioxide (CO_2), or via the introduction into the atmosphere of new GHG, such as halocarbons.

Since 1988, the Intergovernmental Panel on Climate Change (IPCC) summarizes in reports, which are regularly updated, the state of scientific knowledge on climate change, its causes, the processes involved, and the possible consequences. The Fifth assessment report (also known as AR5) was published in 2013–2014 (IPCC, 2014). Its main conclusions are summarized below.

14.1.1 Observed Changes

A mean global warming since the middle of the 19th century is certain and is without any precedent since decades or even millennia. The GHG concentrations have increased, the mean temperatures of the atmosphere and of the ocean have increased, the total amounts of snow and ice at the Earth's surface have decreased, and the sea level has increased. These changes are documented in greater detail below.

The atmospheric concentrations of CO_2, methane (CH_4), and nitrous oxide (N_2O) have increased compared to their concentrations in 1750 (i.e., prior to the industrial era) by 40, 150, and 20 %, respectively. The current concentrations are without any precedent since at least the past 800,000 years.

Each of the past three decades has been warmer for the atmosphere than all previous decades since 1850 (i.e., since about the beginning of the industrial era). The period ranging from 1983 to 2012 was likely the warmest 30-year period over the past 1,400 years. Based on the longest time series of available temperature data, the globally averaged (land and ocean combined) surface temperature has increased by 0.72 to 0.85 °C from the 1850–1900 period to the 2003–2012 period. However, there is a large inter-annual and inter-decadal variability.

These changes in atmospheric temperature have been accompanied by changes in precipitation. These latter changes are more difficult to document and are, therefore, less certain. Nevertheless, it seems that on average precipitations have increased in the mid-latitudes. In addition, changes in terms of extreme events (heat waves, hurricanes, flooding ...) have been observed since about 1950. It is very likely that the number of cold days/nights has decreased and that the number of warm days/nights has increased globally. It is likely that the frequency of heat waves in Europe, Asia, and Australia has increased. It is likely that there are more regions where the number of heavy precipitation events has increased than there are regions where that number has decreased. Finally, it seems that the frequency or the intensity of heavy precipitation events has increased in North America and Europe.

The ocean dominates in terms of the amount of thermal energy being stored in the climate system, since it accounts for more than 90 % of this increased energy. It is virtually certain that the upper part of the ocean (i.e., from the surface to a depth of about 700 m) has warmed up from 1971 to 2010 and it is likely that it has warmed up from 1870 to 1971. For example, the first 75 meters would have warmed up at the rate of 0.09 to 0.13 °C per decade during the period ranging from 1971 to 2010. Because of limited data and analyses, no trend has been observed in the meridional oceanic circulation of the Atlantic Ocean (AMOC for "Atlantic Meridional Overturning Circulation"), which includes the Gulf Stream in the North Atlantic. The increase in the CO_2 concentration (in the atmosphere and, therefore, also in the ocean since it is water soluble as carbonic acid, H_2CO_3) leads to acidification of the ocean: the pH of surface waters has decreased on average by 0.1 since 1850, i.e., an increase in H^+ ions by 25 %.

Concerning the cryosphere (the land surfaces covered with solid water, i.e., ice or snow), the ice sheets of Greenland and Antarctica have lost mass and the glaciers throughout the world, the Arctic sea ice, and the spring snow cover of the northern hemisphere have continued to lose surface area. It is likely, or even very likely, that the loss rates of mass or surface area of the cryosphere have accelerated over the past several years. For example, it is very likely that the annual mass loss rate of the Greenland ice sheet has increased from 1992 to 2011 and it is likely that the annual mass loss rate of the Antarctic ice sheet has increased over the same period.

Sea level has increased by about 19 cm ($\pm$ 2 cm) from 1901 to 2010. In addition, it is very likely that the rate of increase of sea level keeps increasing. It has been on average in the range of 1.5 to 1.9 mm per year during the period ranging from 1901 to 2010, but it was in the range of 2.8 to 3.6 mm per year for the period ranging from 1993 to 2010. The melting of continental glaciers, the melting of polar caps, and the thermal expansion of the ocean contribute in about equal amounts to this increase in sea level (about one-third each).

14.1.2 Causes of Climate Change

Total radiative forcing is defined as the difference between the radiative energy received and the radiative energy emitted by a given system. Since the Earth system tends to be at equilibrium from an energy viewpoint, radiative forcing in the climate change context concerns the different factors contributing to perturbations of the radiative energy budget. The definition provided by the IPCC is as follows.

Radiative forcing measures the impact of perturbations (for example, an increase in the atmospheric concentration of a greenhouse gas) influencing climate on the energy balance of the Earth system. The term radiative is used because these perturbations modify the balance between incoming solar radiation and infrared radiation outgoing from the atmosphere. Since this radiative balance controls the temperature at the Earth's surface, the term forcing is used to indicate that the radiative balance of the Earth is being perturbed. Radiative forcing is generally quantified as the rate of energy transfer per unit surface area of the globe, measured in the upper layers of the atmosphere. It is expressed in watts per square meter (W m^{-2}). The standard definition of radiative forcing is the change due to a perturbation in the downward radiative flux at the tropopause after allowing for stratospheric temperatures to adjust to radiative equilibrium, while holding surface and tropospheric temperatures and state variables fixed at their values before the perturbation. The term effective radiative forcing is also used. It is calculated at the top of the atmosphere (instead of being calculated at the tropopause) and it involves allowing some atmospheric variables (temperature, clouds, etc.) to adjust to the new radiative equilibrium, while holding surface conditions (temperature, ice, etc.) fixed. For well-mixed GHG, both definitions lead to the same results, but the uncertainty range is greater for the effective radiative forcing than for the standard radiative forcing (see Myhre et al., 2013, for details). A radiative forcing caused by one or more perturbations is said to be positive when it leads to an increase in energy for the Earth/atmosphere system and, therefore, warming of the system. In the opposite case, a radiative forcing is said to be negative when the energy decreases, which leads to cooling of the system.

The increase in thermal energy in the climate system is directly linked to an increase in total radiative forcing. The greatest contribution to this radiative forcing is due to the increase in the atmospheric concentration of CO_2.

Figure 14.1 shows the radiative forcing due to emissions from several GHG, GHG precursors, particulate matter (PM), and PM precursors. For example, the radiative forcing due to emissions of CO_2, CH_4, halocarbons, and N_2O is estimated to be 1.68, 0.97, 0.22, and 0.17 W m^{-2}, respectively. Some GHG such as CO_2, halocarbons, and N_2O are chemically inert (or quasi-inert); therefore, their radiative forcing contribution is estimated simply from their atmospheric concentration (balance between their emission and atmospheric removal rates). On the other hand, CH_4 is slowly oxidized to CO_2

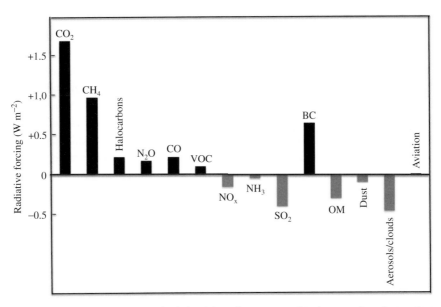

Figure 14.1. Radiative forcing for the 1750–2011 period calculated for pollutants emitted in the atmosphere. Positive forcing is in black and negative forcing is in gray. Net forcing is shown for species that have both positive and negative forcings. VOC do not include CH_4, which is listed separately, nor their effect on secondary organic aerosols (SOA), which is not taken into account here. Halocarbons include here both chlorofluorocarbons and fluorocarbons, which are listed separately in the IPCC report. BC is the black-carbon fraction of atmospheric PM, and OM is the primary fraction of organic compounds present in PM. The effect of these particles on cloud formation is listed as "aerosols/clouds." The effects of NH_3 and SO_2 occur through their formation of particulate ammonium and sulfate, respectively. Dust corresponds to soil dust and the contribution of aviation corresponds to contrails. Data source: Chapter 8, IPCC report (IPCC, 2014).

and leads to ozone (O_3) and water vapor (H_2O) formation, two other GHG. Thus, the contribution of CH_4 emissions to radiative forcing must also take into account the contribution of CH_4 to the whole set of these related chemical species and not be limited to the CH_4 concentration.

Similarly, some pollutants are not GHG, but their chemical reactivity influences some GHG concentrations. For example, carbon monoxide (CO) and volatile organic compounds (VOC) are ultimately oxidized to CO_2 and form O_3 (therefore, a positive radiative forcing). On the other hand, by forming oxidants (particularly the hydroxyl radical, OH), they lead to faster oxidation of CH_4 and, therefore, a slight negative radiative forcing. Nitrogen oxides (NO_x) lead to O_3 formation (positive forcing), but to an increase in the oxidation rate of CH_4 and to the formation of particulate nitrate (i.e., negative forcing). Ozone does not appear in Figure 14.1, because it is not emitted in the atmosphere. It is produced by chemical reactions from VOC, NO_x, CO, and CH_4. Therefore, it appears indirectly via the contributions of these chemical species.

Atmospheric particles (aerosols) lead to absorption of radiation in the case of black carbon (positive forcing), but to scattering of solar radiation in the case of the other components (organic and inorganic) of particulate matter (PM) (negative forcing). These effects are called the direct effects of atmospheric particles on atmospheric radiation. It seems that negative forcing dominates for particles, but there are large associated uncertainties. Particles are also involved in cloud formation and influence precipitation via the size distribution of cloud droplets. This effect is called the indirect effect of atmospheric particles. Clouds have compensating effects. On one hand, they reflect some of the solar radiation back to space and induce some negative forcing. On the other hand, they absorb some of the IR radiation emitted by the Earth, which leads to positive forcing. Overall, clouds are estimated to lead to negative forcing. Therefore, the indirect effect of particles is negative forcing. Furthermore, there is some additional effect due to radiation-absorbing particles. It results from slight associated changes in the atmospheric temperature, which may then decrease the occurrence of low-altitude clouds. If those clouds have a negative forcing contribution, then their decrease would lead to positive forcing. This effect of atmospheric particles on clouds via the perturbation of atmospheric temperatures is called a semi-direct effect.

14.1.3 Future Climate Change

Climate change over the next several years will depend on future emissions of GHG, atmospheric particles, and their precursors. Different scenarios, called representative concentration pathways (RCP), have been used to represent various scenarios of population, energy production and consumption, and technological progress. The conclusions of the IPCC based on scenario calculations are summarized here.

It is likely that in 2100 the mean temperature at the Earth's surface will have increased by 1.5 °C compared to its value during the second half of the 19th century (1850–1900), except for the most optimistic scenario. This temperature increase could exceed 2 °C for the most pessimistic scenarios. This warming would continue during the following century, except for the most optimistic scenario.

Changes in the water cycle, which are due to this global warming, will not be uniform and the wettest regions and seasons will typically become wetter, whereas the driest ones will become drier. Therefore, larger differences between wet and dry regions and between wet and dry seasons are likely to increase.

Changes currently observed for the ocean temperature, ice, sea ice, and snow cover will continue to increase. The increase in sea level will probably accelerate because of the increase in the ocean temperature (dilatation) and the melting of ice and snow. Estimated increases in sea level from the beginning to the end of the century (i.e., from 1986–2005 to 2081–2100) are in the range of 26 to 55 cm for the most optimistic scenario and in the range of 45 to 82 cm for the most pessimistic scenario.

14.2 Effect of Air Pollution on Climate Change

As described in Section 14.1, some air pollutants have a direct or indirect effect on climate change.

Ozone (O_3) is a GHG. It is also an important air pollutant (see Chapter 8). However, its lifetime is short, because it is reactive (destruction by photolysis and reactions with other gases, such as nitrogen oxides and alkenes) and it deposits readily on surfaces. Therefore, any action taken to reduce ozone concentrations has an immediate effect on its radiative forcing. This is a major difference with a GHG such as CO_2, which has a lifetime on the order of one century and for which any action taken now will show some notable effect only decades later.

Other air pollutants are involved in climate change in various ways:

– By leading to the formation of a GHG
– By leading to PM formation
– By leading to oxidant formation, which may affect the lifetime of some GHG

CO, NO_x, and VOC are precursors of O_3. Therefore, their emissions contribute via O_3 formation to radiative forcing.

These O_3 precursors also lead to hydroxyl (OH) radical formation. OH radicals are the oxidant of CH_4. Therefore, an increase in OH radical concentrations due to emissions of CO, NO_x, and/or VOC leads to a decrease in the concentration of CH_4, i.e., less radiative forcing from this GHG. Therefore, the corresponding contribution is negative radiative forcing.

NO_x, VOC, sulfur dioxide (SO_2), and ammonia (NH_3) are precursors of PM (nitrate, organics, sulfate, and ammonium, respectively).

Particles have mostly a negative radiative forcing, with the exception of black carbon (BC), which absorbs radiation (see Chapter 5) and, therefore, contributes to global warming. The effect of BC on atmospheric radiation depends on its particulate state, i.e., whether it is present as individual particles (external aerosol mixture) or whether it is mixed with other chemical species in particles (internal mixture) (Jacobson, 2001). Therefore, the radiative forcing due to BC is uncertain, but it is estimated to be commensurate with that of CH_4 (e.g., Bond et al., 2013). The term "brown carbon" is also used to describe the organic PM fraction that absorbs some radiation (but to a lesser extent than BC). In addition, the absorption of radiation has a semi-direct effect by modifying the thermal budget, which may lead to changes in cloud occurrence (see the semi-direct effect of particles described in Section 14.1.2).

Particles scatter atmospheric radiation and, therefore, they tend to reflect solar radiation back to space, which leads to negative radiative forcing. In addition, given their hygroscopic properties, fine particles act as nuclei for cloud droplets. If there are more particles in the atmosphere, cloud formation will take place on a greater number of condensation nuclei and, for a given amount of liquid water (which depends on the supersaturation of the atmosphere), there will be a greater number of cloud droplets and these droplets will be smaller. Therefore, these clouds will tend to precipitate less and will remain in the atmosphere longer. Since clouds have on average a negative radiative forcing, the result of an increase in the number of particles is negative forcing (indirect cooling effect). The effects of particles on climate are, therefore, complex and require that their physico-chemical and optical properties be well characterized (Boucher, 2015).

Therefore, PM air pollution tends to partially hide global warming. Without any air pollution, global warming would have been greater. However, one should note that the main GHG (CO_2) is the major product of combustion and that combustion processes are also the source of many air pollutants and precursors such as NO_x, CO, VOC, SO_2, and particles. Therefore, one cannot really dissociate the increase of CO_2 emissions from the increase in air pollution. However, one may consider that emission control technologies may be used to reduce air pollution (see Chapter 2) without reducing CO_2 emissions. In that case, a reduction in PM emissions (other than black carbon) will lead to an increase in global warming.

14.3 Effect of Climate Change on Air Pollution

Climate change has various effects on air pollution through the change in atmospheric temperature and precipitation for example, but also more specifically through the change in the occurrence and intensity of weather types (anticyclones, fronts, etc.). It is important to distinguish two kinds of analyses when addressing the effect of climate change on air pollution:

- The effect of a future scenario (such as an RCP scenario) on air pollution;
- The effect of climate change only (i.e., keeping air pollutant emissions constant) on air pollution.

The first analysis consists of a scenario study that treats jointly climate change and air pollution. However, it does not provide any information on the specific relationships that may exist between the change in meteorological conditions and the change in air pollutant concentrations, because the change in air pollution is typically dominated by the change in the air pollutant emissions, as defined by the scenario being studied.

The second analysis does not correspond to a specific change in the future state of the atmosphere, because one is interested in the effect of meteorological conditions corresponding to a future climate (i.e., resulting from different emissions) on current air pollution (i.e., due to current emissions). Nevertheless, such an analysis isolates the effect of climate change on air pollution and evaluates whether this effect is significant or not. We are interested here in this second analysis. The effect of climate change on air pollution has been studied mainly for ozone and fine particles.

The effects of climate change on ozone have been studied for North America and Europe. Overall, an increase in the frequency of occurrence of anticyclonic regimes combined with a slight increase in temperature (which favors chemical kinetics and some emissions such as biogenic emissions and anthropogenic emissions by evaporation) leads to a slight increase in O_3 concentrations. Estimates of the potential increase in O_3 concentrations are on the order of a few ppb (1 to 10 ppb) for North America (Jacob and Winner, 2009). For Europe, a range of −1.7 to +1.6 ppb depending on the regions has been estimated for a climate change corresponding to scenario RCP4.5 (Lacressonnière et al., 2016).

The effects of climate change on fine particles are particularly interesting because of some antagonistic effects. An increase in temperature leads on one hand to an increase in biogenic VOC emissions and anthropogenic VOC emissions by evaporation, as well as faster kinetics for the formation of semi-volatile and non-volatile compounds, which are PM precursors. On the other hand, it may favor the volatilization of semi-volatile particulate-phase compounds, such as SVOC and ammonium nitrate. In addition, the change in the frequency of occurrence of weather types will have an effect since anticyclonic conditions favor air pollution and low-pressure systems favor particle scavenging (i.e., removal from the atmosphere). Studies conducted in North America and Europe show that these effects compensate each other to some extent with the net result being that the effects of climate change on PM air pollution are limited (Lecœur et al., 2014; Lacressonnière et al., 2016; Shen et al., 2017). For a climate change corresponding to RCP4.5, the effect of climate change on annual concentrations of fine particles is estimated to range between about -1 µg m^{-3} and 1 µg m^{-3} depending on the regions. Larger effects for seasonal concentrations have been estimated for North America, with changes reaching up to about 1 µg m^{-3} in Europe and ±3 µg m^{-3} in North America. Nevertheless, it appears that in Europe the annual variations will be less than or similar to the inter-annual variability of PM$_{2.5}$ concentrations except for a very pessimistic climate change scenario (e.g., RCP8.5) (Lecœur et al., 2014).

Problems

Problem 14.1 Climate and meteorology
Explain briefly the difference between climate and meteorology.

Problem 14.2 Greenhouse gases and air pollutants
Give examples of technological changes that are beneficial for air quality, but are detrimental for climate change, or, conversely, that are beneficial for climate change, but are detrimental for air quality.

Problem 14.3 Particles and climate change
What are the effects of atmospheric particles (aerosols) on climate?

References

Bond, T., S.J. Doherty, D.W. Fahey et al., 2013. Bounding the role of black carbon in the climate system: A scientific assessment, *J. Geophys. Res.*, **118**, 5380–5552.

Boucher, O., 2015. *Atmospheric Aerosols – Properties and Climate Impacts*, 248 pp., Springer Atmospheric Sciences, New York.

IPCC, 2014. *Climate Change 2014: Synthesis Report. Contribution of Working Groups I, II and III to the Fifth Assessment Report of the Intergovernmental Panel on Climate Change* [Core Writing Team, R.K. Pachauri and L.A. Meyer (eds.)], IPCC, Geneva.

Jacob, D.J. and D.A. Winner, 2009. Effect of climate change on air quality, *Atmos. Environ.*, **43**, 51–63.

Jacobson, M.Z., 2001. Strong radiative heating due to the mixing state of black carbon in atmospheric aerosols, *Nature*, **409**, 695–697.

Lacressonnière, G., G. Foret, M. Beekmann, G. Siour, M. Engardt, M. Gauss, L. Watson, C. Andersson, A. Colette, B. Josse, V. Marécal, A. Nyiri, and R. Vautard, 2016. Impacts of regional climate change on air quality projections and associated uncertainties, *Climatic Change*, **136**, 309–324.

Lecœur, È, C. Seigneur, C. Pagé, and L. Terray, 2014. A statistical method to estimate $PM_{2.5}$ concentrations from meteorology and its application to the effect of climate change, *J. Geophys. Res.*, **119**, 3537–3585.

Myhre, G., D. Shindell, F.-M. Bréon, W. Collins, J. Fuglestvedt, J. Huang, D. Koch, J.-F. Lamarque, D. Lee, B. Mendoza, T. Nakajima, A. Robock, G. Stephens, T. Takemura, and H. Zhang, 2013. Anthropogenic and natural radiative forcing. In: *Climate Change 2013: The Physical Science Basis. Contribution of Working Group I to the Fifth Assessment Report of the Intergovernmental Panel on Climate Change* [Stocker, T.F., D. Qin, G.-K. Plattner, M. Tignor, S.K. Allen, J. Boschung, A. Nauels, Y. Xia, V. Bex, and P.M. Midgley (eds.)]. Cambridge University Press, Cambridge, UK and New York.

Shen, L., L.J. Mickley, and L.T.L. Murray, 2017. Strong influence of 2000–2050 climate change on particulate matter in the United States: Results from a new statistical model, *Atmos. Chem. Phys.*, **17**, 4355–4367.

15 Regulations and Public Policies

Improving air quality by decreasing air pollutant concentration levels requires promulgating regulations that protect public health, ecosystems, the agriculture, buildings, and atmospheric visibility. Then, public policies must be implemented to design, apply, and evaluate emission control strategies to meet the regulatory standards. In this chapter, the general approach for the development of regulations to protect public health via ambient air quality standards is described first. These regulations and the implementation of the associated public policies differ among countries. Those used in the United States and in France are presented here comparatively to illustrate slightly different approaches. Finally, approaches used at the national and international levels to regulate atmospheric deposition and global atmospheric issues (i.e., destruction of the stratospheric ozone layer and climate change) are presented.

15.1 Regulations for Air Pollutant Concentrations

A regulation of ambient air pollution includes six components:

- The regulated air pollutant, called the indicator species
- The exposure duration
- The regulatory value
- The statistical form of the regulation
- The location of the monitoring stations
- The measurement method

15.1.1 Indicator Species

Regulations are set either for a specific air pollutant (for example, lead or benzene) or for a group of air pollutants. The latter include, for example, photochemical oxidants, nitrogen oxides (NO_x), sulfur oxides (SO_x), and particulate matter (PM). In the case of a group of pollutants, a representative pollutant must be selected. For photochemical oxidants, ozone (O_3) is used because it is the chemical species of that group that is present in the highest concentrations. For NO_x, nitrogen dioxide (NO_2) is used because it shows well-documented adverse effects on the human respiratory system. For SO_x, sulfur dioxide (SO_2) is used because it also shows well-documented adverse effects on the human respiratory system. Sulfur trioxide (SO_3) has a very short atmospheric lifetime, because it is rapidly hydrolyzed to sulfuric acid (H_2SO_4). H_2SO_4 is present in the atmosphere in the particulate phase because of its very low saturation vapor pressure. Particles lead to adverse respiratory and cardio-vascular effects; therefore, particles are regulated separately. Thus, it is appropriate to select SO_2 as the indicator species for SO_x. In the case of particles, the regulations have targeted their size rather than their chemical composition up to now. The reason is that there is a large body of health studies that associate adverse health effects with inhalable particles (PM_{10}) and fine particles ($PM_{2.5}$), whereas there is currently insufficient epidemiological evidence to identify the adverse health effects of individual constituents of PM (one exception could be diesel particles).

15.1.2 Exposure Duration

Health effects may result from acute exposure (ranging from a few minutes to a few hours) and/or chronic exposure (ranging from a few months to a few years). Therefore, the regulation must correspond to the exposure duration that is considered representative of the health effects. For some air pollutants, regulations may be appropriate for both short (acute exposure) and long (chronic exposure) durations. Table 12.2 summarizes the exposure durations relevant to health effects due to air pollutants regulated in North America and in Europe. In some cases, a pollutant may be regulated for an exposure duration in the absence of conclusive evidence of adverse health effects (see Tables 12.2, 15.1, and 15.2). Such cases (e.g., the annual standard for NO_2 in the United States and in Europe and the annual standard for PM_{10} in Europe) generally result from historical reasons (earlier health data led to the regulation, which is then kept by default). The regulation that corresponds to the exposure duration with the best evidence of adverse health effects is typically the most constraining. On the other hand, it may happen that a regulation is not set for both acute and chronic exposures in spite of epidemiological evidence for both. This is the case, for example, for $PM_{2.5}$ in Europe, where only a long-term standard is used (see Tables 12.2 and 15.2). This choice may, however, be appropriate because the long-term exposure is considered to correspond to the most important adverse health effects. In the United States where both short- and long-term exposures to $PM_{2.5}$ are regulated, the long-term regulation (annual concentration averaged over three years) is the most constraining and it is rare that a region would be in attainment of the annual standard, but would not be in attainment of the short-term (daily) standard. It is also the case for ozone (O_3), which is only regulated for short-term exposure (8-hour average concentration), although there is some evidence of likely adverse health effects for chronic exposure to ozone (see Table 12.2). It is considered that an 8-hour average regulatory value also protects against chronic exposure, based on air quality data analyses relating ozone concentrations averaged over short and long periods.

15.1.3 Regulatory Value

The regulatory value (the value of the national ambient air quality standard in the U.S., the limit or target value in Europe) is the best-known aspect of the air pollutant regulation.

Table 15.1. National ambient air quality standards (NAAQS) to protect public health in the United States (Clean Air Act and regulations of the U.S. Environmental Protection Agency).

Pollutant	Concentration[a]	Sampling duration	Statistical form[b] (number of authorized exceedances per year)
Pb	0.15 μg m^{-3}	3 mo[c]	(0)[d]
CO	40 mg m^{-3}, 35 ppm	1 h	99.99th percentile (1)[e]
	10 mg m^{-3}, 9 ppm	8 h[c]	99.99th percentile (1)[e]
SO$_2$	197 μg m^{-3} 75 ppb	1 h	99th percentile (3)[f]
NO$_2$	189 μg m^{-3}, 100 ppb	1 h	98th percentile (7)[g]
	100 μg m^{-3}, 53 ppb	1 year	(0)[d]
O$_3$	137 μg m^{-3}, 70 ppb	8 h[c]	99th percentile (3)[h]
PM$_{10}$	150 μg m^{-3}	24 h	99.7th percentile (1)[i]
PM$_{2.5}$	35 μg m^{-3}	24 h	98th percentile (7)[j]
	12 μg m^{-3}	1 year	(0)[d, k]

(a) regulatory concentrations are expressed in μg m^{-3} for Pb and PM and in ppb or ppm for gaseous pollutants; conversion from ppb (or ppm) to μg m^{-3} (or mg m^{-3}) is at 25 °C and 1 atm. Concentrations are measured at background monitoring stations except for the concentrations of NO$_2$ and PM$_{2.5}$, which must be measured near sources (i.e., mostly near roadways), and for the concentrations of SO$_2$, which must be both measured and modeled near sources.
(b) the statistical form is provided here both as a percentile and a number of exceedances per year; see footnotes for the exact definition.
(c) moving average.
(d) not to be exceeded.
(e) not to be exceeded more than once per year.
(f) 99th percentile of daily maximum 1-hour average concentrations averaged over 3 years; 24-hour average and annual standards were eliminated in 2010 (except for non-attainment areas).
(g) 98th percentile of daily maximum 1-hour average concentrations averaged over 3 years.
(h) annual fourth-highest daily maximum 8-hour average concentrations averaged over 3 years.
(i) not to be exceeded more than once per year averaged over 3 years.
(j) 98th percentile averaged over 3 years.
(k) annual mean averaged over 3 years not to be exceeded.

However, it cannot be dissociated from the exposure (or sampling) duration, the statistical form of the regulation (see Section 15.1.4), or the location of the monitoring stations (see Section 15.1.5). The regulatory value is determined from toxicological and/or epidemiological studies. For standards corresponding to durations of 24 h or more, epidemiological studies are typically used, since human toxicological studies are not conducted for exposure durations exceeding a few hours. For shorter exposure durations, toxicological studies may be favored. However, epidemiological studies add some useful information concerning sensitive populations, which may not be available from toxicological studies. In addition, experimental air quality data may be used to extrapolate ambient pollutant concentrations temporally or spatially, in order to relate results available from toxicological or

Table 15.2. Ambient air quality standards to protect public health in Europe (European Directive).

Pollutant	Concentration[a]	Sampling duration	Statistical form[b] (number of authorized exceedances)
Pb	0.5 µg m^{-3}	1 year	(0)[c]
CO	10 mg m^{-3}, 9 ppm	8 h[d]	(0)[c]
SO$_2$	350 µg m^{-3}, 133 ppb	1 h	99.7th percentile (24)[e]
	125 µg m^{-3}, 47 ppb	24 h	99.2th percentile (3)[f]
NO$_2$	200 µg m^{-3}, 106 ppb	1 h	99.8th percentile (18)[g]
	40 µg m^{-3}, 21 ppb	1 year	(0)[c]
O$_3$[h]	120 µg m^{-3}, 61 ppb	8 h[d]	93th percentile (25)[i]
PM$_{10}$	50 µg m^{-3}	24 h	90.4th percentile (35)[j]
	40 µg m^{-3}	1 year	(0)[c]
PM$_{2.5}$[k]	25 µg m^{-3}	1 year	(0)[c]
C$_6$H$_6$	5 µg m^{-3}, 1.5 ppb	1 year	(0)[c]

(a) all regulatory concentrations are expressed in µg m^{-3} (or mg m^{-3}); conversion of gaseous pollutant concentrations from µg m^{-3} (or mg m^{-3}) to ppb (or ppm) is at 25 °C and 1 atm.
(b) the statistical form is provided here both as a percentile and a number of exceedances per year; see footnotes for the exact definition.
(c) not to be exceeded.
(d) daily maximum 8-hour (moving) average concentrations.
(e) not to be exceeded more than 24 times per year.
(f) not to be exceeded more than 3 times per year.
(g) not to be exceeded more than 18 times per year.
(h) target value.
(i) not to be exceeded more than 25 days per year averaged over three years.
(j) not to be exceeded more than 35 times per year.
(k) limit value of 25 µg m^{-3}; target value of 20 µg m^{-3} (three-year average at urban background monitoring stations) in 2020.

epidemiological studies with some characteristics of the regulation (e.g., sampling duration or location; see examples later in this section). Studies used to set up regulatory values in the United States are summarized in Table 15.3.

Toxicological studies (clinical studies conducted under controlled exposure) have been used for O$_3$ (concentration averaged over eight hours) and CO (concentrations averaged over one and eight hours). On the other hand, epidemiological studies have been used for the NO$_2$ and PM$_{2.5}$ annual values. Epidemiological studies were used in combination with a model representing the relationships between atmospheric concentrations, exposure, blood concentrations, and adverse health effects for Pb (3-month moving-average concentration) in order to better account for various possible exposure pathways. Epidemiological studies have also been used for the 24-hour average values of PM$_{2.5}$ and PM$_{10}$.

Table 15.3. Data used to define the regulatory values of the national ambient air quality standards in the United States. Source: Clean Air Act and regulations of the U.S. Environmental Protection Agency, Federal Register.

Pollutant and sampling duration	Toxicological studies	Epidemiological studies	Ambient concentrations
Pb, 3 months		✓	✓
CO, 1 h	✓		
CO, 8 h	✓		
SO_2, 1 h	✓		✓
NO_2, 1 h	✓	✓	✓
NO_2, 1 year		✓	
O_3, 8 h	✓	✓	
PM_{10}, 24 h		✓	
$PM_{2.5}$, 24 h		✓	
$PM_{2.5}$, 1 year		✓	

In some cases, toxicological and epidemiological studies are used in combination. In the case of the hourly NO_2 value, toxicological studies provided the low value of the range of concentrations corresponding to adverse health symptoms for individuals with moderate asthma. Epidemiological studies were then used to better account for more sensitive individuals (e.g., severe asthma). The regulatory value was to be set for near-source situations. Since the epidemiological studies were mostly based on existing air quality monitoring networks, which only included urban background locations, the analysis was complemented with an analysis of air quality data to extrapolate the urban background data to near-roadway locations. In the case of the hourly SO_2 regulatory value, toxicological studies have also provided the low value of the range of concentrations leading to adverse health effects for individuals with moderate asthma exposed for short periods (5 to 10 minutes). Since the regulatory value was set for a one-hour average concentration to minimize temporal fluctuations in the measured concentrations, ambient concentration data were used to relate the 5-minute average toxicological data and the 1-hour ambient concentrations. In the case of O_3, epidemiological studies have been used only to complement toxicological studies to confirm that significant adverse health effects do not appear below the low value of the range of concentrations proposed for the standard.

In summary, the U.S. regulations are based on epidemiological studies for concentrations averaged over periods ranging from 24 hours to one year. They are mostly based on toxicological studies for concentrations averaged over periods ranging from one to eight hours. However, those latter regulatory values are in some cases also based on epidemiological studies that provide useful information concerning sensitive individuals, who cannot participate in toxicological studies. In addition, air quality data have been used to extrapolate the results of health studies spatially (in the case of NO_2) or temporally (in the case of SO_2).

15.1.4 Statistical Form of the Regulation

A regulation may allow a number of exceedances of the regulatory value. This aspect may be stated either explicitly in terms of the number of allowed exceedances (for SO_2, NO_2, O_3, and PM_{10} in the European regulations and for CO, O_3, and PM_{10} in the U.S regulations) or in the form of a percentile that corresponds to the concentration not to be exceeded (for SO_2, NO_2, and $PM_{2.5}$ in the U.S. regulations). Since a number of allowed exceedances may be converted into a percentile and vice-versa, both are provided in Tables 15.1 and 15.2 to facilitate comparisons between the U.S. and European standards. The ambient concentrations of air pollutants depend strongly on meteorological conditions. Thus, extreme meteorological events may lead to large fluctuations in the highest measured pollutant concentrations from year to year. Therefore, allowing a limited number of exceedances of a regulatory value eliminates those highest measured concentrations from consideration and leads to comparisons of the measured concentrations to the regulatory value that are statistically more robust; i.e., they do not fluctuate as much from one year to the next compared to the highest concentrations. In addition, the meteorological variability may be taken into account for long periods by averaging results over several years. For example, some U.S. regulations, such as $PM_{2.5}$, apply to three-year periods. Averaging over several years may also be applied for short-term regulations. For example, the 99th percentile of the hourly SO_2 concentration is averaged over a three-year period in the U.S. regulation.

15.1.5 Locations of Monitoring Stations

The location of monitoring stations that measure the ambient concentrations of the regulated air pollutants is important for the definition of the air quality standard, particularly in the case of primary pollutants. Urban background stations measure concentrations of primary pollutants that are lower than those measured near the source of those pollutants. For example, an urban background station will measure lower concentrations of CO, NO_2, and PM than those measured near a major roadway, with a difference of a factor of 1.5 to 2, for example, in Paris. In Europe, the air pollutant concentrations are measured at urban background stations as well as at near-source stations (mostly near-roadway stations). In the United States, air pollutant concentrations in urban areas were measured only at background stations until 2009. However, near-source monitoring stations are now required for NO_2 (since 2009) and $PM_{2.5}$ (since 2013). For a primary air pollutant, a given regulatory value (including averaging time and number of allowed exceedances) will correspond to a more constraining standard if a near-source monitoring location is used instead of an urban-background location.

15.1.6 Measurement Method

Some experimental uncertainty is associated with any measurement method. In some cases, there may also be a bias due to artifacts. As a result, different sampling instruments may give different values for a given concentration at a given time and location. Therefore, it is essential to specify which method and sampling instrument must be used to monitor air pollutant concentrations. This measurement method, which is specified in the regulation, is called the reference method. Other methods may be used, if they are shown to give results equivalent to the reference method. For example, the reference method in the United States for PM is a gravimetric method (a filter is weighed under controlled conditions for temperature and relative humidity before and after sampling). Continuous measurements with an instrument such as a tapered element oscillating microbalance (TEOM) are considered equivalent to the reference method. In early TEOM instruments, there were some significant artifacts due to the volatilization of semi-volatile particulate compounds such as ammonium nitrate and some organic compounds. A correction is now performed with a filter dynamics measurement system (FDMS), which minimizes the effect of those artifacts.

15.1.7 Regulations in the United States and in France

Regulations of air pollutants in use in 2018 in the United States and in France are presented in Tables 15.1 and 15.2, respectively. These regulations are based on toxicological and epidemiological studies conducted to identify and quantify the cause-effect relationships for acute and chronic exposure to those pollutants (see Table 15.3). There are several differences. First, some pollutants are regulated in both countries (NO_2, SO_2, O_3, PM, CO, and Pb), whereas others are regulated only in one country. Benzene is regulated in France, but not in the United States and, until 2008, fine particles ($PM_{2.5}$) were regulated in the United States since 1997, but not yet in Europe. Benzene is a carcinogenic pollutant (category 1 according to the IARC) and the European regulation corresponds to an excess cancer risk of about 10^{-5} (see Chapter 12). Benzene is not the only carcinogenic air pollutant and the U.S. Environmental Protection Agency favors treating all carcinogenic air pollutants together in terms of their excess cancer risk, rather than targeting a single pollutant. The sampling times are generally identical, although there are a few differences (one year in Europe for lead and three months in the United States). Statistical forms of the regulatory values are rarely identical and it is difficult to compare two regulations with different statistical forms and regulatory values. For example, the regulatory value for ozone is 120 $\mu g\ m^{-3}$ in France (it is a target value), but this value may be exceeded 25 days per year. The regulatory value is greater in the United States since it is 137 $\mu g\ m^{-3}$, however, the number of allowed exceedances is much lower (three days per year only).

15.2 Public Policies

The implementation of air quality regulations is similar overall in the United States and in France, but there are a few differences. Both approaches are summarized in Sections 15.2.1 and 15.2.2, respectively.

15.2.1 Public Policy in the United States

In the United States, the Clean Air Act (CAA) and its amendments govern the air quality regulations. In the CAA, the major air pollutants are regulated through National Ambient Air Quality Standards (NAAQS, see Table 15.1). These NAAQS are regularly updated as follows. The setup of a standard starts with a review of the scientific literature, targeting primarily the health effects of the pollutant(s) considered. There are six pollutants or pollutant categories that undergo this process of standard setting: lead, carbon monoxide, nitrogen oxides, sulfur oxides, ozone, and particulate matter (PM). These pollutants are called "criteria pollutants" because the product of the scientific literature review was originally called the "criteria document" (it is now called the integrated scientific assessment or ISA). The next step is the preparation of an evaluation of the population exposure and associated health effects. This work is presented in a report titled "Risk and exposure assessment" (REA). The REA documents quantitatively the exposure and health risks of the population for different regulatory scenarios. These two documents constitute the scientific basis for the development of the NAAQS. Both documents are reviewed by a scientific committee, called the Clean Air Scientific Advisory Committee Review Panel (CASAC Review Panel). NAAQS are then proposed by the U.S. EPA and subsequently promulgated by the federal government. Primary NAAQS concern the protection of public health. Secondary NAAQS may be added to protect the environment. At each step of this public process, the stakeholders concerned by the regulation (industry, ecologists, citizens ...) may comment.

Monitoring air quality is essential to ensure that an area of the United States is in attainment or not of the NAAQS. The U.S. EPA sets up a protocol for the air quality monitoring networks, including monitoring sites, measurement instruments, sampling chain procedure (quality assurance/quality control), etc. The air quality monitoring networks are typically operated by the states and measured concentrations are transferred by the state agencies to EPA and incorporated into a database. Based on these air quality data, EPA determines which areas are in attainment of the NAAQS and which ones are not (i.e., non-attainment). In case of non-attainment in a state, the state agency must develop an action plan, called a State Implementation Plan (SIP). A SIP lists all the emission control measures that will be implemented by the state agency to attain the NAAQS within the period allowed by EPA (the length of the period depends on how severe the non-attainment conditions are). The efficiency of these measures is typically evaluated quantitatively with an air quality numerical model (also called chemical-transport model, see Chapter 6). The models that are the most widely used in the United States for this task are CMAQ

(Appel et al., 2017) and CAMx (Vijayaraghavan et al., 2012). EPA may ask for revisions if a SIP seems inappropriate (for example, poor performance of the air quality model against available data).

In most states, the state agency is responsible for stationary sources, whereas mobile sources are under the federal government authority. California is an exception because it regulates its own mobile source emissions. Then, the regulation of stationary sources in California may be delegated to districts. If a state does not attain the objectives listed in its SIP, EPA may implement some sanctions, such as the cancellation of funding for road networks, and/or take control of the SIP management by temporarily cancelling the delegation of the process to that state.

In some cases, it is not possible for a state to attain the NAAQS because long-range air pollution imported from other regions may be a major cause for the exceedance of some NAAQS. This has been the case, for example, in some small states of the northeastern U.S., which are downwind of states that are large emitters of primary air pollutants (in particular NO_x), and, therefore, could not meet the ozone NAAQS. The states with large NO_x emissions experienced little ozone formation, because this secondary pollutant was being formed farther downwind. EPA decided to set up a multi-state approach over the northeastern U.S. to tackle this issue on a regional basis and control NO_x emissions in the states that were the large emitters (Godowitch et al., 2008). This process was called the "SIP Call." This example illustrates the fact that the existing law (here the SIP, which is limited to a state) was not sufficient to attain the regulatory objectives (the NAAQS), but that the federal agency, EPA, did not hesitate to modify the regulatory process to attain the desired objectives. In other words, reaching the objective prevailed over the simple application of the regulatory process, once the latter was shown to be inadequate.

Benzene, which is carcinogenic, is not regulated in the United States, because all carcinogenic pollutants are regulated as a whole as "air toxics." They are subject to a regulatory process that differs from the NAAQS. Initially, their emissions were regulated via National Emission Standards for Hazardous Air Pollutants (NESHAP). This approach implied the concept of zero risk, which in theory could be attained if the emissions complied with the NESHAP values. Although this approach is feasible for non-carcinogenic risks (which involve a health threshold), it is inappropriate for carcinogenic pollutants, which typically do not have such a threshold (see Chapter 12). In the 1990s, a different approach was introduced that is based on the calculation of health risks to evaluate the carcinogenic risks due to air pollutants. If, after implementation of emission control technology required by the regulations (which are defined by source categories), individual excess cancer risks exceed a value in the range of 10^{-6} to 10^{-5}, additional emission controls may be required for specific sources to decrease the risk (Ohshita and Seigneur, 1993).

The temporal trends of the maximum concentrations of three pollutants regulated in the U.S. are shown in Figure 15.1 for the Los Angeles Basin in California. These three pollutants, O_3, NO_2, and $PM_{2.5}$, exceed the NAAQS. The inter-annual variability is mainly due to the meteorological variability. In addition, the NO_2 monitoring network was modified starting in 2014 in response to a revision of the NO_2 NAAQS to include near-roadway locations in addition to the urban background stations (see Section 15.1.5). This change led to an increase in the maximum NO_2 concentration. There are significant

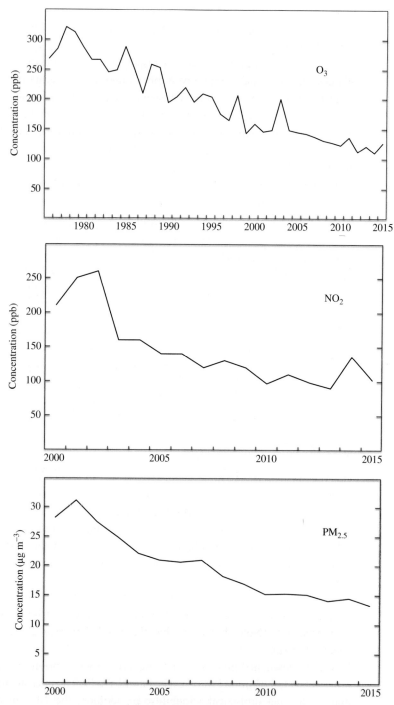

Figure 15.1. Timelines of the concentrations of ozone (O_3), nitrogen dioxide (NO_2), and fine particles ($PM_{2.5}$) in the Los Angeles basin, California. Top figure: maximum 8-hour averaged O_3 concentration from 1976 to 2015; middle figure: maximum hourly NO_2 concentration from 2000 to 2015; bottom figure: maximum $PM_{2.5}$ annual concentration from 2000 to 2015. Source: SCAQMD (2017).

decreases of the concentrations of these three air pollutants over the long term. On average, the decrease is 4 to 5 ppb per year for O_3 over the past 40 years, 9 to 10 ppb per year for NO_2, and about 1 μg m^{-3} per year for $PM_{2.5}$ over the past 15 years. However, these annual decreases may become less important with time, because the sources that are the easiest to control or to eliminate have typically been addressed upfront and, consequently, it becomes harder to control the emissions of primary pollutants and those of precursors of secondary pollutants. In addition, the relative contribution of the long-range transport of air pollution increases when the local emission contribution decreases. In the Los Angeles Basin, this imported pollution originates mostly from Asia. In summary, these figures illustrate the efficiency of regulatory measures that have been implemented to improve air quality in the Los Angeles Basin, while highlighting the efforts that are still pending to attain the NAAQS.

15.2.2 Public Policy in France

In France, an air pollution regulation is based on directives issued by the European Union. The European directive is then converted into French law. A European directive for a given air pollutant takes into account the health effects of that air pollutant. However, there is a major difference between the U.S. approach and the European one in the fact that the European directive also takes into account the time needed to bring the air quality toward the health-based objective. As a result, the limit values (equivalent to the regulatory value of the U.S. NAAQS) may vary over time, starting with a high value for the first few years and decreasing to lower values as time goes on, tending toward a value corresponding to public health protection. It is currently the case for fine particles, $PM_{2.5}$, with an annual limit value of 25 μg m^{-3} in 2015 and a target value of 20 μg m^{-3} for 2020 (for comparison, the U.S. NAAQS annual value for $PM_{2.5}$ is 12 μg m^{-3}).

The European approach uses an ensemble of values that present various levels of constraints:

- Air quality objective: A concentration level of air pollutants to be attained over the long term in order to achieve an efficient protection of human health and the environment, except when it is unachievable through reasonable measures.
- Target value: A concentration level of air pollutants defined in order to prevent, warn or reduce adverse effects on human health and the environment, to be attained to the extent possible by a given deadline.
- Limit value: A concentration level of air pollutants based on scientific knowledge, not to be exceeded, in order to prevent, warn or reduce adverse effects on human health and the environment. (Note, however, that the regulation may allow a limited number of exceedances.)
- Information and recommendation threshold: A concentration level of air pollutants above which a short-term exposure presents a health risk for sensitive individuals, requiring timely and relevant communication to the public.
- Alert threshold: A concentration level of air pollutants above which short-term exposure presents a health risk for the whole population or environmental degradation and justifies emergency measures.

In France, air quality monitoring is conducted by non-governmental organizations (*Associations agréées de surveillance de la qualité de l'air,* i.e., AASQA), which transfer the air quality monitoring data into a national air quality database (*Base de données sur la qualité de l'air,* BDQA). Air quality monitoring must follow standard procedures defined at the European level.

In the U.S., the same entity (the state or, in California, the district) monitors air quality and is responsible for taking actions to meet the NAAQS. In France, the AASQA have no authority on air pollutant emission control. Instead, French government agencies are responsible for regulating stationary sources. Those are the Regional agencies for environment, planning, and housing (*Directions régionales de l'environnement, de l'aménagement et du logement,* DREAL), except in Paris, where it is the Regional agency for environment and energy (*Direction régionale et interdépartementale de l'environnement et de l'énergie,* DRIEE). Each country belonging to the European Union must then demonstrate that it is in attainment of the limit values of the European regulations. If not, an action plan must be developed to reach attainment within a reasonable time. In addition, exceedances of the alert and information thresholds must be communicated to the public and reported to the European Agency. An exceedance of limit values may lead to important fines by the European Union Court of Law (several tens of million euros per year). For example, as of May 2018, six European countries have been referred to the Court of Justice of the European Union because of exceedances of air pollutant limit values: France, Germany, and the United Kingdom for non-attainment of NO_2 limit values and Hungary, Italy, and Romania for non-attainment of PM_{10} limit values.

There are three main programs in France that pertain to air quality protection. They are part of the 1996 Air Law (*Loi sur l'air et l'utilisation rationnelle de l'énergie,* LAURE).

– Urban transportation planning (*Plans de déplacements urbains,* PDU), which are developed by 59 urbanized areas with more than 100,000 inhabitants. They have actually limited effect on air quality, because they mostly impact the spatio-temporal distribution of on-road traffic emissions, without significant impact on the area-wide total mobile source emissions.
– Regional planning for climate, air, and energy (*Schémas régionaux climat air énergie,* SRCAE), which have replaced the earlier regional air quality plans (*Plans régionaux de la qualité de l'air,* PRQA). SRCAE were put into place following the so-called *Grenelle de l'Environnement* meeting in 2007 and were implemented through the Grenelle I (2009) and Grenelle II (2010) laws. They are developed by the administrative regions. They are open for public consultation and are revised every five years. They present recommendations to reduce emissions in order to reach objectives for air quality, while minimizing greenhouse gas emissions and energy consumption. However, they do not lead to specific regulations to reduce air pollutant emissions. Therefore, they are tools for providing recommendations in terms of public policy, rather than decision tools for emission controls.
– The Atmosphere protection plans (*Plans de protection de l'atmosphère,* PPA), which are developed by 24 urbanized areas with more than 250,000 inhabitants. They are developed under the authority of the government representative for the region ("*préfet*").

The Paris PPA is developed by the DRIEE. PPA are developed with the objective of reaching attainment of the European limit values. Therefore, they may include regulations to reduce stationary source emissions. Thus, a PPA is a regulatory tool (actually, it is the only regulatory tool at the regional level). Its development typically involves the government agency (DRIEE or DREAL), technical support from the regional air quality organization (AASQA, e.g., Airparif in the Paris region), as well as stakeholders and experts. In the case of the Paris PPA, the emission control measures have been selected according to a process where all stakeholders participated in the PPA development. A PPA does not always demonstrate conclusively that the measures listed will lead to attainment of the limit values. Nevertheless, a PPA documents, at least partially, which impacts those measures will have on air quality improvement. The list of long-term measures identified to control air pollutant emissions may be complemented with emergency measures in the event of exceedance of alert thresholds.

It is important that these three plans (PDU, SRCAE, and PPA) be consistent among each other. Note that the French PPA have the same limitation as the U.S. SIP, i.e., they are limited to a geographical region and, therefore, may not be sufficient in the case of imported long-range air pollution. A national or European approach may be needed to treat not only regulated air pollutants with long atmospheric lifetimes (O_3 and $PM_{2.5}$), but also their precursors (VOC, NO_x, SO_2, and NH_3). Emission control scenario simulations are conducted to evaluate the impact of emission control measures on future air quality. Numerical air quality models used to that end include, for example, Chimere (Menut et al., 2014) and Polyphemus/Polair3D (Sartelet et al., 2012).

The reduction of stationary source emissions and the regulation of fuels are under the authority of the government. Emission standards for on-road vehicles are set at the European level (see Chapter 2).

Temporal trends of maximum concentrations at urban background stations for three regulated air pollutants are presented in Figure 15.2 for the Paris region. The maximum 8-hour average O_3 concentrations and the maximum annual NO_2 concentrations are shown for the 1994–2015 period. The maximum annual $PM_{2.5}$ concentrations are shown only for the 2007–2015 period, because of a change in monitoring technique in 2007. The inter-annual variability is mostly due to meteorological variability. For example, there is an increase in the maximum O_3 concentration in 2003 due to a heat wave in August 2003. Nevertheless, there are overall decreasing trends over the long term. O_3 concentrations, despite large year-to-year fluctuations, decreased by 23 µg m^{-3} (11.5 ppb), if one compares the averages of the maximum concentrations over the last two decades, i.e., 1996–2005 and 2006–2015). The number of exceedances of the target value remains within the authorized amount (<25 per year); however, the air quality objective is exceeded (8-hour average concentration of 120 µg m^{-3}; Airparif, 2016). The maximum urban background NO_2 concentration decreases annually by about 1.4 µg m^{-3} on average over the period analyzed. The NO_2 concentrations are currently below the annual limit value at urban background stations. However, they exceed this limit value at stations located near roadways (Airparif, 2016). The maximum annual $PM_{2.5}$ concentration has decreased by about 1 µg m^{-3} per year on average over the past eight

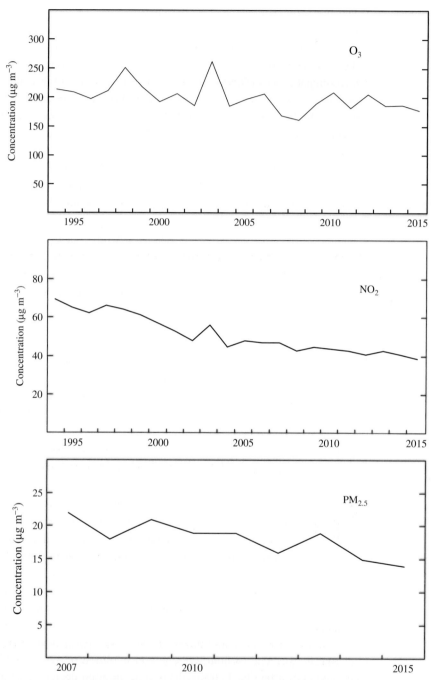

Figure 15.2. Timelines of concentrations of ozone (O_3), nitrogen dioxide (NO_2), and fine particles ($PM_{2.5}$) in the Paris region. Top figure: maximum 8-hour averaged O_3 concentration from 1994 to 2015; middle figure: maximum annual NO_2 concentration at urban background stations from 1994 to 2015; bottom figure: maximum annual $PM_{2.5}$ concentration at urban background stations from 2007 to 2015. Source: Airparif (2016).

years. It is likely that the use of diesel particle filters has contributed to the decrease of fine particle concentrations in Paris. The annual $PM_{2.5}$ concentration is below the target value of 20 μg m^{-3} at urban background stations. At near-roadway stations, the limit value of 25 μg m^{-3} was attained in 2015 (Airparif, 2016). In the case of O_3 and the secondary fraction of $PM_{2.5}$, the contribution of long-range atmospheric transport is important and, therefore, limits the impact of local emission control strategies on the concentration levels of these secondary pollutants in the Paris region.

15.3 Regulations for Atmospheric Deposition and Global Pollution

Regulations for atmospheric deposition of air pollutants concern primarily the protection of ecosystems, agriculture (crops), and buildings, as well as the protection of public health, which can be affected via the contaminated food chain. Similarly to public health protection in terms of inhalation of air pollutants, one may consider establishing concentration levels in ecosystems that are not to be exceeded. This is the case, for example, for mercury concentrations in fish: concentrations less than 0.3 ppm are considered safe for fish consumption. One may also consider establishing atmospheric deposition fluxes to ecosystems that are not to be exceeded. This is the case, for example, with the definition of critical loads, which correspond to levels below which the atmospheric inputs should not have adverse effects on the ecosystem. However, these critical loads may vary significantly among ecosystems because some may be more sensitive to atmospheric inputs than others.

These objectives (pollutant concentrations or deposition fluxes not to be exceeded) must be translated into air pollutant emission limits. In that sense, three main categories of regulations for air pollution emission controls may be identified:

- Total ban of the emissions of a pollutant. This is the case, for example, for some pesticides, which are now banned in terms of manufacturing and use in some countries (e.g., DDT in North America and Europe).
- Emission limits of some pollutants defined by source categories and to be met by each individual source of a given category. This is the case, for example, for emissions of dioxins and furans from incinerators in France and in the U.S., as well as for the recent regulation on mercury emissions from power plants in the U.S. (Mercury and Air Toxics Standards).
- Overall emission limits of some pollutants for a source category, to be met by the sources of that category as a whole. This system is called "cap and trade" in the U.S. and emission trading in Europe. This is the case, for example, for SO_2 and NO_x emissions from coal-fired power plants in the U.S., where caps on the emissions of those pollutants were set that are not to be exceeded by the ensemble of coal-fired power plants in the northeastern U.S. This approach has been shown to work well to reduce acid deposition in North America. Figure 15.3 shows the change in the pH of precipitation and the atmospheric

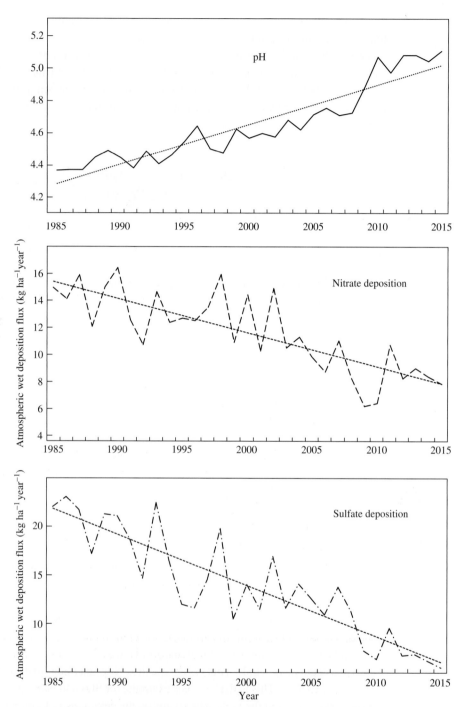

Figure 15.3. Evolution of atmospheric acid deposition from 1985 to 2015 at White Face Mountain, New York State. Top figure: pH of precipitation; middle figure: annual atmospheric wet deposition flux of nitrate; bottom figure: annual atmospheric wet deposition flux of sulfate. Regression lines are shown as dotted lines to illustrate long-term trends. Source: NADP (2017).

wet deposition fluxes of sulfate and nitrate at White Face Mountain in the northern part of New York State. The pH was about 4.4 in 1985 and, 30 years later, it is greater than 5. This significant decrease in the acidity of precipitation results from decreases of about 50 and 75 % in the wet deposition fluxes of nitrate and sulfate, respectively. A regulatory approach based on a cap-and-trade system is best suited for pollutants that have long-range impacts and have negligible local contributions.

In many cases, these regulations are developed at the national or multi-national (e.g., European) level. Nevertheless, there exist some cases of international agreements that are set up to manage air pollution control at the international level. This is the case of course for global environmental problems such as the depletion of the stratospheric ozone layer (see Chapter 7) and climate change (see Chapter 14). In such cases, it is essential that a large majority of countries sign the agreement and meet their commitments. Among the main international agreements pertaining to air pollution, one may mention the following:

- Montreal Protocol of 1987 for the elimination of chlorofluorocarbons and other substances depleting the stratospheric ozone layer (see Chapter 7). This protocol was complemented by the amendments of London (1990), Copenhagen (1992), Vienna (1995), Montreal (1997), Beijing (1999), and Kigali (2016). The last amendment concerns the effect of hydrofluorocarbons on climate.
- Stockholm Convention of 2001, which was elaborated by the United Nations Environmental Program (UNEP) for the reduction of the emissions of persistent organic pollutants (POP) (see Chapter 13).
- Minamata Convention of 2013, which was elaborated by the UNEP for the reduction of mercury emissions (see Chapter 13).
- Göteborg Protocol of 1999, which was elaborated by the Economic Commission for Europe of the United Nations (ECE-UN) for the reduction of the emissions of nitrogen oxides, sulfur oxides, ammonia, and volatile organic compounds to reduce acid deposition, eutrophication, and ozone formation (see Chapter 13).
- Paris Agreement of 2015, which was elaborated during the Conference of the Parties (COP 21) for the reduction of the greenhouse gas emissions (see Chapter 14).

Most international agreements target air pollutants with long atmospheric lifetimes, which, therefore, have health and environmental impacts at a global scale. However, the Göteborg Protocol addresses air pollutants with impacts at regional scales (acid deposition, eutrophication, and tropospheric ozone). This example highlights the fact that an international approach is desirable in terms of harmonization of the regulations, development of common public policies, and monitoring of the impacts of the emission controls on air quality and its environmental impacts. As the previous chapters have shown, air pollution is a complex system to understand and manage in an optimal manner, because it involves complex interactions among a large number of pollutants and across the large spatio-temporal spectrum of atmospheric scales. Therefore, the development of efficient emission control strategies must take into account the complexity of the various multi-pollutant emission/concentration relationships.

Problems

Problem 15.1 Development of a regulation
A regulatory agency is considering setting some air quality standards for an air pollutant emitted from on-road traffic. The agency hesitates between setting up a standard for near-roadway concentrations and urban background concentrations. Give at least one reason for (1) setting a near-roadway standard and (2) setting an urban background standard.

Problem 15.2 Public policy
Alternate driving was introduced in Paris for one day during an air pollution episode in March 2014 (see Chapters 3 and 9). On that day (March 17), only vehicles with an odd license plate number were allowed to drive (except for some categories of vehicles such as taxicabs and emergency vehicles). The effects of this alternate driving measure were estimated by the local air quality organization, Airparif, as follows:

– Decrease of traffic-related PM_{10} emissions by 15 %
– Decrease of PM_{10} ambient concentrations at urban background monitoring stations by 2 %
– Decrease of PM_{10} ambient concentrations at near-roadway monitoring stations by 6 %

Give an argument for and an argument against the use of alternate driving during air pollution episodes in Paris.

References

Airparif, 2016. "*Rapports sur la qualité de l'air en Île-de-France,*" 2011 to 2015, available at: www.airparif.asso.fr/telechargement/telechargement-statistique.

Appel, K.W., S.L. Napelenok, K.M. Foley et al., 2017. Description and evaluation of the Community Multiscale Air Quality (CMAQ) modeling system version 5.1, *Geosci. Model Dev.*, **10**, 1703–1732.

Federal Register, Vol. 75, N° 26, Section II, 9 February 2010, www.federalregister.gov.

Federal Register, Vol. 75, N° 119, Section II, 22 June 2010, www.federalregister.gov.

Federal Register, Vol. 76, N° 169, Section II, 31 August 2011, www.federalregister.gov.

Federal Register, Vol. 78, N° 10, Section III, 15 January 2013, www.federalregister.gov.

Federal Register, Vol. 78, N° 10, Section IV, 15 January 2013, www.federalregister.gov.

Federal Register, Vol. 79, N° 242, Section II, 17 December 2014, www.federalregister.gov.

Federal Register, Vol. 80, N° 2, Section II, 5 January 2015, www.federalregister.gov.

Godowitch, J.M., A.B. Gilliland, R.R. Draxler, and S.T. Rao, 2008. Modeling assessment of point source NO_x emission reductions on ozone air quality in the eastern United States, *Atmos. Environ.*, **42**, 87–100.

Menut, L., B. Bessagnet, D. Khvorostyanov et al., 2014. CHIMERE 2013: A model for regional atmospheric composition modeling, *Geosci. Model Dev.*, **6**, 981–1028.

NADP, 2017. National Atmospheric Deposition Program (NRSP-3), NADP Program Office, Illinois State Water Survey, Champaign, IL.

Ohshita, S.B. and C. Seigneur, 1993. Risk and technology in air toxics control: California and the Clean Air Act amendments, *J. Air Waste Manage. Assoc.*, **43**, 723–728.

Sartelet, K., F. Couvidat, Y. Roustan, and C. Seigneur, 2012. Impact of biogenic emissions on air quality over Europe and North America, *Atmos. Environ.*, **53**, 131–141.

SCAMQD, 2017. "Historical data by year," www.aqmd.gov/home/library/air-quality-data-studies/historical-data-by-year, South Coast Air Quality Management District, Diamond Bar, California.

Vijayaraghavan, K., C. Lindhjem, A. DenBleyker, U. Nopmongcol, J. Grant, E. Tai, and G. Yarwood, 2012. Effects of light duty gasoline vehicle emission standards in the United States on ozone and particulate matter, *Atmos. Environ.*, **60**, 109–120.

Index

Absorption, 13
   of atmospheric radiation, 66, 76
Accumulation. *See* Mode
Acetaldehyde, 157
Acid
   Carbonic, 246, 319
   Carboxylic, 9, 208, 210, 211
   Hydrochloric, 306
   Nitric, 3, 148, 160, 169, 205, 206, 244, 249, 273, 306
   Nitrous, 153
   Sulfuric, 3, 24, 146, 195, 204, 212, 242, 250, 306
Acid rain, 3, 255, 306
Adiabatic
   Gradient, 37
   Transformation, 36
Adsorption, 14, 262
Advection, 100
Aerosol
   Definition of, 190
   Primary organic aerosol (POA), 192, 207
   Secondary organic aerosol (SOA), 192, 207, 209, 211, 212, 231
Agriculture, 6, 204, 305, 308, 309
Albedo, 79, 81, 85
Alcohol, 9, 148, 160, 162, 210, 211, 212
Aldehyde, 9, 148, 153, 156, 162, 166, 168, 208, 211
Aldol condensation, 211
Alkane, 7, 149, 159, 208, 209, 214
Alkene, 7, 146, 160, 208, 210, 214
Alkoxy radical, 160, 162, 169
Alkyne, 148
alpha-Pinene. *See* Pinene
Aluminum, 25, 306
Ammonia, 8, 15, 26, 149, 195, 204, 205, 219, 245, 309, 323
Ammonium, 191, 205, 206, 219, 248
Anticyclone, 324
AOT40, 304
Argon, 33
Aromatic compound, 7, 148, 166, 210, 214
Arrhenius expression, 133

BC. *See* Black carbon
Beer-Lambert, Law of, 83
Benzaldehyde, 166, 167
Benzene, 148, 178, 214, 293, 295, 333

beta-Pinene. *See* Pinene
Bicarbonate, 246
Bioaccumulation, 1, 311, 314
Bioconcentration, 311
Biogenic organic compound. *See* Organic compound
Biomagnification, 311
Biomass, 2, 7, 8, 13
Bisulfate, 195, 204, 205, 206
Bisulfite, 244, 248
Black body, 77, 78
Black carbon, 191, 323
Boltzmann constant, 78
Boussinesq hypotheses, 58, 61
Breeze, 39, 71
Bromine, 141, 142, 313
Brown carbon, 323
Brownian. *See* Diffusion
Building, 37, 63, 111, 119, 307
Butadiene, 148

Cadmium, 311
Calcium, 14, 306, 307
Cancer slope factor, 299
Canopy, Urban, 51, 58, 63, 119
Carbon, 7, 14, 18
Carbon dioxide, 1, 7, 25, 33, 248, 318, 320
Carbon monoxide, 7, 8, 14, 19, 25, 147, 155, 295, 334
Carbonate, 14, 246, 307
Carbon-bond mechanism, 186
Carbonic acid. *See* Acid
Carboxylic acid. *See* Acid
Carcinogenic. *See* Health effects
Carene, 216
Carson, Rachel, 3, 309
Case-control study. *See* Epidemiological study
CFC. *See* Chlorofluorocarbon
Chapman mechanism, 137
Chemical kinetics, 144, 176, 249
Chloride. *See* Sodium
Chlorine, 140, 306, 310
Chlorofluorocarbon, 140, 343
Chromium, 311
Clean Air Act, 256, 334
Climate change, 2, 318
Cloud, 1, 39, 47, 79, 239
Coagulation, 200

Index

Coal, 2
Coefficient
   Absorption, 83
   Activity, 203, 241
   Brownian diffusion, 200, 201, 264, 265
   Coagulation, 200
   Cunningham, 201
   Dispersion, 106, 116, 118
   Drag, 260, 261
   Extinction, 84
   Mass transfer, 120, 196, 271
   Molecular diffusion, 196, 264, 265, 271
   Scattering, 84, 88
   Scavenging, 114, 273
Cohort, 290
Collision, 17, 127, 200, 269, 273
Color of air pollution, 91
Condensation, 43, 49, 196, 248
Conference of the Parties, 343
Constant
   Equilibrium, 136
   Partitioning, 232
   Rate, 128, 132, 133, 134
   Solar, 79
Contrast, 85
Convention
   Minamata, 314, 343
   Stockholm, 343
Coriolis. *See* Force
Coulomb. *See* Force
Criegee radical, 165
Cross-section, Radiative transfer, 128, 130
Crutzen, Paul, 140
Cunningham. *See* Coefficient

Dalton, Law of, 55
DDT, 3, 309
Deliquescence, 206
delta-Carene. *See* Carene
Deposition
   Acid, 3, 306
   Dry, 114, 259
   Regulation for, 341
   Velocity of dry. *See* Velocity
   Wet, 269
Diameter
   Aerodynamic, 190
   Mean, 227
   Median, 228
   Stokes, 190
Diesel. *See* Fuel
Diffusion
   Brownian, 16, 17, 262, 265, 266, 269
   Molecular, 100, 262
   Numerical, 118, 121
   Turbulent, 96, 100
Dimethyl sulfide, 9, 307

Dioxin, 7, 310
Dispersion
   Gaussian, 106
   Horizontal, 118
   Vertical, 116
Distribution
   Gaussian, 96
   Lognormal, 193, 227
   Particle size, 193
   Sectional, 229
DMS. *See* Dimethyl sulfide
Drop, droplet
   Cloud, 240
   Rain, 240
Dry, 120
Dust, 6, 89, 260, 280

Earth
   Radius, 79
   Temperature, 82
EC. *See* Elemental carbon
Efficiency
   Absorption, 83
   Collision, 273
   Emission control, 18
   Scattering, 84
EKMA diagram, 180
Ekman spiral, 57
Electroneutrality, 127, 246
Electrostatic precipitator (ESP), 17
Elemental carbon, 191
Emission
   Aeolian dust, 280
   Anthropogenic, 8, 325
   Biogenic, 11, 29, 324, 325
   Control, 13
   Factor, 10
   Inventory, 10
   Natural, 8
   Resuspension, 277
   Sea salt, 281
Energy
   Activation, 130
   Bond dissociation, 130
   Budget, 82, 320
   Gibbs, 230
   Kinetic, 62
   Mean thermal, 131
   Turbulent kinetic (TKE), 62
Enol, 211
Enthalpy
   of dissolution, 243
   of reaction, 243
Epidemiological study
   Case-control, 290
   Cohort, 290
   Cross-sectional, 290

Epidemiological study (cont.)
   Ecological, 289
   Longitudinal, 289
Epidemiology, 288
Episode, Air pollution, 2, 47, 49, 90, 173, 223, 224
Epoxydiol (IEPOX), 210
Equation
   Atmospheric diffusion, 99
   Continuity, 59
   General dynamic, 195
   Heat transfer, 66
   Hydrostatic, 33
   Navier-Stokes, 58
   Radiative transfer, 85
   RANS, 61
Erodible soil, 280
Erosion by wind, 280
ESP. *See* Electrostatic precipitator
Ester, 210, 211
Ether, 148
Ethylene, 304
Etiologic fraction, 292
Eulerian
   Dispersion, 100
   Model. *See* Model
Eutrophication, 309
Evaporation, 40, 51, 81, 190, 195, 196
Excess risk, 291
Extinction. *See* Coefficient

Ferrel cell, 43
Fick, Law of, 100
Filter
   Fabric, 17
   Particle, 27
Fluorine, 140, 141
Flux
   Actinic, 128, 136
   Dry deposition, 268
   Latent heat, 71, 82
   Radiative, 79, 139, 320
   Sensible heat, 67, 81
   Wet deposition, 272
Fog, 239
Force
   Coriolis, 40, 42
   Coulomb, 201
   Frictional, 259
   Gravitational, 33
   Pressure gradient, 42
   van der Waals, 201
Formaldehyde, 22, 148, 156
Front, 43, 47
Fuchs equation, 201
Fuchs-Sutugin equation, 197
Fuel, 7

Function
   Error, 112
   Phase, 84, 85
Functional group, 186, 233
Functionalization, 207
Furan, 7, 210, 215, 310

Gasoline. *See* Fuel
General atmospheric circulation, 33
GHG. *See* Greenhouse gas
Gibbs. *See* Energy
Glyoxal, 166
Gradient
   Pressure, 42
   Vertical temperature, 57
Grasshopper effect, 276, 310
Gravity, 60, 260
Greenfield gap, 274
Greenhouse gas, 1, 10, 78, 318
Growth law. *See* Rate
Gypsum, 14, 307

Haagen-Smit, Arie, 3, 147
Hadley cell, 43
Half-life, 150
Halocarbon, 141, 320
Halon, 141, 142
Hazard quotient, 299
Haze index, 89
Health effects
   Carcinogenic, 3, 7, 297, 298
   Non-carcinogenic, 297, 298
Health risk, 299
Health risk assessment, 297
Heat capacity
   Molar, 104
   Specific, 36, 37
Heating, 6, 67, 307
Height, Characteristic, 34
Hemiacetal, 210, 211, 212
Henry's law
   Constant, 241, 243
   Definition of, 241
   Effective constant, 244
HRA. *See* Health risk assessment
Humidity
   Absolute, 55
   Relative, 55
   Specific, 56
Humulene, 216
Hydrargyrism, 312
Hydrocarbon, 7, 22, 25, 148, 170
Hydrochloric. *See* Acid
Hydrogen peroxide, 153, 169, 250, 312
Hydroperoxyl radical, 139, 156, 169, 172, 312
Hydrophilic, 209, 233
Hydrophobic, 89, 209

Hydrostatic. *See* Equation
Hydroxyl radical, 138, 149, 153, 212, 249, 312
Hygroscopic, 89, 248, 323
Hysteresis, 206

IARC. *See* International Agency for Research on Cancer
Ice, 40, 80, 319, 322
Ideal gas law constant, 33
Ideal gas law, 33
IEPOX. *See* Epoxydiol
Incidence, 291
Incineration, 7, 13, 15, 27
Inertia, Deposition via, 261, 265, 273
Infrared. *See* Radiation
Interception, Deposition via, 261, 265, 273
International Agency for Research on Cancer, 3, 297
Inversion, Temperature, 40, 43, 54
Ionic dissociation, 241
Ionosphere, 38
Iron, 248, 251
Irradiance, 78, 79
Isomerization, 160, 210
Isoprene, 8, 168, 210

Johnston, Harold, 139

Kelvin effect, 197
Ketone, 9, 162, 165, 168, 208, 210
Knudsen number, 197
Koschmieder relationship, 86
K-theory, 116, 117

Lagrangian
    Dispersion, 96, 102
    Model. *See* Model
Laplace, Law of, 52
Layer
    Mixing, 53, 54
    Planetary boundary (PBL), 37, 51
    Quasi-laminar, 262, 264, 265
    Residual, 54
    Surface, 51, 63, 67, 264
Lead, 18, 24, 295, 311, 328, 334
Leighton photostationary state, 154
Length
    Mixing, 61, 63
    Monin-Obukhov, 65, 68, 117
    Roughness, 63, 263
Letovicite, 205
Lifetime
    Chemical, 150
    Cloud droplet, 254
    Raindrop, 272
Limit value, 337
Limonene, 211, 216
Lognormal. *See* Distribution

Loi sur l'air et l'utilisation rationnelle de l'énergie (LAURE), 338
Longifolene, 216
Longitude, Ecliptic, 129

Magnesium, 306
Manganese, 248, 251
Mass action, Law of, 133
Maxwell-Boltzmann distribution, 131
MBO, 211, 216
Mean free path, 197, 201
Mechanism
    Carbon-bond. *See* Carbon-bond mechanism
    Chromium. *See* Chromium
    Mercury. *See* Mercury
    Photochemical pollution, 169, 170
    Surrogate molecule, 185, 233
Mercury, 3, 8, 18, 312, 341, 343
Mesosphere, 38
Metals, Heavy, 311
Methacrolein (MACR), 168, 210
Methane, 149, 159, 319, 320
Methyl vinyl ketone (MVK), 168, 210
Methylbutenol. *See* MBO
Methylperoxyl radical, 158
Microreversibility, Principle of, 135
Mie theory, 83
Mixing ratio, 56, 125
Mode
    Accumulation, 193, 227
    Coarse, 193, 227
    Nucleation, 193, 227
Model
    Chemical-transport, 116, 121, 187, 307, 314, 334
    Eulerian, 116, 118
    Gaussian, 103
    Lagrangian, 103
    Meteorological, 72
    Plume, 108, 109, 121
    Puff, 108, 121
    Street-canyon, 119
    Trajectory, 181
Modeling
    Aqueous-phase chemistry, 257
    Deposition, 283
    Emissions, 28
    Gas-phase chemistry, 185
    Meteorology, 72
    Particles, 227, 233
    Radiative transfer, 92
    Transport and dispersion, 121
Molina, Mario, 140
Monin-Obukhov. *See* Length
Monoterpene, 9, 168, 211, 216
Montreal. *See* Protocol
Morbidity, 291, 295
Mortality, 291, 296

National Ambient Air Quality Standards (NAAQS), 334
National Emission Standards for Hazardous Air Pollutants (NESHAP), 335
Navier-Stokes. *See* Equation
Newton, Law of, 58
Nitrate
  Inorganic, 191, 204, 205, 206, 219, 256, 308, 343
  Ion, 248
  Organic, 148, 159, 162, 169, 207, 213, 214, 309
  Peroxyacetyl. *See* Peroxyacetyl nitrate (PAN)
  Radical, 148, 149, 153, 207, 212, 249
Nitric acid. *See* Acid
Nitric oxide, 7, 18, 109, 139, 148
Nitrogen dioxide, 7, 18, 91, 146, 148, 154, 173, 249, 295
Nitrogen oxides, 8
Nitrogen pentoxide, 148, 249, 250
Nitrogen, Molecular, 1, 7, 14, 15, 77, 131, 134
Nitrous acid. *See* Acid
Nitrous oxide, 148, 276, 319, 320
Non-carcinogenic, see Health effects
NO_x. *See* Nitrogen oxides
NO_y. *See* Nitrogen oxides
NO_z. *See* Nitrogen oxides
Nucleation, 195

Octane, 8
Odds ratio, 293
Odum two-compound model, 231
Oligomerization, 212
Optical thickness, 84
Organic compound
  Aqueous-phase, 212, 255
  Biogenic, 162, 168, 216
  Emissions, 6, 7, 8, 19, 113
  Semi-volatile (SVOC), 18, 27, 148, 207
  Volatile (VOC), 18, 148
Oxygen, Molecular, 1, 40, 77, 128, 137
Ozone
  Depletion potential, 141
  Formation potential, 182
  Greenhouse gas, 323
  Stratospheric, 77, 125, 137, 139
  Tropospheric, 146, 147, 149, 154, 173, 175, 185, 212, 251, 295, 304, 324, 334
Ozonide, 162

PAH. *See* Polycyclic aromatic hydrocarbon
PAN. *See* Peroxyacetyl nitrate (PAN)
Particle
  Coarse, 191
  Definition of, 190
  Diesel, 27
  Emissions, 18, 27, 280, 282
  Fine, 191
  Inorganic, 230
  Organic. *See* Aerosol
  Size distribution, 227
  Ultrafine, 191
Particulate matter (PM). *See* Particle
PBL. *See* Layer
PCB. *See* Polychlorobiphenyl
Pentane, 160
Peroxyacetyl nitrate (PAN), 130, 158
Persistent organic pollutant (POP), 3, 14, 276, 309
pH, 246
Phase
  Aqueous, 204, 212, 240
  Gas, 153, 204, 205, 209
  Organic, 223, 232
  Particulate, 211
Photochemical. *See* Reaction, Smog
Photolysis, 125, 151, 153, 156
Photon, 76, 93, 128
Pinene
  alpha-Pinene, 211, 216
  beta-Pinene, 216
Plan de protection de l'atmosphère (PPA), 338
Planck
  Constant, 76
  Law of, 78
Plume
  Color, 91
  Dispersion, 96
  Height, 105
  Rise, 104
  Slender plume approximation, 103
PM_{10}, 9, 191, 328
$PM_{2.5}$, 9, 191, 291, 328
POA. *See* Primary organic aerosol (POA)
Polar
  Cell, 43, 82
  Stratospheric clouds, 141
Pollutant
  Precursor, 2, 146, 190, 255, 320
  Primary, 2, 6, 146, 190
  Secondary, 2, 6, 146, 190, 307
Polychlorobiphenyl (PCB), 310
Polycyclic aromatic hydrocarbon (PAH), 7, 213, 310
POP. *See* Persistent organic pollutant (POP)
Prandtl
  Mixing length. *See* Length
  Number, 70
Precipitation, 33, 240, 269
Premature death, 301
Pressure
  Atmospheric, 33, 35
  Partial, 56, 203, 204
  Saturation vapor, 203
Propene, 162
Propylene. *See* Propene

Protocol
  Göteborg, 307, 343
  Montreal, 143, 343
Puff, 91, 101

Quasi-laminar. *See* Layer

Radiance, 78, 83
Radiation
  Earth's, 76, 80
  Infrared (IR), 1, 77, 318, 320
  Solar, 1, 40, 66, 76, 128, 320
  Ultraviolet (UV), 1, 77, 125, 139
  Visible, 1, 83
Radiative forcing
  Direct effect of particles, 322
  Effective, 320
  Indirect effect of particles, 322
  Negative, 320
  Positive, 320
  Semi-direct effect of particles, 322
  Total, 320
Radiative transfer. *See* Equation, Modeling
Rain, 240, 246, 269, 270
Rain, Acid. *See* Acid rain
Rainout, 269, 270
RANS (Reynolds-averaged Navier-Stokes), 60
Raoult, Law of, 204, 231
Rate
  Chemical reaction, 132
  Growth law, 196
  Photolysis, 128, 184
  Resuspension, 277
Rayleigh scattering, 83
Reaction
  Bimolecular, 130, 133
  Elementary, 131, 135
  Heterogeneous, 14, 141, 241
  Photochemical, 127, 130, 147, 153
  Photolytic, 127
  Thermal, 127, 130, 131
  Titration, 154, 250
  Trimolecular, 130, 134
  Unimolecular, 130, 131
Reduction of air pollution
  Acid rain, 255, 256
  Mercury, 316
  Ozone, 179
  Particles, 217
  Pollutant content in fuels, 24
Reemission, 276
Reference
  Concentration, 298
  Dose, 298
Regime
  Ammonia-poor, 220

Ammonia-rich, 220
Continuous, 200
High-NO_x, 176, 177, 213
High-pressure, 132, 134
Low-NO_x, 176, 177, 213
Low-pressure, 132, 134
Meteorological, 45
Molecular, 200
Transient, 102
Relative risk, 292
Reservoir species, 130
Resistance to dry deposition
  Aerodynamic, 262
  Diffusion, 264
  Surface, 266
  Total, 266
Resuspension. *See* Emission
Reynolds
  Decomposition, 60
  Number, 60
  Stresses, 61
Richardson number, 68
Roughness. *See* Length
Rowland, Sherwood, 140

Salt
  Ammonium. *See* Ammonium
  Nitrate. *See* Nitrate
  Sea, 89, 281
  Sulfate. *See* Sulfate
Scale
  Atmospheric, 39
  Lagrangian time, 98
  Length. *See* Length, Monin-Obukhov
  Temperature, 67, 69
Scattering, 83
Scavenging, 114, 269, 273
Schmidt number, 265
Scrubber, 14, 17
Sedimentation, 16, 259, 269
Sesquiterpene, 9, 168, 216
Sherwood number, 271
Similarity theory, 65, 107, 117
Simulation, Numerical. *See* Modeling
Smagorinsky dispersion coefficient, 118
Smog
  London, 2
  Photochemical, 3, 146, 147, 192
Smog chamber, 184, 186, 212
Smoke, 2, 147
Smolarkiewicz algorithm, 121
Snow, 40, 80, 319, 322
SOA. *See* Secondary organic aerosol (SOA)
Sodium, 192
Solubility, 233, 244, 269
Soot, 4, 7, 191

Source
  Anthropogenic, 6
  Area, 112
  Continuous point, 102
  Instantaneous point, 101
  Line, 111
  Natural, 6
  Volume, 113
Spectrum, Atmospheric radiation, 77, 81
Stability, Atmospheric, 68, 105, 114
Standard deviation
  Dispersion, 96, 101
  Horizontal wind speed, 98
  Lognormal distribution, 227
  Vertical wind speed, 120
Standard, Air quality, 328
State Implementation Plan (SIP), 334
Stefan-Boltzmann
  Constant, 79
  Law of, 79
Stokes
  Diameter. *See* Diameter
  Law of, 260
  Number, 265
Stratopause, 38
Stratosphere, 37, 38
Sulfate, 89, 191, 195, 204, 205, 206, 208, 219, 250, 255, 307, 343
Sulfite, 14, 244, 248
Sulfur dioxide, 8, 18, 24, 146, 243, 295
Sulfur trioxide, 327
Sulfuric acid. *See* Acid
SUM60, 305
Sun, 76, 79
Surface tension, 197
SVOC. *See* Organic compound

Target value, 337
Taylor, Theorem of, 97
Temperature
  Ambient, 33
  Potential, 52
  Scale. *See* Scale
  Vertical profile, 36
  Virtual, 56
  Virtual potential, 57
Terpene, 92, 168, 213
Terpenoid, 92, 168
Tetrol, 211, 212
Thermosphere, 38
Titration. *See* Reaction
Toluene, 24, 166, 186, 214
Toxicological study
  Definition of, 286
  In vitro, 287
  In vivo, 287
Toxicology, 286
Traffic, On-road, 18, 112, 146, 277
Transfer
  Heat, 37, 51, 66, 67, 71
  Mass, 51, 116, 196, 234, 241, 262, 270
  Momentum, 51, 59, 60
Troe correction, 132, 135
Tropopause, 37
Troposphere, 37
Turbulence, 61, 65, 67, 95, 96, 103

Ultraviolet. *See* Radiation
Unit risk factor, 299

van der Waals. *See* Force
van't Hoff equation, 214, 243
VBS. *See* Volatility Basis Set (VBS)
Vegetation, 6, 8, 71, 92, 120, 259, 304, 306, 308
Vehicle, 19, 24, 217, 277
Velocity
  Convective, 69
  Dry deposition, 259, 266, 267
  Fall, 240, 259, 260
  Friction, 61, 117, 280
  Horizontal wind, 42, 66
  Mean thermal, 197, 201
  Sedimentation, 260, 267
  Vertical wind, 42, 64
Viscosity
  Dilatational, 58
  Dynamic, 58
  Kinematic, 59
  Turbulent kinematic, 61
Visibility, 89, 93, 190, 192
Visual range, 85
VOC. *See* Organic compound
Volatility, 195, 206, 208, 255, 276
Volatility Basis Set (VBS), 233
von Kármán constant, 62, 70, 117
Vorticity, Potential, 37

W126, 305
Washout. *See* Scavenging
Water
  Content, 54, 206, 240
  Liquid, 240
  Vapor, 1, 7, 33, 37, 55, 71, 239
Wien, Law of, 78

Wind
    Direction, 57, 64
    Geostrophic, 42
    Speed. *See* Velocity
    Vertical profile, 63

Xylene, 214

Yield
    Ozone, 182
    Quantum, 128
    Secondary organic aerosol (SOA), 214, 232

Zenith angle, 82